In a continuation of Simon Cotton's 2012 *Every Molecule* ...., his new book, *Every Molecule Matters*, covers in great detail the story of Food, Vitamins, Hot and Cold, Abused Pain Killers and Other Drugs of Abuse, Nasty Smells, War and Peace, Organochlorine Compounds, Organofluorine Compounds, Smoking and Vaping, Isotopes and Methane. Each chapter is replete with extensive historical facts, chemical structures and origins, and present-day situations. Each chapter has a detailed bibliography at the end of the volume.

The chapter, **Spices, 'Hot' and 'Cold,'** explains in fascinating detail the chemistry and biology of capsaicin ("Hot") and menthol ("Cold"), and the evolution of these chemicals from antiquity to modern times. Also covered are the lesser known "spice chemicals" such as nutmeg, black pepper, cinnamon and Szechuan peppers.

Nasty Smelling Molecules is a fascinating exploration of the chemicals with which every human and animal species is familiar! In addition to the ubiquitous indole and hydrogen sulfide, Cotton presents a myriad of other sulfur compounds from a variety of sources and discusses in detail the origin of the 'Scent of Death' heterocyclic amines.

Abused Painkillers and Other Drugs of Abuse reveals the evolution of the widely abused heroin, fentanyl, and oxycontin and presents the lesser known spice and nitazenes. The extraordinary toxicity of fentanyl analogues, such as sufentanil and carfentanil, is outlined in alarming detail. The nitazenes came to prominence in 2019 as new analgesics, more potent than morphine, but now recognised to have fatal toxicities comparable to fentanyl.

Vitamins is a marvellous presentation of the myriad vitamins that humans ingest daily. From the connection of scurvy to the lack of vitamin C, to rickets to the lack of vitamin D, and pellagra to the absence of vitamin B_3 (niacin), this chapter is a marvellous presentation of the history, sources, and function of all 18 vitamins.

Food is a wonderful summary of food types, from their constituents (protein, lipids, carbohydrates) to both fresh and cooked foods, fruits and vegetables, cheeses, and their myriad aromas. All of this is combined with enough organic chemistry to satisfy the professional (e.g., aroma structures, thermal reactions, oxidation and enzymatic chemistry).

War and Peace in Nature concisely summarises the extraordinary chemistry utilised by plants and insects in chemical defence against potential predators (man, birds, animals, and other insects). These repellent "allomones," "alarm pheromones" and natural insecticides comprise an array of volatile organic compounds, some of which are well known (benzoquinone, hydrogen peroxide, acetic acid, hydrogen cyanide, volatile aldehydes, phenols) and others such as piperidines, pyrethrins and the "blister beetle" cantharidin are new to science. Interestingly, one of these piperidines is a close analogue to the poison hemlock alkaloid that killed Socrates. The well-known plant alkaloids caffeine, cocaine, morphine, and nicotine are toxic to insects. Organic chemistry is limited to the structures of these compounds and some of the fascinating routes to their formation.

Organochlorine Compounds is a chapter dear to me! I've been following and documenting organochlorine compounds found in nature for the past 45 years. Dr. Cotton concisely presents an account of both man-made and natural organochlorine compounds. From the Civil War anaesthetic chloroform and the life-saving vancomycin to the pesticides DDT and Dieldrin, this short chapter succinctly covers this enormous field. A highlight is the sex pheromone 2,6-dichlorophenol produced by species of female ticks. Interestingly, this isomer is essentially impossible to prepare in the laboratory because chlorination of phenol gives the 2,4-isomer! Other relevant examples are the chemical defensive secretion epibatidine, which is a rare chloropyridine, and the soil antibiotic chlorotetracycline. The presentation of organic structures is minimal and appropriate.

Organofluorine Compounds is arguably the most important chapter in this book. It deals with the two current "hot" topics of CFCs (chlorofluorocarbons) and PFAS (polyfluorinated alkyl substances) – the former in great detail with respect to the ozone layer and the latter fleetingly. Coverage of CFCs (historical, evolution, chemistry, consequences) is the most detailed and comprehensive that I have ever seen! The rare natural organofluorine and highly toxic compound, fluoroacetic acid, and some fluorinated pharmaceuticals are covered briefly.

Smoking and Vaping. Following an introduction to the use of tobacco dating back to perhaps 18,000 years ago and certainly used by people in the Andes during 3000–5000 BC, this chapter focusses on the adverse health effects of tobacco smoking. These include formation of the ubiquitous polycyclic aromatic hydrocarbons and N-nitrosoamines. The section on the relatively new addiction of Vaping (via e-cigarettes) includes the fascinating chemical formation of various toxic compounds (acrolein, other aldehydes) and other compounds from the flavouring additives such as diacetyl ("popcorn aroma"). Presented also is the surprising generation of higher reactive (and presumed toxic) ketene from Vitamin E acetate. The virtually unknown effects of metals and alcohols, which are present in e-cigarettes, are disclosed. This chapter closes with an illuminating "Conclusions and Summary".

Isotopes at Work. In what is an enormous area comprising hundreds of isotopes, Dr. Cotton concentrates this chapter on a few of the most relevant isotopes. Beginning with definitions and familiar examples ("heavy water," radioactive vs. non-radioactive, uranium and carbon isotopes), a major attraction of this chapter is the use of isotopes in food fraud, carbon dating, criminology and forensic science, one case of which analysed the skeleton of King Richard III. Also presented are "isoscapes," the new study of lighter isotopes.

Methane. The simplest organic molecule is methane (one carbon atom bonded to three hydrogen atoms). In this concise chapter, Dr. Cotton presents the astonishing natural occurrences of methane in our universe, and its role in the "Greenhouse Effect" leading to global warming. No chemistry is explicitly depicted.

Bibliography. Dr. Cotton has assembled a bibliography of nearly 60 pages that covers each book chapter. From book references to primary literature citations, this is a wealth of information for the reader. References include both classic reviews and modern journal citations. Several references to the 2024 literature are listed. For example, the Food chapter lists nearly 170 references. This document alone is worth the price of this book!

Gordon W. Gribble, *Professor of Chemistry Emeritus,*
Department of Chemistry, Dartmouth College,
Hanover, New Hampshire, USA

Every Molecule Matters

Every Molecule Matters is a successor to the author's earlier *Every Molecule Tells a Story* and tells the story of a wide range of molecules.

These range from the familiar odours that enhance the pleasure of eating (and the spices that add piquancy) to the vitamins vital to our survival, as well as the ways in which insects and plants use chemicals to protect themselves. There's controversy surrounding the compounds of chlorine, which encompass life-supporting anaesthetics and natural antibiotics, as well as insecticides like DDT, which saved innumerable lives but became an environmental *cause célèbre*. Through the addictive power of nicotine, smoking tobacco transformed from a curiosity imported from the Americas into a megapound industry that has caused significant human illness and death. And how safe is vaping? Discover the painkillers that have become drugs of abuse, and smile at the smelly sulfur compounds that serve as unpleasant human odorants (and defence molecules for skunks), control natural cycles in the environment or act as flavourings in wine. You will discover them all here.

This book showcases the structures of hundreds of compounds used by humans, animals and plants. Some are beneficial; some are not. Find out here why you should be better informed about them.

- This collection of molecules includes human issues, such as the chemistry of vaping, and drugs of abuse, including 'spice', nitazenes and fentanyl.
- 'Chemistry of Everyday' includes vitamins and the molecules that give foods their aromas and appetizing appeal.
- The chemistry of nature – how plants and insects use chemicals to defend themselves against potential predators, whether humans, birds, animals or other insects.
- Organohalogen compounds, encompassing the atmosphere-damaging CFCs and their replacements, and the chlorine compounds that are important medicines (e.g. vancomycin).
- Using isotopes, from archaeologists faced by mysteries of ancient Rome and silver from Spanish conquistadors, to tracking down the origin of South American cocaine and solving the 'Scissor Sisters' murder case.

Every Molecule Matters

Simon Cotton

CRC Press
Taylor & Francis Group
Boca Raton London New York

CRC Press is an imprint of the
Taylor & Francis Group, an **informa** business

First edition published 2026
by CRC Press
2385 NW Executive Center Drive, Suite 320, Boca Raton FL 33431

and by CRC Press
4 Park Square, Milton Park, Abingdon, Oxon, OX14 4RN

CRC Press is an imprint of Taylor & Francis Group, LLC

© 2026 Taylor & Francis Group, LLC

ISBN: 9781041110637 (hbk)
ISBN: 9781041110620 (pbk)
ISBN: 9781003658115 (ebk)

DOI 10.1201/9781003658115

Typeset in Times
by codeMantra

Dedication

This book is offered for the well-being of May Nisbet and Lisa Duffin and for the repose of the soul of María de los Ángeles Santiago.

Contents

Preface xiii
Overview xv
Acknowledgements xix
About the Author xx

Chapter 1   Food 1

Introduction 1
Carbohydrates 2
    Polysaccharides 6
Amino Acids and Proteins 8
    Amino Acids 8
    Polypeptides and Proteins 10
Lipids 11
Taste Sensations 12
Meat 12
Cheese 15
    Blue Cheeses 17
    Lactose and Camembert 18
    Cheddar 21
Bread 22
    Baking 24
Potato 28
    Boiled Potatoes 30
    Baked Potatoes 31
    Potato Chips 32
Mushrooms 33
    Puffball Mushrooms 36
Onions 37
    Fresh Onions 38
    Cooked Onions 39
Tomatoes 39
Strawberries 45
    Wild Strawberries 48
Oranges and Lemons 51
    Orange Juice and Limonene Isomer Smells 53

Chapter 2   Vitamins 56

Introduction 56
Vitamin A 56
Vitamin B_1, Thiamin 58
Vitamin B_2, Riboflavin 60

Vitamin B_3, Niacin 61
Vitamin B_5, Pantothenic Acid 64
Vitamin B_6, Pyridoxine 64
Vitamin B_7, Biotin 66
Vitamin B_9, Folic Acid 66
Vitamin B_{12}, Cobalamin 68
Vitamin C, L-Ascorbic Acid 69
Vitamin D, Calciferol 70
Vitamin E 72
Vitamin K 73

Chapter 3   'Hot' and 'Cold' 75

Introduction 75
Spices and 'Hot' 75
    Capsaicin 76
    The TRPV1 Receptor 77
    Black pepper 79
    Ginger 79
    Clove 80
    Nutmeg 81
    Cinnamon 82
    Szechuan Peppers 82
Cool and Menthol 84
    How It Works? 87
    Carvone 87
    A New Minty Molecule 88
Mustard 89

Chapter 4   Abused Painkillers and Other Drugs of Abuse 92

Opium 92
Morphine 94
Heroin 95
Fentanyl 97
    Carfentanil and Other Powerful Fentanyls 100
Oxycodone and Oxycontin 102
    Oxycodone 102
Krokodil 105
Spice 108
Nitazenes 112

Chapter 5   Nasty-Smelling Molecules 115

Introduction 115
Hydrogen Sulfide 115
Dimethyl Sulfide 116

Dimethyldisulfide and the Titan Arum 119
Thiols... 120
Skunks.. 123
Personal Hygiene – and Wines... 125
Trimethylamine.. 128
The Smell of the Living and the Dead 129
The Scent of Death... 132
Euglossine Bees.. 136

Chapter 6   War and Peace in Nature... 137

Cut Grass.. 137
Plant Defence 141
Weaponised Insects.. 145
    Out of Detective Novels.. 150
    Blister Beetles... 151
    Plants Attracting Insects... 152
    Insect Repellents... 155

Chapter 7   Organochlorine Compounds 159

Organic Chlorine Compounds 159
Pesticides.. 161
Dieldrin and Aldrin.. 163
Anaesthetics – and Others.. 164
Chlorinated Phenols 165
Natural Germ-Killers from the Earth 166
Teicoplanin 169
Chloramphenicol 170
Halogenated Compounds from Marine Fungi 171
Another Killer 171

Chapter 8   Organofluorine Compounds 173

CFCs, Serendipity and a Serious Problem 174
Other Organofluorine Compounds.. 181
Other Problematic Organofluorine Compounds 182
A Natural Problem 182
Fluorinated Pharmaceuticals... 183
Anaesthetics 183
Celecoxib.. 187
Blood Substitutes.. 188
And the Latest News?.. 189

Chapter 9   Smoking and Vaping 190

Smoking 190
Nicotine 191

Health Effects of Smoking 194
Toxic Chemicals in Tobacco 195
Vaping 198
Issues with Nicotine 198
Metal Heating Elements and Metal Pieces 199
Solvents 200
Flavourings.. 201
    Diacetyl... 202
    Benzaldehyde, Cinnamaldehyde and Vanillin 203
Vitamin E Acetate.. 204
Nicotine Analogues 205
Student Vaping 206
Conclusions and Summary.. 206

Chapter 10 Isotopes at Work... 208

Isotopes.. 208
What's the Difference between H_2O and D_2O? 210
Lead Isotopes 211
Coinage and Isotopes 213
Uranium Isotopes and Their Separation 213
Carbon Isotopes... 214
Detecting Food Fraud.. 215
Radiocarbon 217
Fraudulent White Truffles 218
Isoscapes.. 218
Carbon in Foods 219
Strontium Isotopes... 219
Isotopes and Drugs 219
King Richard III... 220
Isotopes in Solving Crimes 221
    The Welsh Case 221
    The 'Scissor Sisters' Case 221
    A Unsolved Crime.... Five Thousand Years Ago............. 222

Chapter 11 Methane... 224

Introduction and Historical Background............................... 224
Wood, Coal and Coal Gas.. 225
Coal, the Environment and Smogs.. 226
Natural Gas and Methane... 226
Methane, a Greenhouse Gas... 227

Bibliography 229

Index.. 287

Preface

Back in 2009, I began writing a book that was published by CRC Press in 2012 under the title *Every Molecule Tells a Story*. It detailed, in 14 chapters, aspects of several areas of chemistry, including 'Atmosphere and Water', 'Hydrocarbons', 'Acids and Alkalis', 'Steroids and Sex', 'Cosmetics and Perfumes', as well as toxic and explosive molecules, medicinal molecules and polymers. It was aimed at students on either side of the school–university divide in Europe and the USA, as well as at their teachers and other interested persons.

The present book is intended for broadly the same audience. I have chosen a number of topics that interest me and that I hope will also interest the reader. Part of it will, I hope, correct some misapprehensions. For example, the use of the word 'chemical' is often used by many people, including some in the media, as an insult. Chemicals are assumed to be dangerous. Well, dear reader, we are all made of chemicals, so what does that make us? Take organochlorine compounds (Chapter 7). Yes, the misuse of DDT has caused real environmental problems; yet, Rachel Carson did not advocate its banning but its sensible use. Without chlorine, how do we make our water supply safe to drink? Do we want to return to early Victorian levels of water-borne disease? In the case of organofluorine compounds (Chapter 8), CFCs caused environmental problems that could not have been foreseen when they came into use in the 1930s; yet, once their effects upon the ozone layer were appreciated, there was quick action to eliminate them and to replace them with safer chemicals to carry out their roles. Organic compounds of chlorine and fluorine are widely used in medicine, not least in the world of anaesthetics.

Morphine has been used as a medicinal molecule for much longer; it was the active ingredient of opium obtained from poppies which was used as a painkiller as far back as prehistoric times. Once morphine was isolated two centuries ago, its addictive power was soon discovered. The section on painkillers (Chapter 4) relates how the ingenuity of chemists has enabled the synthesis of new painkillers, starting with diacetylmorphine and continuing through substances like the fentanyls and the nitazenes, only for them to be abused by drug users, with all the problems that implies and, of course, to be exploited by criminals. The problems created by painkiller addictions in the USA are well documented – but remember, these molecules have their proper purposes.

Addiction is not confined to painkillers; another kind is provided by tobacco (Chapter 9). This was not such a problem until the widespread adoption of cigarette smoking, with smoking-related diseases leading to considerable mortality from cancer and other causes. The addiction here is caused by nicotine, of course. As evidence accumulated linking smoking with disease in the period after World War II, there was a steady drop in the number of smokers; but it was only in the 21st century that vaping emerged as a 'safer alternative'. As many vaping liquids contain substances known to be irritants of the human respiratory system, it remains to be seen how safe they are, as discussed here; it is too soon for their possible long-term effects on health to be known.

Violence and sex tend to be associated with humans, rather than other inhabitants of the planet. Thus, some of the ways in which plants and insects use chemicals, discussed in Chapter 6 (War and Peace in Nature), may come as a surprise to some readers.

Rather than just thorns and spines, plants use chemicals to deter predatory insects, with nicotine and caffeine being two 'everyday' examples. Less well known are the plant-produced pyrethrin insecticides. Similarly, while formic acid is a well-known defensive chemical weaponised by ants (also known as formica), hydrogen peroxide is one of a number of other liquids fired by insects in their defence. The 'Green Leaf Volatiles', with their inimitable odour produced when grass is cut, seem to have a defensive function, with bactericidal and antifungal properties.

On the contrary, there are many areas where chemicals make more innocent contributions. The longest chapter of this book is devoted to certain chemicals in food. People know something of the proteins, carbohydrates and lipids which supply the energy we need to live and the chemicals to build and repair the body, but are less familiar with the molecules that contribute to the smell of the dishes we eat and how these chemicals are generated. Thus, Chapter 1 is mainly concerned with these. Chemicals are indispensable in the area of vitamins too, and a section is devoted to how they came to be discovered and their nature determined (Chapter 2). Many foods test our senses in other ways. Spices make notable contributions to our enjoyment of our meals. They have been known for many years, of course, but it is only very recently that we have found out why chilli peppers are 'hot' or why menthol is 'cool', not to mention why mustard has its peculiar effects (Chapter 3).

Of course, most of the students who attend chemistry lessons at school do not pursue their studies any further. If you mention 'chemistry' to them later in life, they will probably reminisce the smelly chemicals in the lab. Some influential smelly chemicals emerge in Chapter 5, whether the role of dimethyl sulfide in the global sulfur cycle or the mercaptans used to provide the odour in natural gas supplies that warns of leaks. Mercaptans make up a family of sulfur compounds with very unpleasant smells when concentrated but in trace amounts lend a pleasant bouquet to certain wines. Then there are the smelly chemicals that crop up in foods, notably durian. The smell associated with the human body also enters into the discussion, whether of the living or the dead.

The existence of isotopes is now taken for granted, but some of them are used in unexpected ways (Chapter 10), particularly in the application of spectroscopic probes to authenticate foods, such as the flavourings associated with truffles or to determine whether vanilla extract is 'natural' or petroleum-derived, with the former bearing a considerable premium. More surprising is their role in forensic science, as demonstrated in the investigation of the remains of King Richard III, whose burial place was unexpectedly discovered, and in solving modern murder mysteries.

And last of all, there's methane, which has changed from being a smokeless fuel that solved a pressing pollution problem (i.e. smogs) to an environmental problem when it was recognised as a greenhouse gas, as discussed in Chapter 11.

Taken together, then, here are hundreds of molecules for you to read about. Some are controversial; some are not. Some will be familiar to you, but I hope this book will introduce you to some new molecules and also to new ways of thinking about them.

Overview

This compilation is an essential reference work for every chemist – particularly organic chemists. It is a handbook of up-to-date knowledge on many of the chemical compounds in our chemical world, and their contexts. Not only is each chapter full of fascinating facts, but each chapter examines the molecules from a different perspective. Here, this reviewer provides commentaries on each of the chapters.

CHAPTER 1: FOOD

What differentiates this book from others that cover the topic of food is that, upfront Dr. Cotton discusses the taste of foods: "So the core of this chapter is to consider molecules responsible for flavour sensations in a range of foods, both cooked and uncooked". This topic is especially relevant to those who, through chemotherapy, or through viral infections, lost their ability to experience the food sensations.

According to Dr. Cotton, roast beef has been the most studied, the delightful smell (to most carnivores) coming primarily from: methional, 2-acetylthiazole, 2-acetylthiazolidine, 2-ethyl-3,5-dimethylpyrazine, and 2,3-diethyl-5-methylpyrazine. On the other hand, fried chicken exudes a completely different array of complex molecules. Eye-opener for this reviewer!

The diversity of odiferous molecules from different cheeses was also fascinating, as were those from the baking process. Vegetarians and vegans need not despair as there are many, many pages on odours from raw and cooked vegetables and fruit.

CHAPTER 2: VITAMINS

As Dr. Cotton notes in the introduction, vitamins are "are an untidy collection of complex organic nutrients". This chapter is built around the discovery of vitamins, particularly through the diseases they prevented. In the descriptions, exceptional care has been taken to accord the correct attributions of discovery and the discovery· process. Then follows a detailed discussion of every vitamin and its chemistry.

CHAPTER 3: SPICES, 'HOT' AND 'COLD'

After the usual outstanding historical background to the discovery and spread of spices around the world, Dr. Cotton focusses upon their mode of action. This includes very useful facts, including: "Much of a capsaicinoid molecule is hydrophobic and so not water-soluble, which is why reaching for the water jug or a beer is not the answer to a curry that is too hot for you; it is thought that milk is the best option, as it contains the lipophilic casein, which is better at removing the lipid-like capsaicins".

CHAPTER 4: ABUSED PAINKILLERS AND OTHER DRUGS OF ABUSE

This chapter commences with a lengthy – and fascinating – discourse on the history of opium ["a mixture that involves over 20 different alkaloid molecules"] and its wide acceptance: "As an over-the-counter cure-all, it was the equivalent of aspirin a century later for many people – but better. It was a painkiller, a sedative and a specific treatment for diarrhoea".

Dr. Cotton has a superlative coverage of fentanyl and its many derivatives. He describes the incredible toxicity of the fentanyl family requiring the wearing of hazmat suits when handling even the smallest dose. He also describes its probable use to break the 2002 siege by Chichen terrorists of 900 people in the Dubrovka Theatre in Moscow, by pumping an aerial suspension into the ventilation system of the Theatre and then treating as many as possible of the hostages with fentanyl antagonists. It is such unusual facts and applications which make this compilation so interesting.

There is also a lengthy coverage of oxycodone and its relatives, including "Krokodil", and many other addictive drugs.

CHAPTER 5: NASTY SMELLING MOLECULES

What an unusual – but very appropriate – chapter title for an academic book! It is, of course, a focus upon molecules with high vapour pressure containing nitrogen and/ or sulfur atoms. Again, Dr. Cotton surprises the reader by a lengthy – and absorbing – discourse on truffles: "The black truffle produces some 80 volatile molecules, including a lot of aldehydes, ketones and esters". Then, a digression on: the smelliest plant in the world, the titan arum *Amorphophallus titanum*. It flowers irregularly, every few years, and then just for two or three days. ... The [rotten-flesh] smell helps it to attract the kind of insects, which like to feed on decaying flesh – flies and carrion beetles – whilst its deep red inflorescence looks like meat. The titan arum and several others that produce the rotting flesh smell owe their odour to mixtures of dimethyl disulphide and dimethyl trisulphide."

CHAPTER 6: WAR AND PEACE IN NATURE

And what curiosities does the reader find in this chapter? Surely the section on plant defences is the most fascinating. As Dr. Cotton remarks: "Plants are at the bottom of the food chain. Though it sounds like confused biology to use the expression, they seem like sitting ducks. Vulnerable, yes, but they have a range of tricks up their sleeves to defend themselves against predators" – many of which Dr. Cotton describes.

Of course, insects with their mobility, have an amazing array of defensive and offensive options. One of the many amazing examples is: "Soldier termites of the Australian species *Nasutitermes exitiosus* do something even more complicated; they have been described by the authors of the book 'Secret Weapons' as 'mobile artillery units'. They fire their weapon from a gland on their heads, not their abdomens, but again it can be directed, ahead; to the sides; and even behind them".

CHAPTER 7: ORGANOCHLORINE COMPOUNDS

Dr. Cotton ends this chapter with what might be more appropriate as an opening statement: "The world contains an amazing variety of organochlorine compounds – some are 'natural', some wholly synthetic. Some of these compounds are toxic or harmful in other ways, but others are not just useful substances but quite safe into the bargain. Molecules are 'morally neutral'; they do not display their good or bad sides until they come into contact with people". This chapter opens with a lengthy and detailed coverage of the history of DDT's "rise and fall". Then amongst other compounds, he describes a large number of the fungal-derived chloro-antibiotics, where they were found, and their amazing complex structures.

CHAPTER 8: ORGANOFLUORINE COMPOUNDS

To begin this chapter, Dr. Cotton discusses in depth how, though fluorine and chlorine are in the same Periodic Table Group, difference in bond energies and electronegativities leads to some very different behaviours and properties in analogous compounds. About half the chapter is consumed by an in-context discussion of chlorofluorocarbons. Then PFOS (perfluorooctane sulfonic acid) and its family are covered, followed by fluoro-anaesthetics, and fluoro-pharmaceuticals.

CHAPTER 9: SMOKING AND VAPING

Opening this chapter is a fascinating detailed history of tobacco smoking around the world. As usual, the account is sprinkled with asides that make this book such a fascinating read: "But the practice of cigarette smoking did not catch on at once [in Britain], until the American invention (Virginia, again) of the cigarette making machine in 1880, which chopped cigarettes from a tube of paper-wrapped tobacco, and which could make up to 212 cigarettes per minute". Dr. Cotton then enters the incredibly complex world of vaping, which he tells us at the beginning: "A Chinese pharmacist named Hon Lik is often given the credit for inventing the precursor of the modern devices in 2003, as an aid to stopping smoking, after his father died of lung cancer (Spoiler: Hon Lik still smokes)".

CHAPTER 10: ISOTOPES AT WORK

Commencing with a review of the history of isotope coverage, Dr. Cotton then reviews some of the isotope relevance of hydrogen, lead, uranium, and others. Of course, we can rely on Dr. Cotton to find some fascinating information which few would know. In this case, the selling of synthetic vanilla as expensive "natural vanilla." He explains: Plants make vanillin via a biochemical pathway that results in a higher $^{13}C/^{12}C$ ratio than that found in synthetic vanillin, ... But the counterfeiters got round this by putting vanillin molecules with extra ^{13}C into their fraudulent "vanilla extract", so that their vanillin samples matched the "natural ratio". Nevertheless, the fake vanilla extract can still be identified, as Dr. Cotton explains: "Carbons outside the aromatic ring are easier to introduce; because of this chemical inequivalence between the carbon atoms in the aromatic ring and those carbon atoms that are substituent groups, the distribution of the ^{13}C atoms in 'faked' vanillin is non-uniform, with greater numbers in the aldehyde and methoxy substituent positions".

SUMMARY

This book represents a life-time of knowledge of chemistry accumulated by Dr. Cotton in the real world. This includes many more anecdotes and asides than are selected here. Sprinkled through the chapters are the chemical structures and many reaction mechanisms. For an organic chemist, these provide a greater depth of comprehension. Should the reader not be so inclined, the text alone is worth the cost of this book. The reader cannot claim to be truly knowledgeable about the chemistry of the world we live in, unless they have read this book and retained it as an ever-ready reference source.

Geoff Rayner-Canham, *F.R.S.C., F.C.I.C., Grenfell Campus,*
Memorial University Corner Brook, Newfoundland, Canada

Acknowledgements

I am very grateful to all those people who have discussed chemistry with me for many years and to those who have provided copies of papers to which I have not had access. In particular, I thank Hilary LaFoe for believing in this project and Varalika Kathuria for her skilled editorial help.

About the Author

Simon Cotton earned his BSc and PhD in chemistry from Imperial College London, followed by research and teaching appointments at Queen Mary College, London, and the University of East Anglia. He subsequently taught chemistry in both state and independent schools for over 30 years. In 2011, he became an honorary senior lecturer in chemistry at the University of Birmingham, where he taught inorganic and organic chemistry for 5 years. He has published research on the chemistry of iron, cobalt, scandium, yttrium and the lanthanide elements.

His 'Soundbite Molecules' feature ran as a regular column in the magazine *Education in Chemistry* from 1996 to 2012, reaching every secondary school in the UK. He has written over 100 'Molecules of the Month' articles, which are featured online at http://www.chm.bris.ac.uk/motm/motm.htm and recognised globally. Additionally, he has delivered over 40 'Chemistry in Its Element' podcasts for the Royal Society of Chemistry's *Chemistry World* website at http://www.rsc.org/chemistryworld/.

In 2005, he shared the Royal Society of Chemistry Schools Education Award, and in 2014, he was awarded the British Empire Medal for his work in chemistry and education.

He was the editor of 'Lanthanide and Actinide Compounds' for the *Dictionary of Organometallic Compounds* and the *Dictionary of Inorganic Compounds* between 1984 and 1997. He wrote the section on lanthanide coordination chemistry for the second edition of *Comprehensive Coordination Chemistry* and the sections on lanthanide inorganic and coordination chemistry for the first and second editions of the *Encyclopedia of Inorganic Chemistry*.

This is his ninth book. His previous books are given as follows:

D. J. Cardin, S. A. Cotton, M. Green, and J. A. Labinger, *Organometallic Compounds of the Lanthanides, Actinides and Early Transition Metals,* Chapman and Hall, 1985

S. A. Cotton, *Building the Late Mediaeval Suffolk Parish Church,* SIAH, 2019

S. A. Cotton, *Chemistry of Precious Metals*, London, Blackie, 1997

S. A. Cotton, *Every Molecule Tells a Story*, CRC Press, Boca Raton, FL, 2012

S. A. Cotton, *Lanthanide and Actinide Chemistry,* John Wiley, 2006

S. A. Cotton, *Lanthanides and Actinides,* Macmillan, 1991

S. A. Cotton and F. A. Hart, *The Heavy Transition Elements,* Macmillan, 1975

P. May and S. A. Cotton, *Molecules That Amaze Us,* CRC Press, Boca Raton, FL, 2015

1 Food

INTRODUCTION

Much of what is written about food is concerned with nutritional value, but food is, of course, much more than this. Taste sensations matter in our enjoyment of a meal. So, the core of this chapter is to consider the molecules responsible for flavour sensations in a range of foods, both cooked and uncooked.

Times have changed since the days of hunter-gatherers and in more recent times when growing and preparing food were exceptionally time-consuming.

As Harold McGee points out, until around 2 million years ago, our primate ancestors in Africa lived on plant-based diets, but a combination of changing climate and a decrease in vegetation led them to start obtaining nutrition from animal carcasses. These provided a more concentrated source of food than plants and contributed to the enlargement of the brain that formed early humans. Furthermore, meat was the concentrated foodstuff that enabled humans to spread to colder regions of Europe or Asia, where plant foods were only sporadically available – or absent altogether. As humans became active hunters, meat became the centrepiece of meals that it remains today. Some 10,000 years ago, agriculture became possible with the domestication of animals and the cultivation of grasses. Domesticated crops like wheat and barley led to population growth, as they were a very efficient way of obtaining protein for feeding people, more so than meat-based diets, which even today are the prerogative of the wealthy.

Our enjoyment of food is driven by the molecules responsible for their aroma, the compounds that stimulate our gastric juices. This chapter briefly covers key types of chemicals in foodstuffs – proteins, carbohydrates and lipids, which are the sources of the molecules that are degraded to produce the molecules that we smell. It then moves on to a selection of foods to explore those molecules that enhance the eating experience.

Two key reactions for generating flavourings in cooked food are the Maillard and Strecker degradation reactions. A French chemist named Louis Camille Maillard discovered the first-named reaction in 1912. It is responsible for the browned colouring of cooked foods as well as for flavouring molecules and involves the reaction upon heating between a reducing sugar and an amino acid. The Strecker degradation is a condensation reaction between α-amino acids and α-dicarbonyl compounds. Both are non-enzymatic reactions. The Strecker degradation breaks down amino acids into small aldehydes, such as 2-methylpropanal, 2-methylbutanal and 3-methylbutanal, as well as sulfur- and nitrogen-containing heterocyclic compounds that often contribute to flavouring. Thus, the reaction between glucose and a primary amine proceeds through many stages (1.1). Two possible α-dicarbonyl compounds are shown as products.

DOI: 10.1201/9781003658115-1

(1.1)

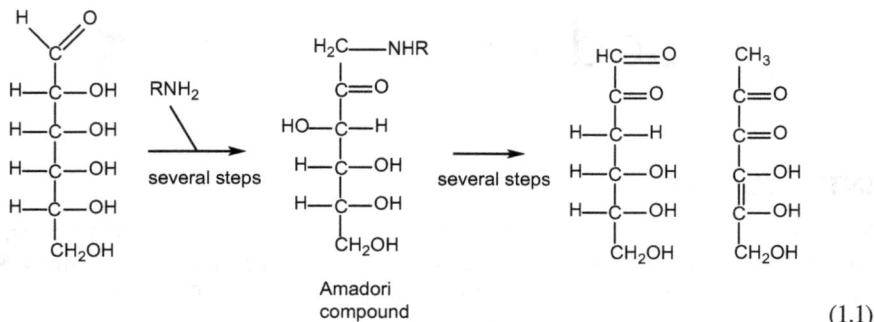

An α-dicarbonyl compound can react with an amino acid such as valine, proceeding through several steps to form a substituted pyrazine and 2-methylpropanal (isobutanal), which is one example of a Strecker degradation (1.2).

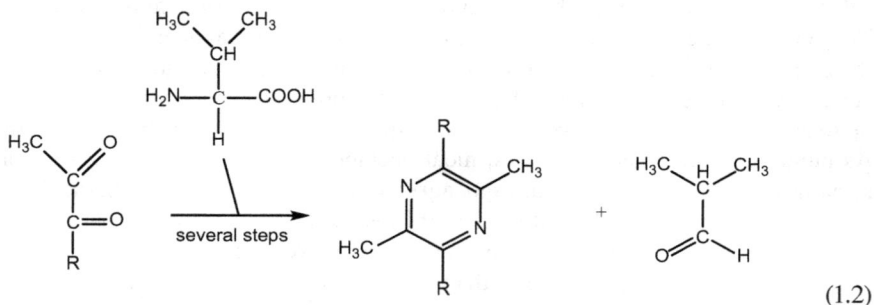

(1.2)

So where do these reactions find their reactants?

CARBOHYDRATES

Carbohydrates mean much more than 'sugar'. Besides the most important ones – glucose, sucrose, starch and cellulose – there are other important compounds. Carbohydrates are the most important energy source quantitatively through their oxidation by oxygen, directly or indirectly. Besides being energy sources, carbohydrates also serve other roles, such as sweetening and thickening agents, as well as precursors for aromas through heat-driven processes like the Maillard reaction. Non-digestible carbohydrates, such as starch and cellulose, play an important role in diet, particularly in storage.

Many important foods that contain carbohydrates are not sweet. The sugar content of foods varies considerably (data from Coultate): hard cheese 0.1%, new potatoes 1.3%, beef sausages 1.8%, white bread 2.6%, cow's milk 4.8%, onions 5.6%, apples 11.8%, bananas 20.9%, dairy ice cream 22.1%, plain chocolate 59.5%, jam 69.0%, raisins 69.3% and honey 76.4%. The large amount of D-glucose in raisins permitted its isolation by Marggraf in 1747. The amount of individual sugars varies from food to food.

Through the photosynthetic synthesis of glucose, carbohydrate biosynthesis driven by light energy supplies other food-dependent organisms with energy:

$$6\,CO_2 + 6H_2O \rightarrow C_6H_{12}O_6 + 6O_2$$

Glucose (1.3) is particularly important due to its role in both photosynthesis and respiration. Beyond that, the glucose ring is incorporated into numerous other carbohydrates; thus, the disaccharide maltose has two glucose units: sucrose combines glucose with fructose, and lactose consists of glucose and galactose joined together. Amongst polysaccharides, prominent examples include starch – both amylose and amylopectin – as well as glycogen and cellulose, which have structures based on glucose units linked together.

The body does not have an immediate use for most of the glucose it produces, so most of it is stored as glycogen, about 350 g of which is contained in the body. This is readily converted into glucose when required. The small amount of glucose in the blood (5 or 6 g) supplies the body with about 15 minutes' worth of energy through cellular respiration.

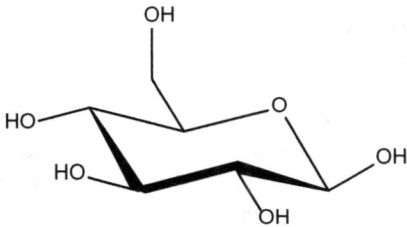

(1.3)

The glucose molecule is not flat (1.3) but is most often shown as Haworth projections, which represent the stereochemistry of the hydroxyl groups clearly. The two forms of D-glucose differ solely in the orientation of the hydrogen and hydroxy groups at carbon 1 (1.4) and (1.5).

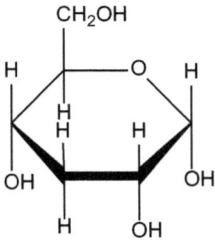

α-D-glucose

(1.4)

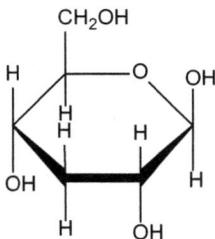

β-D-glucose

(1.5)

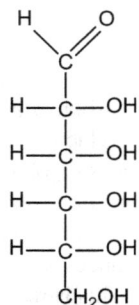

$$(1.6)$$

In solution, the two cyclic forms of D-glucose are in equilibrium with an open-chain isomer (1.6), which allows the α- and β-forms to interconvert. Its carbonyl group is in a terminal position, making it an aldose, a reducing sugar.

Sucrose is the carbohydrate that makes table sugar sweet. It is a disaccharide with an α-D-glucose molecule joined to a β-D-fructose molecule by an α-1,2-glycosidic linkage (1.7). Sugarcane, with a sucrose content of 12%–26%, is a perennial grass of the genus *Saccharum*, native to the Southern Pacific and Southeast Asia; it subsequently spread to the Indian subcontinent, where sugar was extracted from it over 2,000 years ago. Persians took it to Arabia, and returning crusaders brought it to Europe around 1100 AD. Later, Columbus brought sugarcane to the Caribbean in 1493, from where it spread into South America, leading to the growth of a sizeable industry. A sugar beet industry emerged during the Napoleonic Wars when the British blockade of the Continent restricted access to sugar cane, providing a source of sugar that could be harvested in temperate climates.

The presence of all those –OH groups in sucrose enables its water solubility and hence extraction from macerated sugar cane and sliced beet. The introduction of artificial sweeteners in recent years – including the heat-stable sucralose, which can be used in cooking – has restricted the use of sucrose; however, on the other hand, sucrose has tended to replace fats in low-fat food.

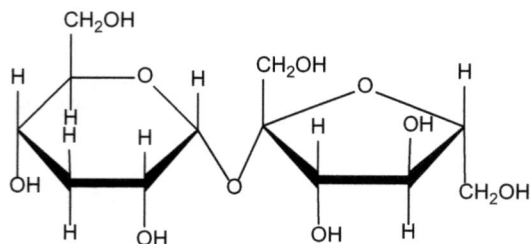

sucrose

$$(1.7)$$

Fructose is important for several reasons. It is found in many fruits, being the most abundant in some, including pears, grapes and apples. It is the principal sugar present in honey, closely followed by glucose. Fructose is around 50% sweeter than sucrose.

In crystalline form, fructose exists in a six-membered ring form (pyranose), similar to that found in glucose (1.8). Intermolecular O–H ... O hydrogen bonds between hydroxyl groups in neighbouring molecules play a role in stabilising this form in the crystal. However, when it bonds to glucose to form the disaccharide sucrose, fructose adopts a form with a five-membered furanose ring (1.9). Aqueous solutions of fructose contain a mixture of the pyranose and furanose forms.

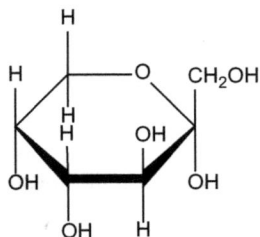

α-D-fructopyranose

(1.8)

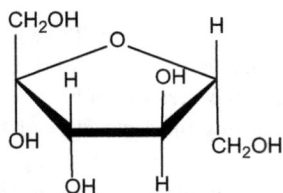

β-D-fructofuranose

(1.9)

Maltose, also known as malt sugar, is a disaccharide formed from two molecules of glucose linked by an α-1,4-glycosidic bond (1.10). Enzymes generate it in germinating cereals (such as barley), and the process is followed, in brewing, by fermentation.

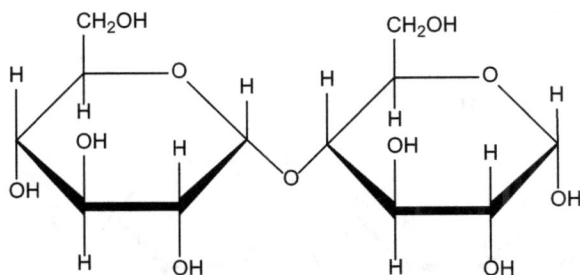

(1.10)

The disaccharide **lactose** (1.11) contains the monosaccharides galactose and glucose, joined through a β-1,4-glycosidic linkage (galactose is a diastereoisomer of glucose, differing in the configuration at carbon 4). The importance of lactose in food is due to its widespread occurrence in milk, comprising around 5% of cow's milk (the percentage in human milk is around 7%).

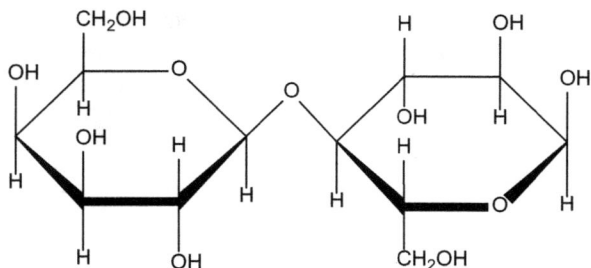

$$(1.11)$$

Most humans secrete the enzyme lactase (β-galactosidase) in the small intestine, which hydrolyses lactose to glucose and galactose, enabling its absorption. Humans who do not possess the lactase enzyme cannot digest lactose and are thus 'lactose intolerant'.

POLYSACCHARIDES

Cellulose, starch and glycogen are all polysaccharides based on glucose. They have insolubility in common but differ structurally. Polysaccharides have two roles. One is as an energy store (notably starch); the other is to provide the skeleton of both plant cells and whole plants. Starch occurs widely in foods, making up some 70% of the food consumed. Some starch contents of some key foods (Coultate) are boiled potatoes 16.7%, cassava (dry) 22.0%, white bread 46.7%, white flour 61.8%, white rice (uncooked) 73.8%, cornflakes 77.7% and maize flour 92.0%.

Starch and glycogen are based on α-glucose, with α-1,4-glycosidic links joining the glucose units. Humans secrete an amylase enzyme that can hydrolyse these links, releasing the glucose monomers.

Starch is composed of two forms of polymer. Amylose makes up about 20% of starch and has 'straight' (unbranched) chains that contain several hundred glucose units linked together. In contrast, amylopectin has branched chains with 1,6-α-glycosidic bonds at the branches.

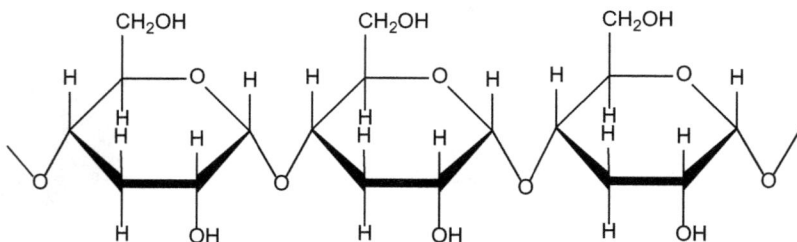

amylose

$$(1.12)$$

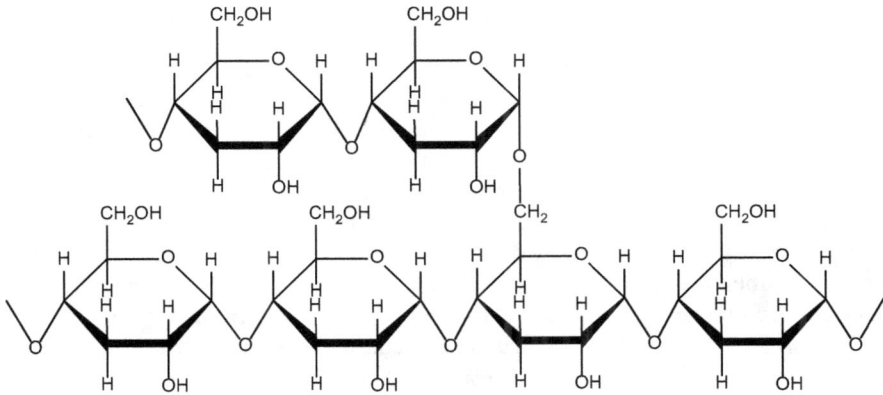

amylopectin

(1.13)

Glycogen is a polysaccharide that acts as 'stored glucose' in the human body. Its structure resembles that of amylopectin, but with more frequent branching every 8–12 glucose units, rather than every 30 units in amylopectin. The glycogen structure is therefore more readily broken down.

Cellulose, the most abundant polysaccharide, is an important ingredient of food, as it makes up the structure of plant cell walls and comprises around a third of plant matter.

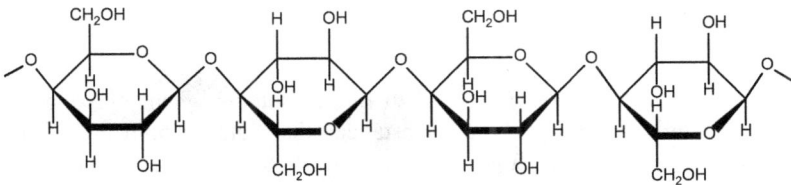

cellulose

(1.14)

Cellulose is composed of glucose units that are linked through β-1,4-glycosidic bonds, whereas the polysaccharides starch and glycogen have α-1,4-glycosidic bonds. Every other glucose molecule in the chain is flipped over, leading to 'linear' unbranched chains, which are quite different from what is found in starch or glycogen. There are hydrogen bonds between neighbouring chains, resulting in a strengthening of the structure.

Humans and some other mammals do not possess the cellulase [*sic*] enzymes that can hydrolyse the β-1,4-glycosidic links in cellulose and thus cannot break down cellulose, but ruminants such as cows (and sheep) have bacteria that contain cellulase in the rumen (part of their intestinal system) that hydrolyse cellulose to glucose (which they can then use as an energy-generating food).

Humans, however, can use cellulose in fruit, vegetables and nuts as dietary fibre (roughage), facilitating the cleaning of the large intestine. Thus, as far as humans are concerned, cellulose has dietary value but no nutritional value.

AMINO ACIDS AND PROTEINS

Proteins are another key component of foods, essential for the growth and repair of the body. The body uses dietary proteins as a source of α-amino acids, which it reassembles differently to make the specific proteins required by the body. Like other living systems, the human body has to synthesise new protein all the time, both for its growth and to replace those molecules that have been broken down, by bacteria, for example. Animals eat protein in their diet, which they break down in their stomachs by hydrolysis reactions; they do this to obtain the amino acids required for *their* protein synthesis. These amino acids are converted into other amino acids and then reassembled into new proteins in the liver, the body's chemical factory.

The protein content of foods varies over a considerable range, from 1.3% in milk, 2.6% in rice and ~9% in bread, to 17.4% in cod, 20.3% in beef and 20.8% in chicken (Coultate). The content in cheese depends upon the type, with examples being 19.3% (Brie), 25.5% (Cheddar) and 39.4% (Parmesan).

Around 15% of the human body is composed of protein, whether in the skin, nails, hair, muscle or haemoglobin (the oxygen carrier in red blood cells) or insulin, the hormone found in the blood. Proteins are large molecules, with molecular masses ranging between around 10,000 and several million. Proteins are assembled from α-amino acids, which are the 'building blocks of proteins'. They contain carbon, hydrogen, nitrogen and oxygen atoms; some amino acids also contain sulfur.

Plants synthesise all the amino acids they need. Animals need plants to survive, as plants supply them with amino acids, several of which they cannot produce themselves.

Amino Acids

These are used as the "building blocks" to make proteins when joined by peptide links. Generally, amino acids are substances that contain both NH_2 (amine) and $-CO_2H$ (carboxylic acid) functional groups. Proteins are constructed from alpha-amino acids, which have the general structure $H_2N-CH(R)-CO_2H$. With the exception of glycine (where $R = H$), they contain an asymmetric carbon atom, allowing them to exist as two enantiomers. In practice, the amino acids that occur in nature are obtained as just one isomer, as shown below for α-alanine (1.15).

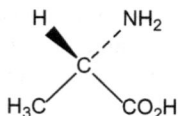

(1.15)

About 21 of these are 'natural' amino acids used in making proteins (over 500 amino acids are known). Nine of these cannot be made by organisms: valine, leucine, isoleucine, methionine, phenylalanine, tryptophan, threonine, histidine and lysine, and thus, they are 'essential' amino acids. A further six cannot *always* be made: arginine, cysteine, proline, glycine, glutamine and tyrosine; whilst six more can be made in the body: alanine, aspartic acid, asparagine, glutamic acid, serine and selenocysteine. Only the natural amino acids are found in the genetic code.

Table 1.1 shows 19 naturally occurring amino acids, with three further structures shown in (1.16)–(1.18).

Amino acid	R	Symbol
Alanine	CH_3	Ala
Arginine	$(CH_2)_3NHC(=NH)NH_2$	Arg
Asparagine	CH_2CONH_2	Asn
Aspartic acid	CH_2CO_2H	Asp
Cysteine	CH_2SH	Cys
Glutamic acid	$CH_2CH_2CO_2H$	Glu
Glutamine	$CH_2CH_2CONH_2$	Gln
Glycine	H	Gly
Isoleucine	$CH(CH_3)(CH_2CH_3)$	Ileu
Leucine	$CH_2CH(CH_3)_2$	Leu
Lysine	$(CH_2)_4NH_2$	Lys
Methionine	$(CH_2)_2SCH_3$	Met
Phenylalanine	$CH_2C_6H_5$	Phe
Serine	CH_2OH	Ser
Threonine	$CH(OH)CH_3$	Thr
Tyrosine	$CH_2C_6H_4OH$	Tyr
Valine	$CH(CH_3)_2$	Val
Histidine	1.16	His
Proline	1.17	Pro
Tryptophan	1.18	Try

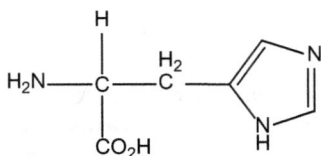

Histidine

(1.16)

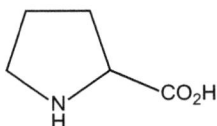

Proline

(1.17)

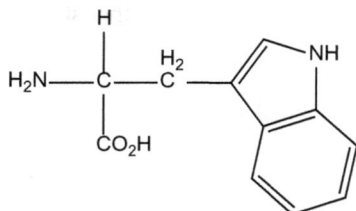

Tryptophan   (1.18)

In proline, the amine group is a secondary group, not primary, as it is in the other alpha-amino acids.

POLYPEPTIDES AND PROTEINS

With the help of an enzyme catalyst, the $-NH_2$ group of one amino acid molecule can react with the $-COOH$ group of another amino acid in a condensation reaction (eliminating a water molecule), forming a dipeptide (1.19) via an amide or peptide link ($-CO-NH-$).

$$H_2N-\square-COOH + H_2N-\blacksquare-COOH \rightarrow H_2N-\square-CO.NH-\blacksquare-COOH + H_2O$$

$\square$ or $\blacksquare$ = 'skeleton' of an amino acid

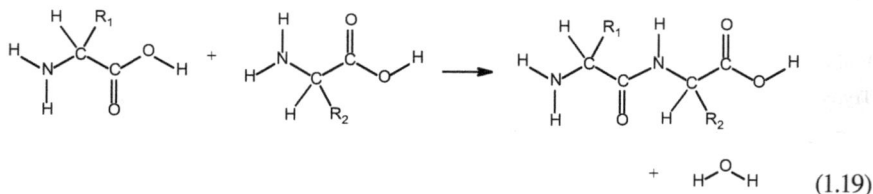

(1.19)

The process can be repeated at each end of the molecule; since there are some 20 natural amino acids, there is in principle a virtually infinite range of possible combinations and sequences. Join three amino acids together, a tripeptide is formed; join four amino acids together, a tetrapeptide is formed; and so on. Thus, a nonapeptide is made by joining nine amino acids together, the best example of a nonapeptide being the peptide hormone oxytocin, which is involved in childbirth (see *Every Molecule Tells A Story* pp. 51–52). Once assembled into a peptide, amino acids are now referred to as 'amino acid residues'.

There comes a point when the molecule is referred to as a polypeptide, though there is no precise definition of the molecular size involved; polypeptides are usually taken to be made of somewhere between 20 and 50 amino acids, all joined together by peptide links. Above that, the molecule is described as a protein.

LIPIDS

Lipids are a diverse group of substances comprising oils and fats, as well as the phospholipids found in cell membranes. They are insoluble in water but soluble in non-polar solvents. The most common type of lipid, for example, in 'body fat', is triglycerides, triesters of glycerol with long-chain carboxylic acids, in which all three of the –OH (alcohol) groups have been esterified. A triester of glycerol with stearic (C_{18}) acid, which is a saturated fat, is shown in (1. 20).

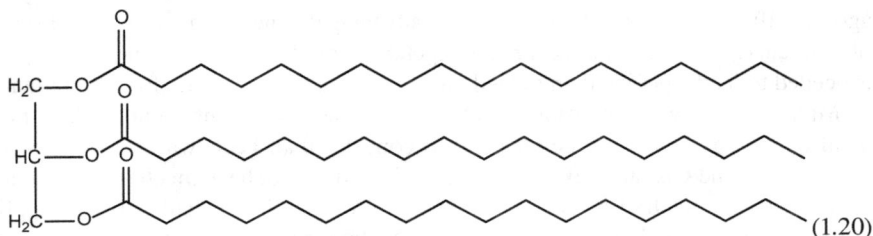

(1.20)

A triester of glycerol with three different acids – from top to bottom, palmitic acid (saturated), oleic acid (unsaturated) and α-linolenic acid (unsaturated) – is shown in (1.21). This is an unsaturated fat.

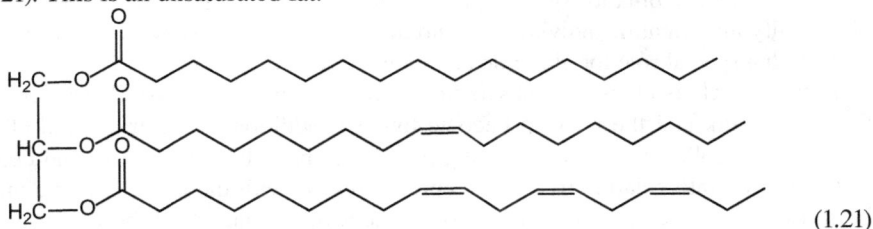

(1.21)

Whilst saturated fats are often solids at room temperature, unsaturated fats have lower melting points than saturated fats of similar molecular size and are frequently liquids.

A third kind of lipid with a similar structure is phospholipids (also known as phosphoglycerides); they are found particularly in cell membranes. The molecules are based upon a glycerol molecule, in which two hydroxyl groups have been esterified by long-chain carboxylic acids (just as in triglycerides), but the third OH group is esterified with the ester of phosphoric acid (1.22). This affects the properties of the phospholipids, as the phosphate ester group is polar. The phospholipid can be described as having a polar head (the phosphate ester part) and a non-polar tail (the fatty acid ester part).

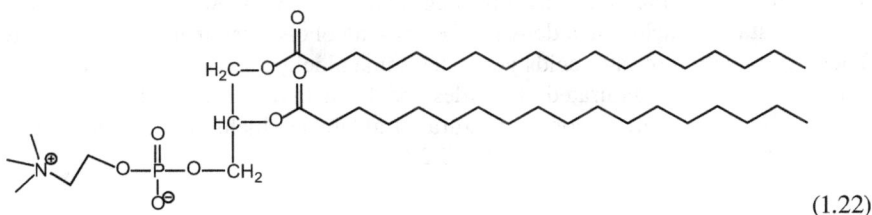

(1.22)

TASTE SENSATIONS

Food is, of course, much more than its nutritional value. Taste sensations matter in our enjoyment of a meal. The core of this chapter considers the molecules responsible for our sensations of a range of foods, both cooked and uncooked.

MEAT

As Harold McGee pointed out, primates were vegetarians until around 2 million years ago. A shift to a meat-based diet gave them advantages, since, as a concentrated source of both energy and protein, meat beats vegetables; on the other hand, much less grain is needed to feed a person than to feed an animal that then becomes food for humans.

Many people would consider meat to be an essential component of their main meal of the day, and what better meat to consider than beef, with the odours that contribute to the gustatory experience? Raw beef does not have much of a smell (nor, for that matter, much of a taste) as it contains few volatile molecules, but it supplies the large molecules that act as precursors for the small odorant molecules, namely proteins and carbohydrates. After the animal has been killed, enzymes are involved in processes like the breakdown of protein into peptides and individual amino acids. Glycogen gets degraded to glucose.

Flavour is a combination of taste plus aroma, with taste composed of sweet, sour, bitter, salty and umami. Individual amino acids contribute to sweetness and bitterness, with a special role for glutamic acid in umami (savouriness).

Proteins such as myosin in muscle fibres are denatured above 40°C, causing the fibres to shrink and the meat to become tougher; additionally, from around 60°C upwards, the collagen in tissues joining muscles to bone becomes softer and denatures, turning into gelatin. To achieve optimal results with the meat, cooking must be carried out at a temperature that balances the relative amounts of these two types. Over 1,000 different molecules have been identified as arising from cooked meat, but only a few of them have a major "impact" and are responsible for a beefy smell. Differing reaction conditions lead to different combinations of reaction products.

Two main reactions lead to the smell of cooked meat. Above 140°C, Maillard reactions involve a series of steps that begin with the condensation between the carbonyl group of a reducing sugar and a free amino group, leading to organic volatiles containing nitrogen, sulfur and oxygen. These reactions also turn the outside of the meat brown. Cysteine degradation is important in contributing to sulfur compounds in meat flavour, whilst methionine is the source of methional (1.23). Some of these compounds appear elsewhere. For example, 2-acetylpyrroline (1.24), with its sweet roasted smell, is also found in popcorn and cooked rice and is formed by Maillard degradation of proline. Secondly, lipid degradation, from below 100°C upwards, produces substances including aldehydes, ketones, alcohols, carboxylic acids and esters. Thermal oxidation of fatty acids generates some aldehydes contributing to meat flavour, such as C_6–C_{10} saturated aldehydes and 12-methyltridecanal (1.25), which has a beefy smell, as well as some unsaturated aldehydes including 2-nonenal (1.26), 2-undecenal and (E,E)-2,4-decadienal (1.27).

Well over a dozen compounds have been described as odour-active compounds in cooked beef. They include aldehydes including the saturated aldehydes 2- and 3-methylbutanal (pungent, green, sweet, roasty), hexanal, heptanal (green) and octanal (fruity, green), methional (1.23), 3-(methylthio)propanal (cooked potato), 2-acetylthiazole ((1.28), roast), 2-octenal (fruity, fatty, tallowy), (E)-2-nonenal (tallowy, fatty), (E, E)-2,4-decadienal (deep fried, fatty), 3-mercaptopentan-2-one ((1.29), beefy), 2,3-diethyl-5-methylpyrazine ((1.30), earthy, roasty, meaty) and 2,4-nonadienal (fatty). These contribute 'notes' including 'green', 'earthy', 'tallowy', 'fatty', 'meaty' and 'roast' to the overall 'cooked beef' smell. Many of them also contribute to the smell of cooked pork.

Roast beef seems to be the cooked meat that has been most investigated. Key odorants contributing to the 'roast beef' aroma include methional, 2-acetyl-thiazole, 2-acetylthiazolidine (1.31), 2-ethyl-3,5-dimethylpyrazine (1.32) and 2,3-diethyl-5-methylpyrazine (1.30). Compounds identified as significant contributors to the smell of fried beef include methional, phenylacetaldehyde, 2-ethyl-3,6-dimethylpyrazine, 2-ethyl-3,5-dimethylpyrazine and 2-propyl-3-methylpyrazine.

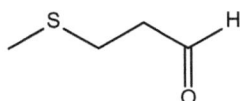

(1.23)

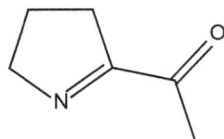

(1.24)

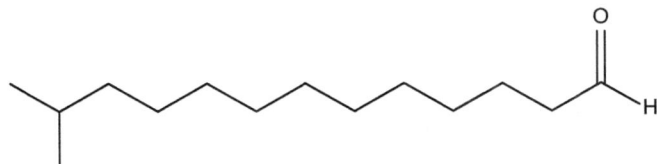

(1.25)

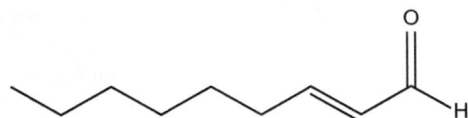

(1.26)

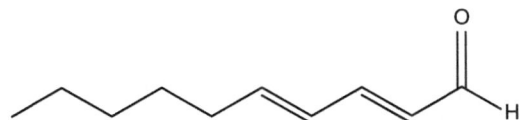

(1.27)

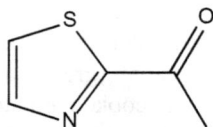

(1.28)

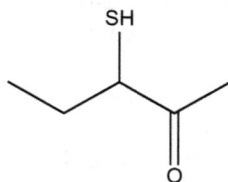

(1.29)

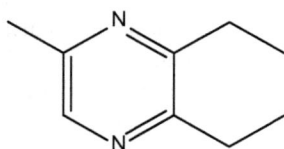

(1.30)

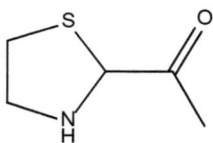

(1.31)

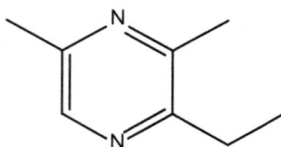

(1.32)

Less study has been made of other meats. 2-methyl-3-furanthiol (1.33), generated by the reaction of ribose with cysteine, is believed to be the most important compound in chicken flavour. Amongst a number of aldehydes, (E, E)-2,4-decadienal (1.27) is considered an important odorant in cooked chicken because of its low odour threshold.

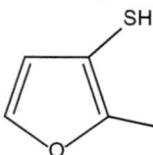

(1.33)

Because of the higher cooking temperature, the flavour of fried chicken brings in other molecules, including trithiolanes such as 3,5-dimethyl-1,2,4-trithiolane (1.34), 3,5-diisobutyl-1,2,4-trithiolane (1.35), 3-methyl-5-butyl-1,2,4-trithiolane (1.36) and 3-methyl-5-pentyl-1,2,4-trithiolane. These result from processes such as the thermal degradation of cysteine. Alkyl pyrazines like 2,6-dimethylpyrazine (1.37) contribute nutty and earthy notes to fried and roasted chicken flavours, as do substituted pyridines. They provide notes such as 2-pentylpyridine, which has a strong fatty and tallow-like odour, and 2-isobutyl-3,5-diisopropylpyridine (1.38), described as having a roasted cocoa-like aroma. These heterocycles result from Maillard reactions.

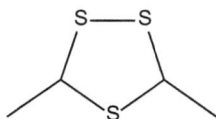

(1.34)

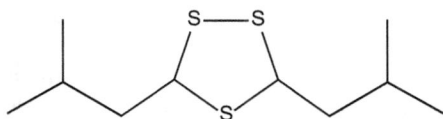

(1.35)

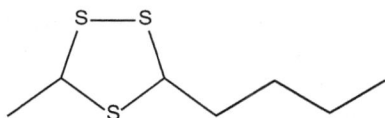

(1.36)

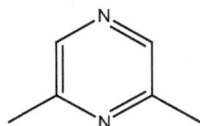

(1.37)

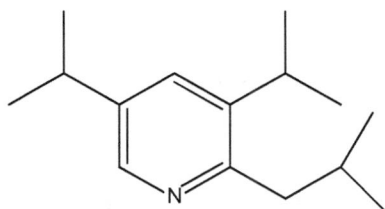

(1.38)

CHEESE

Cheese is loved all over the world. Harold McGee said: "Cheese is one of the great achievements of humankind". Everyone has their favourite. I can recall a memorable Saint-Nectaire one Parisian August afternoon in a café in the Louvre, though my tastes normally lean toward blue cheese, whether from the limestone caves of Roquefort or an English Stilton.

For thousands of years, cheese has provided humans with a foodstuff rich in proteins and fat, which supplies vitamins and minerals like calcium, as well as essential amino acids. Cheese was popular in the time of the ancient Romans, whilst around 800 AD, Emperor Charlemagne is said to have been partial to a blue cheese resembling Roquefort.

Each cheese has a particular flavour, and its odorants depend on the chemistry occurring as it matures. Most cheeses are made starting with milk from cows, which is usually first pasteurised by heating (e.g. 70°C for half an hour), killing off any dangerous pathogenic bacteria present, and then cooled. Now "starter" bacteria (usually from the *Lactococcus, Streptococcus* and *Lactobacillus* families) and rennet are added, after which the mixture is digested for around an hour at 30°C–40°C. During this time, the starter bacteria ferment lactose, generating lactic acid and causing the pH to fall to around 4.6, at which point the protease enzyme chymosin (also known as rennin) coagulates the casein, the principal protein in the milk, forming curds. After setting, the curds are separated from the liquid (whey).

Cottage cheese is made from freshly drained cheese curds. These are quite large molecules, and there has not been enough time for bacteria to break them into smaller molecules that are small enough to be smelled, which is why cottage cheese is tasteless. So where do these flavours come from?

Cheese is made from milk, which contains three classes of chemicals: the protein casein, built from amino acids; lipids present in milk fat; and the carbohydrate (sugar) lactose. It is their decomposition that creates the smaller molecules responsible for the odour of the cheese made from that milk – to be smelled, molecules have to have a molecular mass of around 300 or less. The breakdown of the molecules is the work of enzymes in bacteria – different enzymes produce different odorants. Some of these are the 'starter' bacteria added to initiate fermentation, which are particularly responsible for creating the flavours of hard cheeses like Cheddar, Cantal and Salers as they ripen by breaking down proteins.

However, many other cheeses have different bacteria added. In addition to the starter bacteria, the great blue cheeses such as Gorgonzola, Stilton and the French blues from the Auvergne and Roquefort have the blue–green *Penicillium roqueforti* mould added; after ripening for a few days, their sides are pierced to facilitate the access of oxygen, which enables the β-oxidation of the carboxylic acid side-chains (see below). To facilitate the ripening process on the surface, cheeses like Camembert are given a coating of *Penicillium camemberti* mould; similarly, a smear of *Brevibacterium linens* bacteria is applied to the surface of cheeses such as Limburger and Brick.

The addition of salt draws water out of cells, controlling the growth of microorganisms and thus helping to preserve the cheese – as well as serving as a flavouring.

As mentioned earlier, bacteria decompose three 'starting materials' in milk to create the flavour molecules:

1. Lipids in milk fat.
2. A protein named casein
3. The carbohydrate lactose

In lipolysis, esters, which are formed from long-chain carboxylic acids and the triol glycerol (propane-1,2,3-triol), are decomposed back into their constituents by the lipase enzyme (1.39). This is a hydrolysis reaction known as saponification.

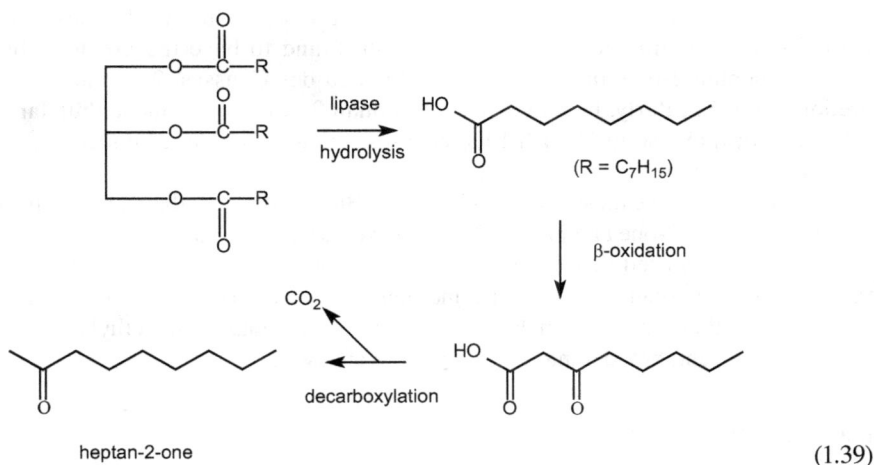

heptan-2-one   (1.39)

The carboxylic acid molecules can then undergo oxidation at the next-but-one carbon to the carboxylate group (β-oxidation), followed by the loss of a CO_2 molecule (decarboxylation) to form alkan-2-ones (1.39). The acids can also react with small alcohol molecules (esterification), forming pleasantly smelling esters.

BLUE CHEESES

Lipolysis and decarboxylation are most noticeable in forming the important odorants in blue cheeses, the alkan-2-ones. They are often called 'methyl ketones' because the methyl group is next to the carbonyl functional group. Heptan-2-one (1.40) and nonan-2-one (1.41) are the ketones that have a 'blue cheese' smell. So, how do they come about and why do they generally have an odd number of carbon atoms?

(1.40)

(1.41)

The natural carboxylic acids used to make the glyceryl triesters found in milk have an *even* number of carbon atoms; the reason for that is that they are created by biosynthesis from CH_3CO units, from acetyl coenzyme A. The free acids are liberated by lipolysis, and then, they undergo decarboxylation, *losing a CO_2 molecule*, which forms an *odd-carbon molecule*. (In fact, the R groups in the lipids generally have a lot more than 10 carbon atoms, so they have to undergo multiple oxidations to produce the C_7 heptan-2-one and C_9 nonan-2-one.)

The different blue cheeses vary in their flavour makeup. In French blues, whilst methyl ketones dominate, Roquefort has been found to have the greatest abundance of heptan-2-one and nonan-2-one. In Bleu des Causses, 2-pentanone was predominant. Similarly, heptan-2-one and nonan-2-one are the most abundant in Gorgonzola; likewise in Danish Blue cheeses, though sometimes undecan-2-one is equally abundant.

Heptan-2-one is the most abundant ketone in Blue Stilton, but there are significant amounts of butan-2-one and pentan-2-one, as well as nonan-2-one.

It should be realised that these alkanones are not the only important odorants in these cheeses; although heptan-2-one and nonan-2-one are key impact molecules in Gorgonzola, other molecules including 1-octen-3-ol, 2-heptanol and ethyl hexanoate are also important, broadening the smell of the cheese.

LACTOSE AND CAMEMBERT

Another key material involved in generating flavourings is the carbohydrate lactose, particularly in Camembert and Brie, which are surface mould-ripened cheeses. First, the disaccharide lactose is split into the monosaccharides glucose and galactose, which in turn are decomposed into lactate (1.42). The surface mould of *Penicillium camemberti* is responsible for further breaking down lactic acid into carbon dioxide and water. The pH on the surface of the cheese increases from 4.6 to 7 as the acid is removed, leading to the migration of calcium phosphate from the interior of the cheese to its surface, whilst the micelles of casein separate, causing the centre of the cheese to soften.

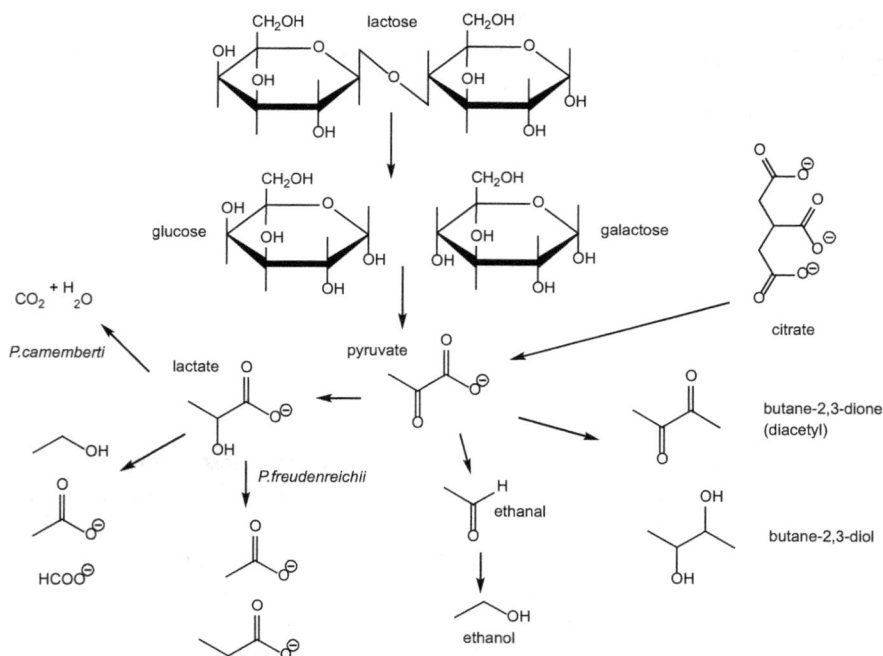

(1.42)

Gruyère and Emmental cheeses owe their holes to the fermentation of lactate into ethanoate, propanoate and gaseous carbon dioxide, the latter migrating and accumulating to form the holes or 'eyes' in the cheese.

$$3 \text{ lactate} \rightarrow 2 \text{ propanoate} + 1 \text{ ethanoate} + 1 \text{ carbon dioxide}$$

Camembert cheese contains a wide range of odorants. The buttery-flavoured butane-1,3-dione (diacetyl) is formed from citrate, as does ethanol. Carboxylic acids such as 3-methylbutanoic acid are derived from amino acids and also from the lipolysis of esters. Acids and alcohols can then react to form esters, which have stronger odours. Hydroxy acids can undergo intramolecular esterification to form internal esters or lactones, which is how molecules such as 2-undecalactone and γ-decalactone become Camembert flavourings. Ammonia is another product of the degradation of amino acids, a reaction known as the deamination of amino acids during proteolysis. This ammonia can sometimes be smelled on a ripe Camembert cheese.

(1.43)

An important breakdown process of proteins is known as proteolysis. This process breaks down proteins (like casein) into smaller chains, peptides and then into individual amino acids. These amino acids contribute their individual tastes to cheeses, but they also serve as starting points for other molecules. This is achieved through several reactions: decarboxylation (loss of a CO_2 from a CO_2H group), deamination (loss of a $-NH_2$), oxidation and reduction. These reactions generate small volatile molecules.

The preceding scheme uses valine as an example (1.43). A transaminase enzyme catalyses the loss of an *amine* group, generating a keto acid; this, in turn, can undergo further change to 2-methylpropanal (in this example), which can either be reduced or oxidised to the corresponding alcohol or carboxylic acid, respectively. Other enzymes convert valine into an amine or a different carboxylic acid. All of these products have their own smells.

Such reactions occur with other amino acids. Thus, another example of an amino acid producing these small odorant molecules is leucine (1.44). Leucine can be converted into a keto acid by transamination, using an aminotransferase enzyme; then the keto acid can be decarboxylated, using a different enzyme, forming the aldehyde 3-methylbutanal, which can be reduced to 3-methylbutan-1-ol or oxidised to 3-methylbutanoic acid. 3-methylbutanal supplies malty and nutty notes to several cheeses. It is amongst the odorants in Gruyère cheese, along with other aldehydes, including methional and 2-methylbutanal, as well as the carboxylic acids butanoic, 2- and 3-methylbutanoic and phenylethanoic acid.

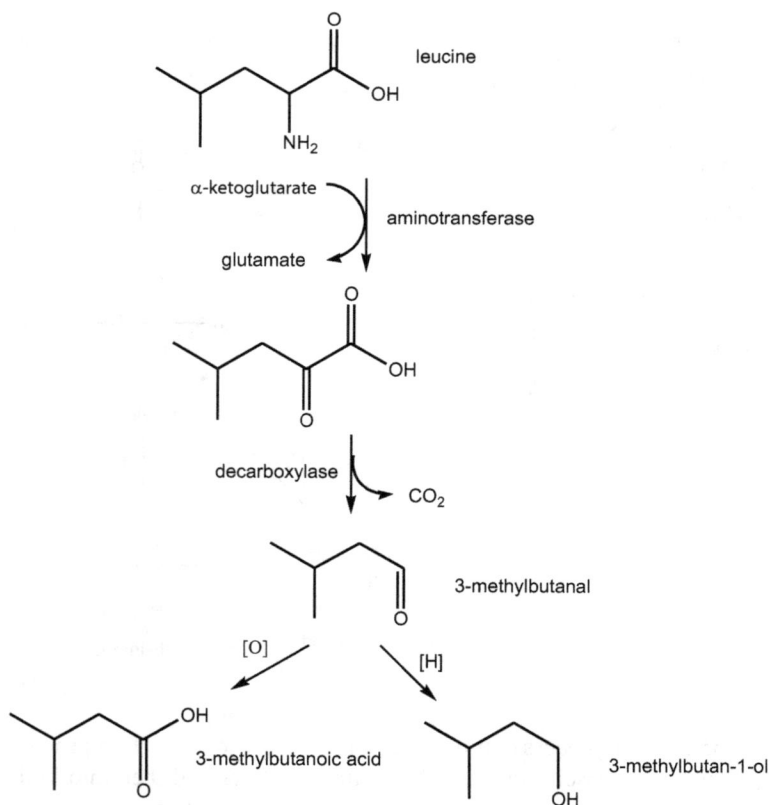

(1.44)

Of the two natural sulfur-containing amino acids, methionine acts as a source of several sulfur-containing flavour molecules in cheese; they are formed through catabolism by enzymes like *Geotrichum candidum* and *Penicillium camemberti*. Methional (1.23) is best known for giving its smell to boiled potatoes (page 30) but also contributes to the smell of both Cheddar and Camembert. Methional can also be converted into methanethiol, CH_3SH, which can be further transformed into other smelly sulfur compounds like $(CH_3)_2S$, $(CH_3)_2S_2$ and $(CH_3)_2S_3$. One particular sulfur compound, the thioester ethyl 3-mercaptopropionate, $CH_3CH_2C=O(SCH_3)$, has a distinct Camembert smell.

CHEDDAR

Cheddar is the most popular cheese in the world. In common with some other hard cheeses such as Cantal and Salers, its flavour molecules are created over several months, largely by enzymes, mainly in *Lactobacilli* that they contain. Its flavour is a blend of numerous molecules; there is no single kind of 'impact' molecule as there is in blue cheeses. The main odorants belong to families whose formation has already been covered, such as carboxylic acids, aldehydes and esters (1.45).

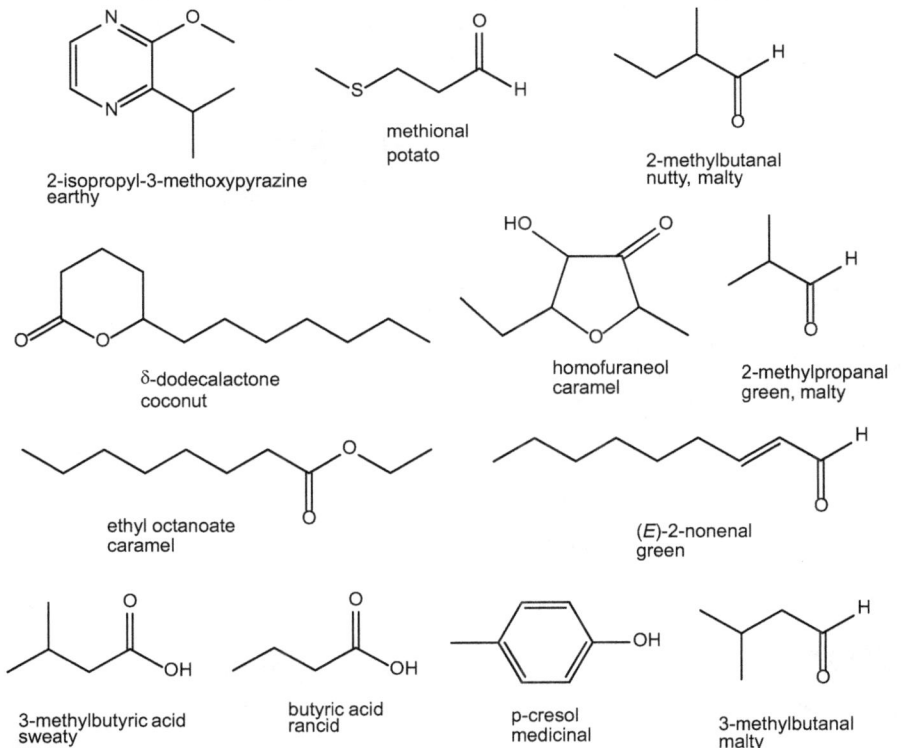

2-isopropyl-3-methoxypyrazine
earthy

methional
potato

2-methylbutanal
nutty, malty

δ-dodecalactone
coconut

homofuraneol
caramel

2-methylpropanal
green, malty

ethyl octanoate
caramel

(E)-2-nonenal
green

3-methylbutyric acid
sweaty

butyric acid
rancid

p-cresol
medicinal

3-methylbutanal
malty

(1.45)

In summary, although there are just three types of starting material from which they are made, there is a wide variety of flavours amongst different cheeses, due to the large number of volatile odorants formed by friendly bacteria and the wide range of reactions that occur.

BREAD

Cereal crops – notably wheat, rye, barley, oats, maize and rice – have been a basis of the human diet worldwide for thousands of years, but wheat, notably *Triticum aestivum*, stands out as the basis of flour for bread-making. After milling the wheat to convert it into flour, water, salt and the indispensable yeast are added; the mixture is kneaded, and the dough develops; yeast gets to work, with fermentation forming carbon dioxide; then, the 'expanded' dough is baked.

The protein essential for forming dough is 'gluten'; it is made up of several molecules collectively known as gliadins and glutenins. Once the flour has had water added, the gluten is hydrated by the added water, forming the dough. At the molecular level, the polypeptide chains in gliadin and glutenin align, forming a hydrogen-bonded network that stiffens the dough. When the dough is kneaded, pressed, stretched and folded, the network is strengthened. Uncoiling of proteins helps align the large molecules into parallel sheets, facilitating the trapping of CO_2 gas, which is important to the texture of the baked loaf (Figure 1.1)

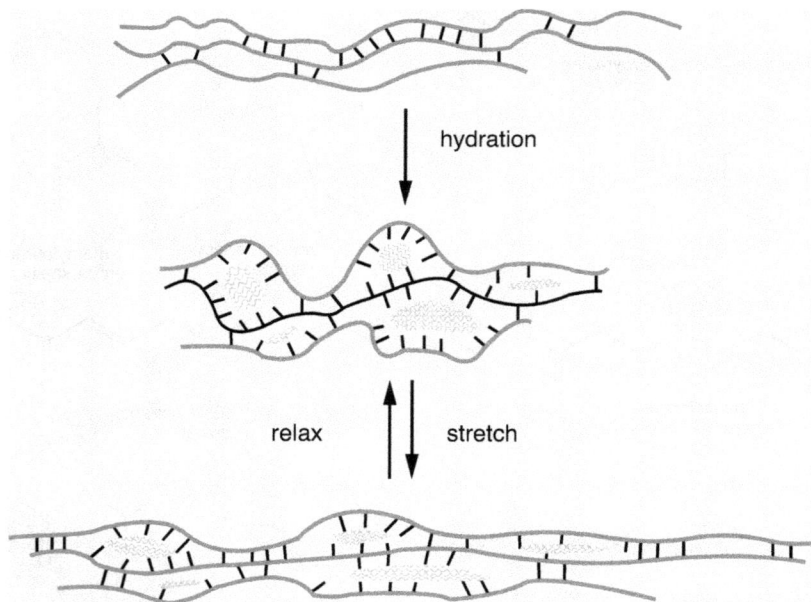

FIGURE 1.1 Role of water in the alignment of polypeptide chains in the formation of dough. From Bryan Reuben and Tom Coultate, "On the rise", *Chemistry World*, October 2009, 54-57.

Starch has more than one role beyond being a source of energy. For one thing, it helps break up the gluten network, tenderising the material.

Starch is made of two different giant carbohydrates: amylose, with its unbranched chain structure, and amylopectin, with a branched-chain structure. Both need to be broken down before they can be fermented, and these reactions require several enzyme catalysts. Both amylose and amylopectin are hydrolysed using amylase (found in wheat kernels) with water as the other reactant, generating the disaccharide maltose (1.46).

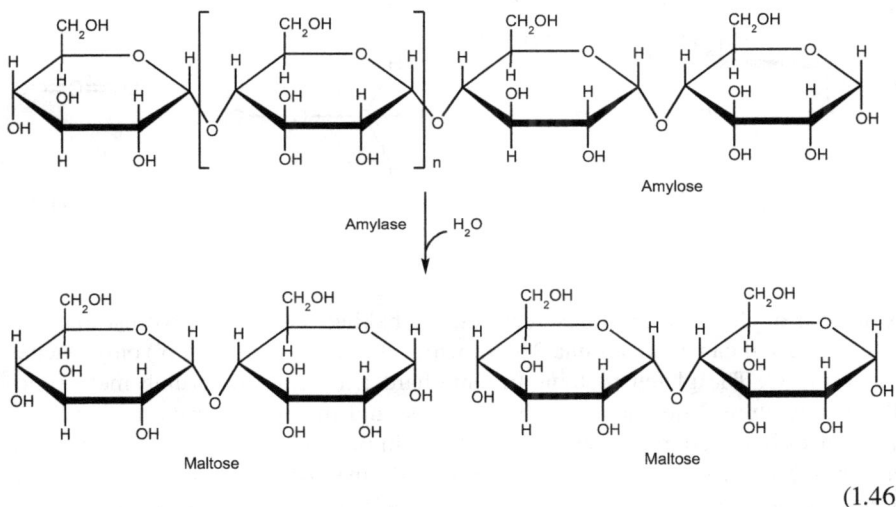

Amylose

Amylase | H_2O

Maltose

Maltose

(1.46)

Next, a different enzyme, maltase (from yeast), catalyses the hydrolysis of maltose into two molecules of the monosaccharide glucose (1.47).

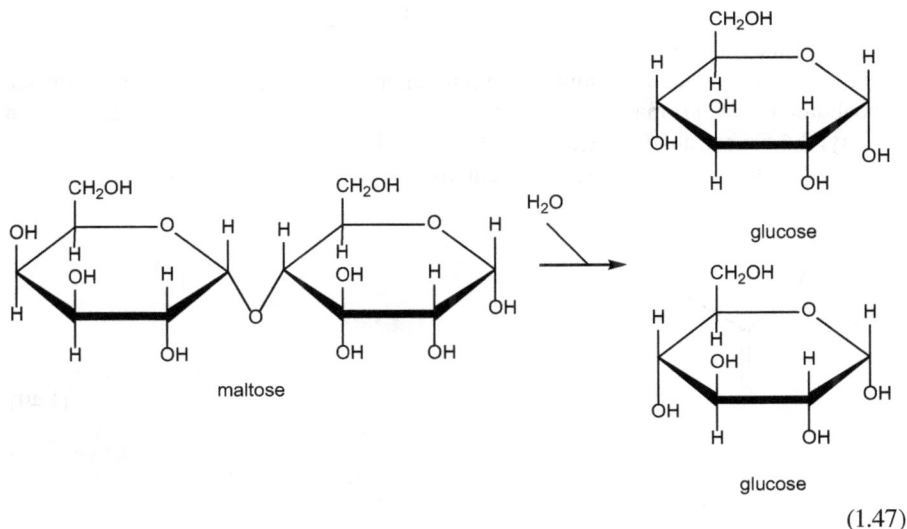

glucose

H_2O

maltose

glucose

(1.47)

Lastly, a third enzyme in yeast called zymase catalyses the fermentation of glucose; each glucose molecule yields two molecules of ethanol and two of carbon dioxide, the gas that causes the 'rise' in the bread (1. 48). Glucose also acts as a source of flavour molecules during baking and participates in the caramelisation of the crust.

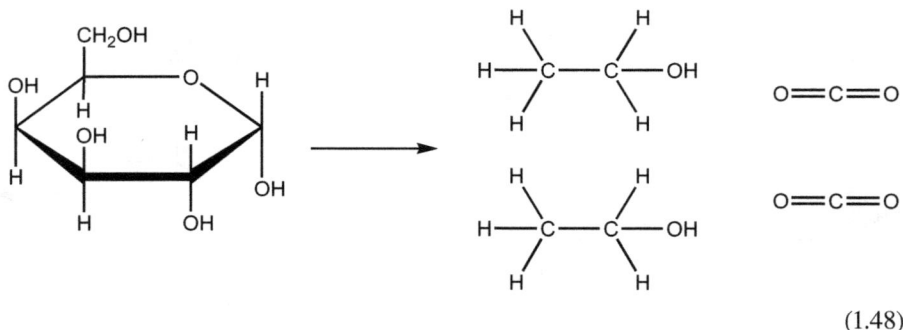

(1.48)

BAKING

A number of chemical reactions occur during baking. Typically, the crust (the outside) of the loaf can attain around 230°C, whilst the centre (the 'crumb') only reaches about 100°C. The gluten proteins become harder, making the crumb more rigid. The temperature difference between the crust and the crumb leads to significant differences in the odorant molecules produced in the different regions. In the cooler crumb, the main reactions taking place are enzyme-controlled, such as fermentation and oxidation of lipids. This leads to most of the small molecules formed being oxygen-containing molecules like alcohols, aldehydes, ketones, carboxylic acids and esters. In the crust, higher temperatures, when enzymes do not work, cause other pathways, such as Maillard reactions, thermal degradation and caramelisation. The reaction products here are often nitrogen-containing heterocycles, such as pyrrolines, pyridines and pyrazines.

The smell of bread is a composite of a number of molecules. The most important contributor to the aroma of the crust is 2-acetylpyrroline (1.49), with the similar 2-acetyl-3,4,5,6-tetrahydropyridine (1.50) also being important. 3-methyl-butan-1-ol (1.51) is thought to be the most important volatile compound in the crumb.

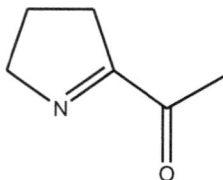

(1.49)

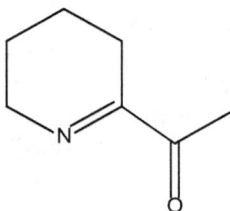

(1.50)

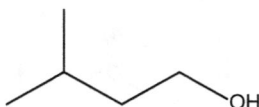

(1.51)

The smell of 2-acetylpyrroline has been variously described as corn chip, roasty, popcorn and bread. Apart from white bread and white bread toast, as well as baguettes, 2-acetylpyrroline is a key contributor to the smell of cooked rice and is also found in popcorn. It also contributes to the smell of roast beef and cooked tails of lobster and crayfish. Also found in popcorn as well as bread, 2-acetyl-3,4,5,6-tetrahydropyridine is a homologue of 2-acetylpyrroline and has a similar smell to 2-acetylpyrroline; both these molecules have very low odour thresholds.

2-acetylpyrroline is formed from the α-amino acid proline through a reaction with 2-oxopropanal, which is derived from the reducing sugar sucrose through a number of steps (1.52).

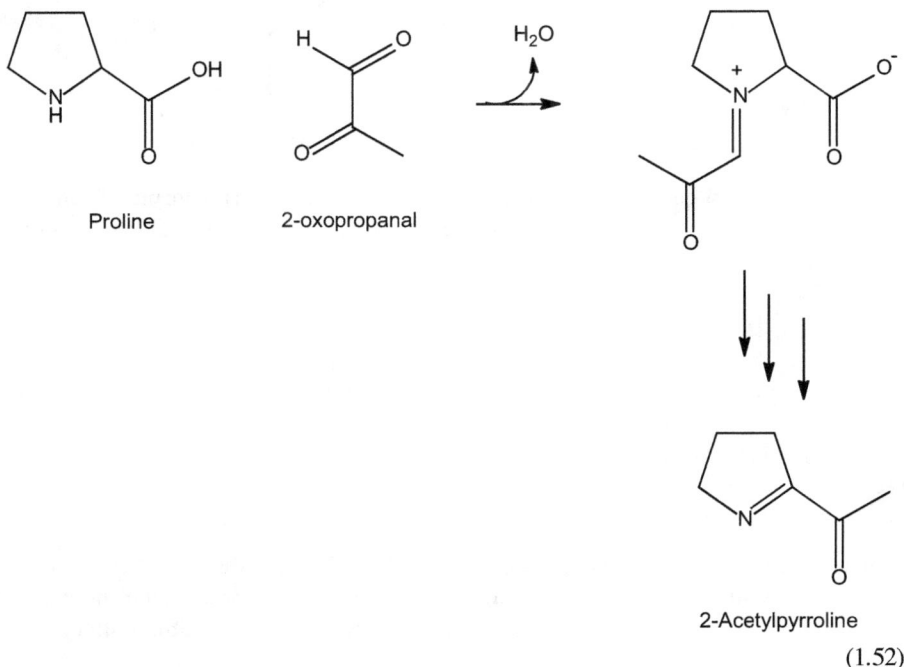

Proline   2-oxopropanal

2-Acetylpyrroline

(1.52)

A number of aldehydes found in bread originate in -amino acids. An amine transferase enzyme changes L-isoleucine into a keto acid through a transamination reaction, where the C–NH$_2$ is swapped with a C=O group in another acid. The keto acid loses a CO_2 in a decarboxylation, forming the aldehyde 2-methylbutanal (1.53). Similar reactions convert leucine into 3-methylbutanal and valine into 2-methylpropanal; all three aldehydes by themselves have 'malty' smells.

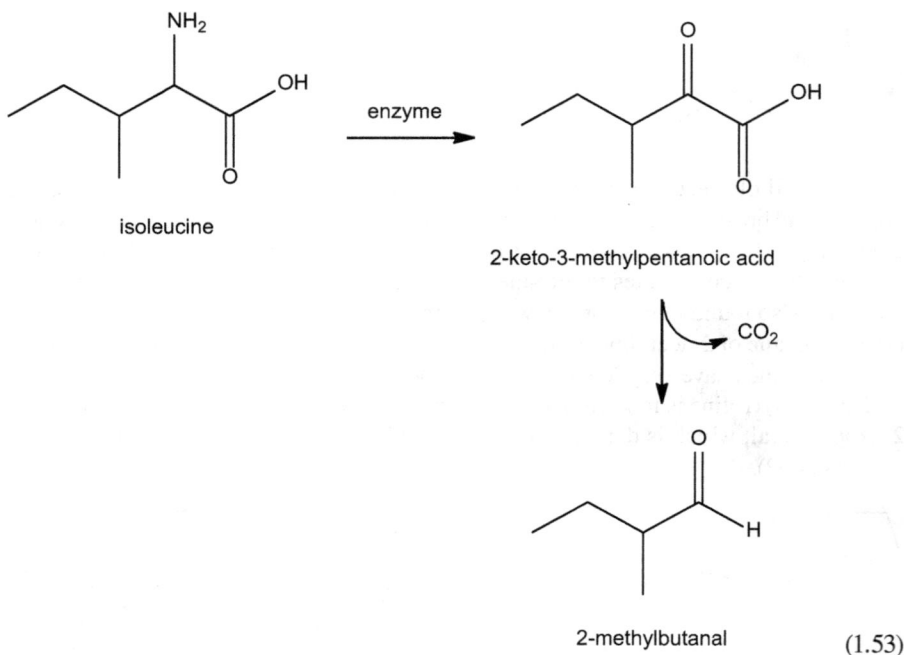

isoleucine

2-keto-3-methylpentanoic acid

CO_2

2-methylbutanal   (1.53)

Using similar sequences of reactions, 3-methylbutanal (1.54) is formed from leucine, and 2-methylpropanal (1.55) is formed from valine. All these aldehydes contribute 'malty' notes to the smell of bread.

(1.54)

(1.55)

The oxidation of lipids yields more carbonyl compounds that significantly contribute to the smell of the crumb, notably (E)-2-nonenal ((1.56), 'green' note), (E, E)-2,4-decadienal ((1.57), 'fatty', 'waxy') and butane-2,3-dione ((1.58), 'buttery').

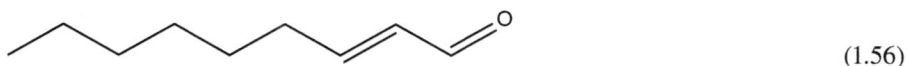

(1.56)

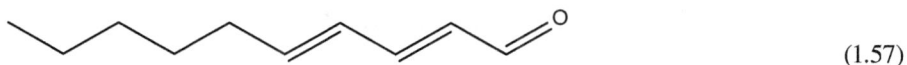

(1.57)

(1.58)

2-phenylethanol and 2-phenylethanal, both of which contribute a 'honey' note to the crumb, are derived from the amino acid phenylalanine through a sequence of enzyme-catalysed reactions. First, a transamination reaction replaces the amino group with an oxo-group, forming the keto acid phenylpyruvic acid. On decarboxylation with the loss of carbon dioxide, the aldehyde 2-phenylethanal is produced, which can be reduced to 2-phenylethanol (1.59).

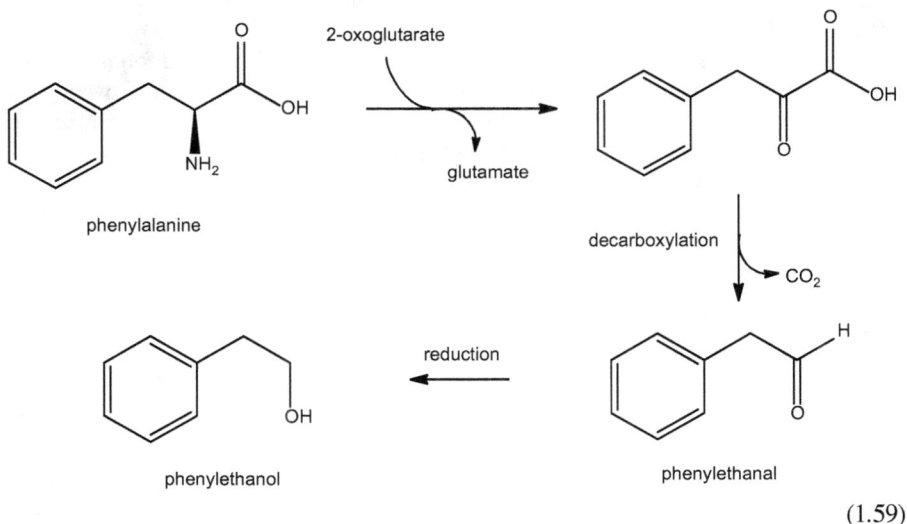

(1.59)

The sulfur-containing aldehyde 3-methylthio-1-propanal, commonly known as methional, also makes an important contribution to the smell of potatoes and certain cheeses. It originates from the sulfur-containing amino acid methionine through a similar sequence of enzyme-controlled reactions: a transamination followed by a decarboxylation (1.60).

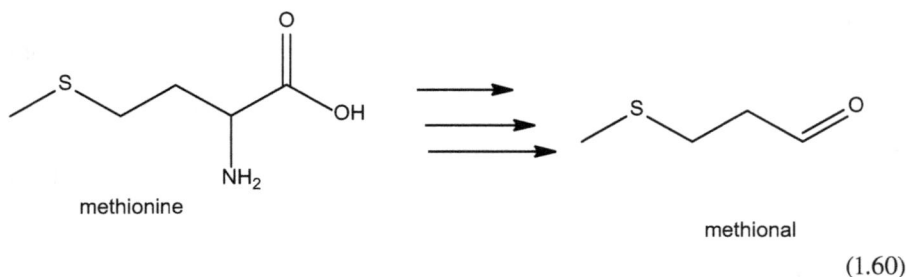

methionine

methional

(1.60)

Enzymes do not function at higher temperatures, where browning occurs. The key reactions then are caramelisation and the Maillard reaction; these primarily occur on the crust, which is about 100°C hotter than the inside of the bread (the crumb). Heat facilitates the reaction between the carbonyl groups in reducing sugars and the $-NH_2$ groups of amino acids, which requires basic conditions to generate the free amine functions from the zwitterions normally present. This is the reaction that causes baked bread to turn brown and also applies to toasting; it is not confined to baking, as it also causes browning in roasted foods such as coffee (and turkey). It is distinct from the caramelisation that occurs when sugars are heated alone. Brown polymers called melanoidins are also products of Maillard reactions, which are major contributors to the colour of such cooked foods; their structure is not well understood at present.

Pyrazines are heterocyclic compounds that are important in roasted foods, not just bread. They result from ring-forming reactions, such as condensation between two α-aminoketones. Although pyrazines have a high odour threshold, they are present in relatively high concentrations in bread, so that molecules such as 2-ethyl-3,5-dimethylpyrazine make a significant contribution to bread's smell (1.61).

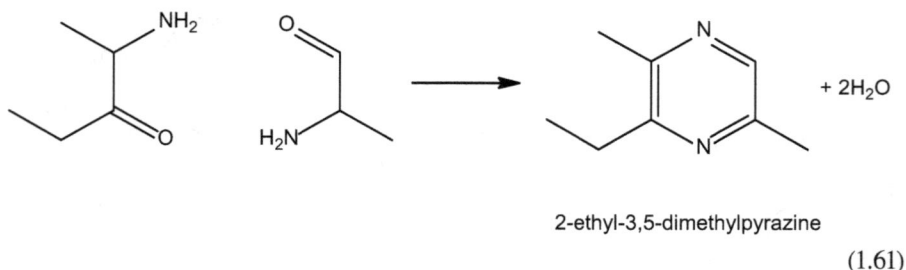

$+ 2H_2O$

2-ethyl-3,5-dimethylpyrazine

(1.61)

POTATO

The potato was first brought to Europe from South America, where it had been domesticated, and *Solanum tuberosum* spread to other countries. It was a hardy and inexpensive crop that was easy to grow, leading to significant production expansion, especially in the 18th century, as it was ideal for feeding growing populations. It could be prepared quickly, was easy to digest and was filling. We now know that potatoes are a good source of minerals, proteins, carbohydrates (notably starch) and dietary fibre. More people could be fed per acre with potato than with cereals. By the 19th century, it became the second staple food – after bread – in Central Europe.

Apart from raw potatoes, they usually come in one of three types of cooked potato: boiled, baked and fried (potato chips in the UK, French fries to our Colonial readership); these look different and have a number of odorants, some found in more than one form of cooked potato.

Unlike fruit, potato tubers don't emit flavours as they ripen. Individually, raw potatoes have a weak vegetable-like smell. It is hard to say which molecules are responsible, as enzymatic activity begins as soon as the potatoes are peeled. The lipids in the potato are hydrolysed by hydrolase enzymes, forming polyunsaturated fatty acids, notably linoleic (1.62) and linolenic (1.63) acids.

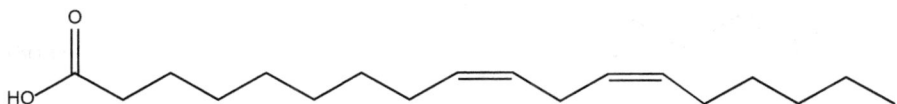

(1.62)

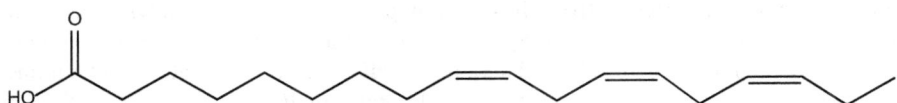

(1.63)

These acids are oxidised by lipoxygenase enzymes, generating molecules like hexanal (1.64) and 2,4-decadienal (1.65).

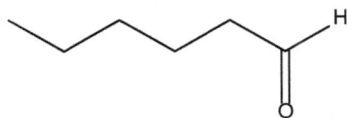

(1.64)

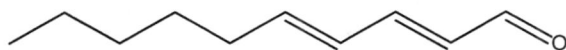

(1.65)

Substituted pyrazines, like 3-isopropyl-2-methoxypyrazine (1.66), 3-isobutyl-2-methoxypyrazine (1.67) and 2,3-diethyl-5-methylpyrazine (1.68), have also been identified as contributors to the aroma of raw potatoes.

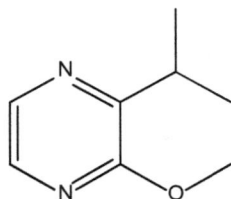

(1.66)

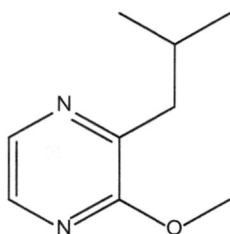

(1.67)

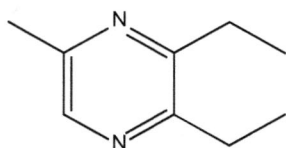

(1.68)

BOILED POTATOES

Heating potatoes in boiling water deactivates some enzymes, so the odorants in boiled potatoes are different from those in raw potatoes. The best-known ingredient of 'boiled potato' smell is the aldehyde methional, formed in the Strecker degradation of methionine (1.60). Methional actually smells like boiled potatoes, but a number of other compounds are part of the boiled potato smell.

In the Strecker degradation, an α-amino acid loses a carboxylate group (decarboxylation) and is converted into an aldehyde (a 'Strecker aldehyde') with one less carbon atom than the parent amino acid. Usually in the Strecker degradation, the α-amino acid reacts with an α-dicarbonyl compound, generating the aldehyde, together with another fragment, as shown in (1.69).

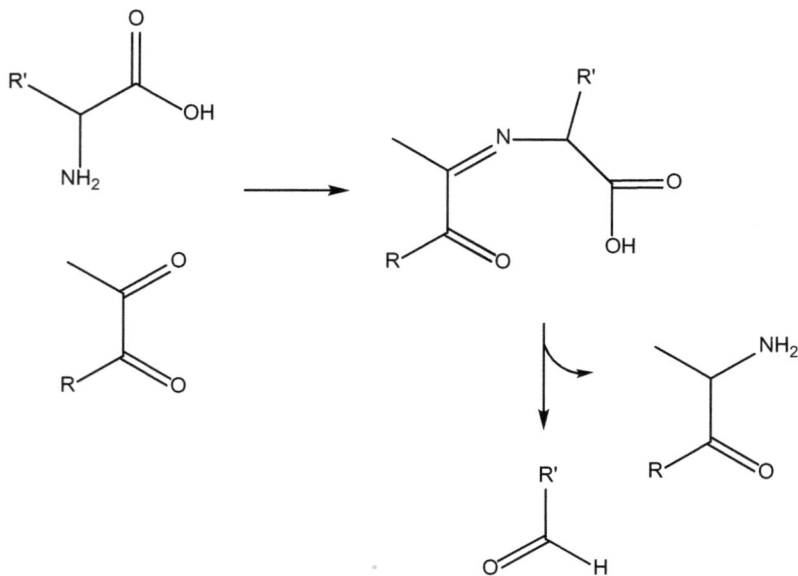

(1.69)

This latter molecule can condense with a similar one, ultimately forming a pyrazine in a Maillard reaction, a likely source of the pyrazines often found in potatoes (1.70).

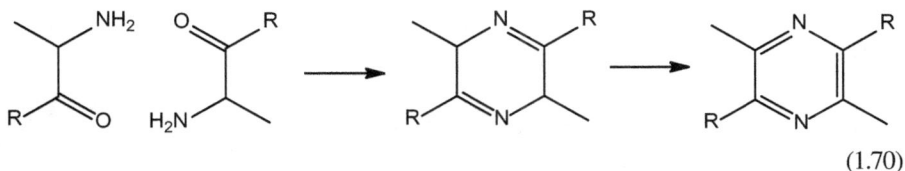

(1.70)

Amongst the other compounds present in boiled potatoes are some carbonyl compounds, the products of degradation of lipid-derived carboxylic acid, including (E)-2-octenal (1.71), 2,4-decadienal (1.65), (Z)-4-heptanal (1.72), 1-octen-3-one (1.73) (which is also a noted ingredient of mushrooms) and (E)-2-nonenal (1.56).

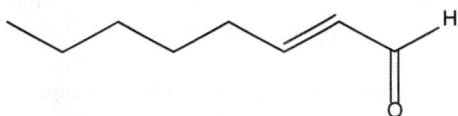

(1.71)

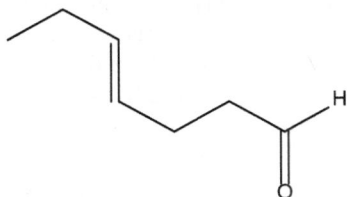

(1.72

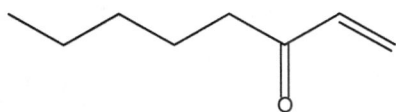

(1.73)

Other important odorants include two alkoxypyrazines, 3-isobutyl-2-methoxypyr-azine (1.67) and 2,3-diethyl-5-methylpyrazine (1.30), already noted as products of the Maillard reaction.

BAKED POTATOES

When the potato is baked, the reaction occurs at a higher temperature than boiling. A mixture of many molecules appears to be generated. Methoxypyrazines are produced particularly in the potato skin, with 3-isopropyl-2-methoxypyrazine (1.66) identified as a key odorant. The aldehydes 2,4-decadienal (1.27), methional (1.23), 2-nonenal (1.26) and 3-methylbutanal (1.54) have been reported as important contributors to flesh aroma, with hexanal (1.64), nonanal (1.74) and decanal (1.75) also present.

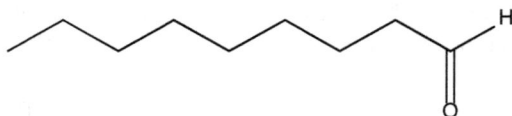

(1.74)

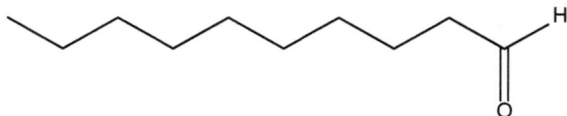

(1.75)

POTATO CHIPS

The high temperatures involved in drying cause oxidation of the fatty acids from the frying oils, whilst sugars and amino acids are also sources of odorants. An initial gentle frying generates surface layers of starch that reinforce the outer cell walls into a robust crust; then, frying at a higher temperature crisps and browns the exterior. Lipids are degraded by frying in oil, and the high temperatures form large amounts of pyrazines, notably 2,3-diethyl-5-methylpyrazine (1.30), 3-ethyl-2,5-dimethylpyr-azine (1.76) and 2-ethyl-3,5-dimethylpyrazine (1.77), which supply earthy and nutty notes. Again, Strecker aldehydes 2-methylpropanal (1.55), 2-methylbutanal (1.53) and 3-methylbutanal (1.54) contribute a malty, sweet character; also, the oxidation of lipids contributes unsaturated aldehydes like (E,E)-2,4-decadienal (1. 27), which has a deep-fried, fatty character.

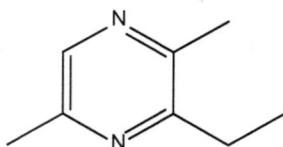

(1.76)

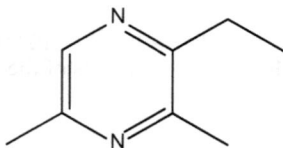

(1.77)

Overall, the different forms of cooked potatoes – boiled, baked and chipped – smell and taste different because of their varying cooking conditions. Key odorants such as pyrazines and aldehydes are produced via Strecker degrada-tions and the Maillard reaction. No single compound is responsible for the smell of any type of cooked potato; some odorants are found in more than one form of cooked potato, with the flavour being the result of the relative amounts and balance of the different chemicals.

MUSHROOMS

Mushrooms are a popular food, known for their characteristic smell and taste. The two leading varieties are *Agaricus bisporus* (essentially the only cultivated mushroom in the UK) and the shiitake mushroom, *Lentinula edodes*, which is most popular in the Far East.

More than a hundred volatiles have been identified in mushrooms, many of which contain eight carbon atoms, especially 1-octen-3-ol ((1.78), widely known as 'mushroom alcohol') and 1-octen-3-one (1.79). Amongst the others are E-2-octen-1-ol (1.80), E-2-octenal (1.81), octanal (1.82), octan-3-ol (1.83) and octan-3-one (1.84).

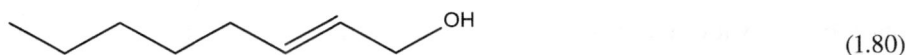

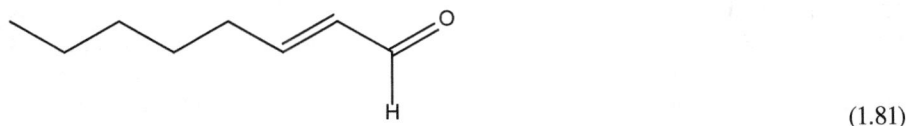

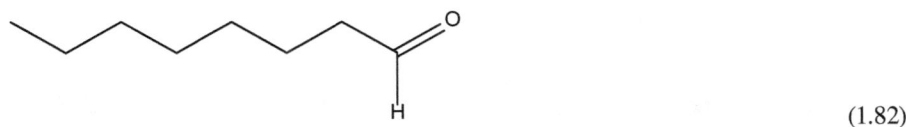

(1.78)

(1.79)

(1.80)

(1.81)

(1.82)

(1.83)

(1.84)

1-octen-3-ol is the best known of these. The presence of a chiral carbon, with four different groups attached, means that it exists as two optical isomers. The (*R*-)-(−) isomer (1.85) is produced in mushrooms to the almost complete (99%) exclusion of the (*S*)-(+)- isomer (1.86); the (*R*-)-(−) isomer has a 'mushroom' smell, whilst the (*S*)-(+)- isomer has a somewhat mouldy, grassy smell. Bodily receptors, e.g. for smell or taste, are 'handed', as they are proteins, made from amino acids, all of which, except glycine, are 'handed'. This means that the bodily response to optical isomers often differs, hence the different smells.

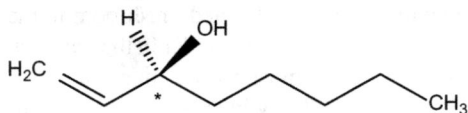

(S)-(+)-1-octen-3-ol

(1.85)

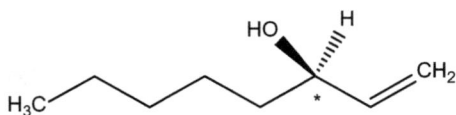

(R)-(-)-1-octen-3-ol

(1.86)

Biosynthesis of 1-octen-3-ol in *Agaricus bisporus* mushrooms involves the enzymatic breakdown of the most abundant fatty acid in the mushroom, linoleic acid (1.87). Linoleic acid is first oxidised by a lipoxygenase enzyme to 10-hydroperoxy-8*E*, 12*Z*-octadecenoic acid, which is then split at a C=C bond by a lyase enzyme, thus forming 1-octen-3-ol as well as 10-oxo-*trans*-8-decenoic acid (ODA). This process is used to produce 1-octen-3-ol in industrial bioreactors. It should be noted that in plants, another enzymatic breakdown occurs at a different point in the carbon chain of linoleic acid, generating C_6 molecules like hexenal; this causes the familiar smell of freshly cut grass (pages 139–141).

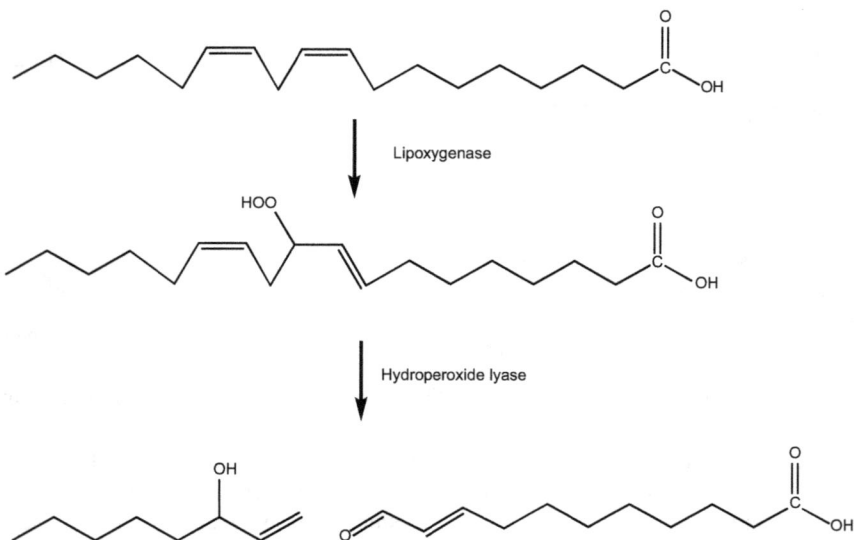

Lipoxygenase

Hydroperoxide lyase

(1.87)

The overall smell of mushrooms is due to a number of compounds, not just 1-octen-3-ol. 1-octen-3-one also has a 'mushroom' smell and is thought to contribute more to the overall aroma of mushrooms. Two other contributors have more nuanced smells – (E)-2-octen-1-ol (1.80) has a mushroom smell with a 'grassy' note, whilst octan-3-ol (1.83) has a mushroom smell with a 'nutty' note; the smells of octan-1-ol and (E)-2-octenal (1.83) do not resemble mushrooms. Other molecules that contribute to the aroma but do not have 'mushroom' notes include the aldehydes benzaldehyde ((1.88), almond note) and phenylethanal ((1.89), honey note).

Another frequent contributor to mushroom smell is another aldehyde, methional (3-(methylthio)propanal), better known for its 'boiled potato' smell (see page 30). Methional (1.23) is generated by enzymatic action on the amino acid methionine. Thus, the overall smell of mushrooms is the result of a blend of several molecules with different smells – dominated, of course, by a couple of molecules with 'mushroom' smells: 1-octen-3-ol and 1-octen-3-one.

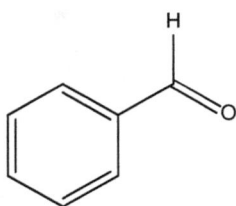

(1.88)

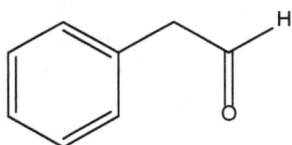

(1.89)

Frying *Agaricus bisporus* mushrooms causes some changes in the odorants that are released. First, there is less of the two important C_8 molecules, 1-octen-3-one and, in particular, 1-octen-3-ol. On the other hand, some other odorants become more noticeable. One of these is 2-acetyl-1-pyrroline (1.24), formed from the amino acid proline; it is found in a number of other foods, including fresh bread and basmati rice, to which it contributes a 'popcorn' note.

A second type of mushroom is the most popular in the Far East, the *shiitake* mushroom (*Lentinula edodes*). It has a special niche in Oriental medicine, credited with lowering blood pressure, strengthening the immune system and being active against microbes and tumours. It is low-fat and contains many vitamins and minerals. In addition to an earthy flavour, raw shiitake mushrooms are noted for their strong and intense mushroom aroma. Key odorants, as in European mushrooms, are 1-octen-3-one and 1-octen-3-ol, whilst other C_8 molecules such as octan-3-ol, octan-3-one, (E)-2-octenol, (E)-2-octenal and octan-1-ol are also present.

Of these, recent research has shown that 1-octen-3-one is the dominant odorant in raw shiitake. 1-octen-3-one has a much lower odour threshold than 1-octen-3-ol, which more than compensates for the alcohol having a higher concentration in the mushroom. Another recent discovery is that 1-octen-3-one is formed directly from linoleic acid, rather than being made by oxidising 1-octen-3-ol (a natural assumption).

When shiitake mushrooms are fried, their odour changes significantly. As is the case with *Agaricus bisporus*, the amounts of both 1-octen-3-one and 1-octen-3-ol drop significantly. This is compensated by the appearance of new compounds, notably sulfur-containing compounds like 1,2,4,5-tetrathiane (1.90) and 1,2,3,5,6-penta-thiepane (1.91), which contribute the 'sulfury' notes expected.

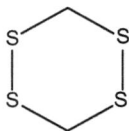

(1.90)

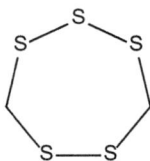

(1.91)

PUFFBALL MUSHROOMS

Capable of reaching the size of a football, giant puffball mushrooms (*Calvatia gigantea*) must be eaten fresh, usually after frying. Amongst the many volatiles they produce, the characteristic C_8 mushroom volatiles, 1-octen-3-one and 1-octen-3-ol, are both present, though in small amounts, along with 3-octanone and 3-octanol. Also found are the aldehydes 2-methylbutanal (1.53) and 3-methylbutanal (1.54), as well as some esters, including the fruity-smelling methyl anthranilate ((1.92), methyl 2-aminobenzoate).

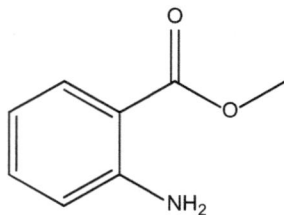

(1.92

As already mentioned, puffball mushrooms are consumed within 24 hours of picking, as there is a significant change in the volatiles – the amount of the C_8 odorants decreases sharply, and there is a rapid increase in the concentrations of disagreeably smelling carboxylic acids 2-methylbutanoic acid (1.93) and 3-methylbutanoic acid (1.94) (see also Chapter 5, p. 130).

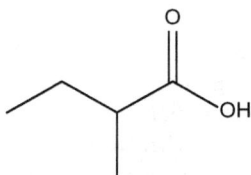

(1.93)

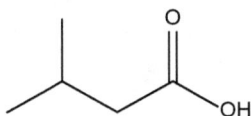

(1.94)

ONIONS

Onions (*Allium cepa*) have been cultivated since the Bronze Age and are used in cookery, whether raw or cooked, in many contexts. Raw onions are odourless, but when cut, they develop a strong smell. Like onions, garlic also belongs to the genus *Allium*, but they have a number of differences, not least in smell and lachrymatory activity. Why is this?

Let's start with garlic (*Allium sativum*). It does not have a smell until it is crushed, when a rather pungent odour rapidly develops. Garlic contains an odour-less amino acid that also has a sulfoxide group; it is called alliin, also known as S-allylcysteine sulfoxide. Tissue damage in garlic results in the alliin coming into contact with an enzyme called alliinase, which cleaves the C–S bond, forming allylsulfenic acid and dehydroalanine. Two molecules of allylsulfenic acid (2-pro-penesulfenic acid) immediately come together in a condensation reaction (1.95), eliminating a water molecule and forming allicin (diallyl thiosulfinate). Allicin is the molecule largely responsible for the smell of crushed garlic. It has antibacte-rial properties, and that is probably why garlic produces it, as a protection against pathogens and pests.

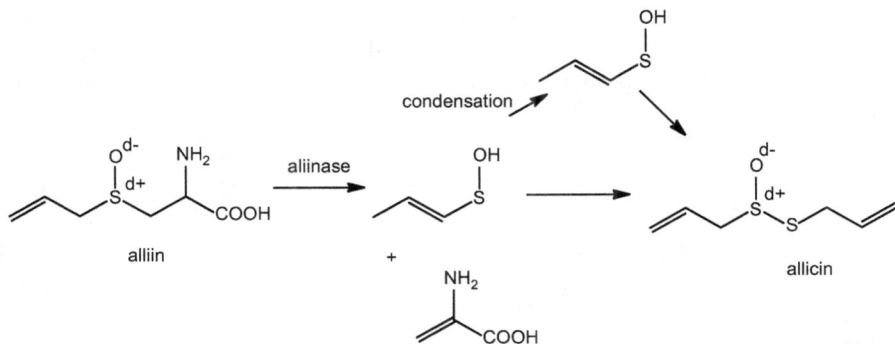

(1.95)

Onions do not contain alliin; instead, its isomer isoalliin (*trans*-(+)-*S*-(1-propenyl)-L-cysteine sulfoxide) is present. As with garlic, cutting onions brings isoalliin into contact with the alliinase enzyme, which again breaks the C–S bond, generating dehydroalanine (as before) and a new compound, 1-propenesulfenic acid (1.96). However, there is another enzyme present in onions that is not found in garlic, lachrymatory factor synthase, which converts 1-propenesulfenic acid to propanethial S-oxide; this is the volatile molecule that causes irritation to your eyes and tears when you peel an onion. This molecule is known as the onion lachrymatory factor – a very rapid reaction, occurring within seconds. The tears are due to your eye trying to wash away the source of the irritation. As in the case of garlic, some of the sulfenic acid undergoes condensation, forming bis(1-propenyl) thiosulfinate (which then undergoes rearrangement to molecules known as zwiebelanes).

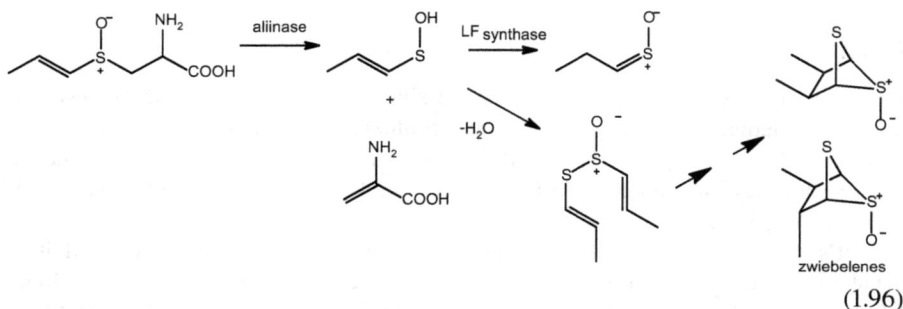

zwiebelenes

(1.96)

Fresh Onions

A group of Danish scientists monitored the volatiles from onions after they were cut, using mass spectrometry. To begin with, for about 10 minutes, the emissions were dominated by propanethial S-oxide (the lachrymatory factor) and its decomposition products like propanal; the amount of propanethial S-oxide dwindled, and it could not be detected after about half an hour. After the initial 10-minute period, the most abundant emissions were propanethiol (1.97) and dipropyl disulfide (1.98), with the former predominating, and traces of thiosulfinates; these reached a maximum after about an hour. The researchers suggested that propanethiol could be the main source of onion odours, and others have suggested that compounds with a propyl thiol group in their structure generally have an onion-like flavour.

(1.97)

(1.98)

Other studies on cut onions indicate a wide range of mainly C_3–C_6 odorants are present, including alcohols, esters and both saturated and unsaturated aldehydes, plus sulfur compounds such as dipropyl disulfide, allylpropyl disulfide (1.99) and dipropyl trisulfide. This study has reported that dipropyl disulfide and dipropyl trisulfide are the main volatile compounds present in fresh onions.

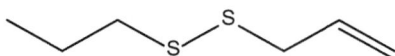

(1.99)

COOKED ONIONS

Cooked onions are another matter, with quite a characteristic smell – sweet, meaty and savoury. One important chemical in the odour from cooked onion is 3-mercapto-2-methylpentan-1-ol (1.100). It was first detected in raw onion, but subsequently, it was discovered that it was more concentrated in cooked onion. Its aroma has been described as a pleasant meat broth, sweaty, onion and leek-like. It is believed to be formed by the action of heat upon propanethial S-oxide.

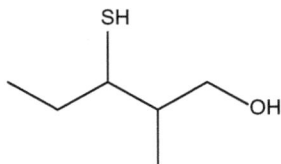

(1.100)

Subsequently, it has been found that the human OR2M3 odorant receptor responds uniquely to 3-mercapto-2-methylpentan-1-ol, out of 190 receptors tested. The extremely low odour threshold for this compound means that OR2M3 responds to the very low levels present.

Studies of cooked onions have again detected a wide range of molecules present, including both saturated and unsaturated aldehydes as well as disulfides (such as dimethyl disulfide, diallyl disulfide and methyl allyl disulfide) and trisulfides (e.g. diallyl trisulfide, methyl propyl trisulfide and allyl propyl trisulfide).

TOMATOES

You see tomatoes before you smell or taste them. Their chlorophyll content makes unripe tomatoes green, but when they ripen, they turn red because of the hydrocarbon lycopene, $C_{40}H_{56}$. Compared to saturated hydrocarbons, this is very hydrogen-deficient; lycopene has 13 double bonds in the molecule, 11 of these being conjugated double bonds, alternating with single bonds. Such compounds are usually coloured, as lycopene is; it absorbs light principally in the blue–green part of the visible spectrum ($\lambda_{max} = 470$ nm), thus its red–orange colour. Contrast this with β-carotene, the molecule that is responsible for the colour of carrots; also $C_{40}H_{56}$, it is thus an isomer of lycopene. Beta-carotene has 11 conjugated double bonds, with two rings in place of the extra two double bonds in lycopene; it absorbs light in a slightly different region of the spectrum ($\lambda_{max} = 448$ nm), which is why β-carotene and carrots are orange, not red.

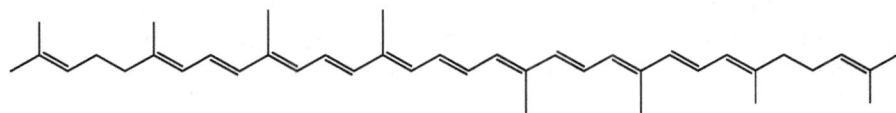

(1.101)

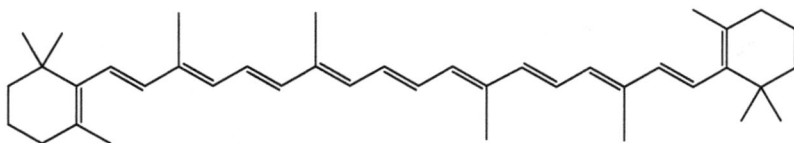

(1.102

The taste of a tomato is the result of a combination of sugars – fructose (1.103) and glucose (1.4) – and acids – weak acids, like citric acid (1.104) and malic acid (1.105). The desirable tomato varieties have the right combination of sweetness and acidity.

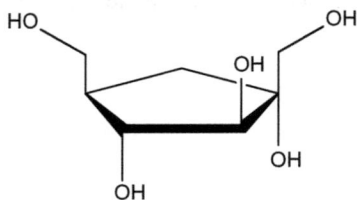

(1.103)

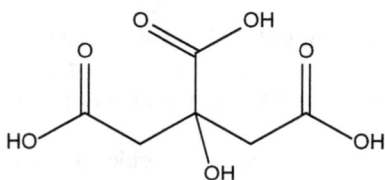

(1.104)

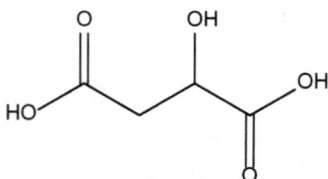

(1.105)

And then, the volatiles, of which there are many. No one molecule smells 'tomato' – flavour is a composite of the efforts of many.

The moment that the tissues of a tomato are disturbed, whether by ripening, biting or attack by a predator, enzymatic reactions begin (1.106). 13-lipoxygenase enzymes target linolenic acid and linoleic acid, generating 13-hydroperoxides, which undergo a further attack by lyases, cutting the carbon backbone and forming C_6 aldehydes, notably (Z)-hex-3-enal (cis-3-hexenal). Amongst some 400 odorants produced by tomatoes, (Z)-hex-3-enal stands out (1.106). One of only about 20 molecules that really contribute to tomato flavour, it has the greatest impact, with its very low odour threshold. It is an important product when plant tissue is damaged and is also the key molecule in 'fresh grass smell' (page 140). Very reactive, it is easily transformed into other C_6 molecules, including (Z)-hex-3-enol (cis-3-hexenol), (E)-2-hexenal (1.107), trans-2-hexenal and hexanal (1.108). Other important contributors to tomato smell include 2-methylbutanal (1.53) and 3-methylbutanal (1.54), β-ionone (1.109), hexanal (1.64), β-damascenone (1.110) and 1-penten-3-one (1.111).

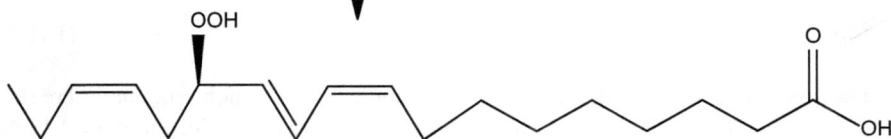

(Z)-hex-3-enal

(1.106)

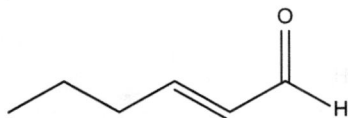

(1.107)

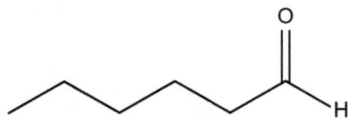

(1.108)

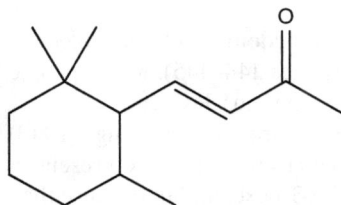

(1.109)

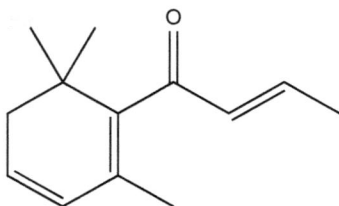

(1.110)

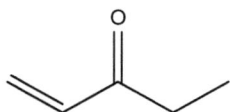

(1.111)

The ripening process adds several of these molecules through enzymatic break-down of amino acids, like leucine, isoleucine and phenylalanine. This process (1.112) creates molecules like 2- and 3-methylbutanal, 2-phenylethanol and methyl salicylate.

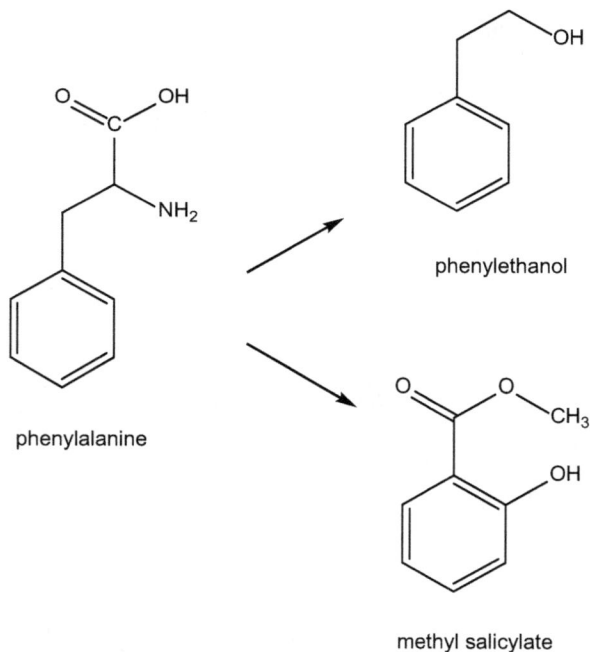

phenylethanol

phenylalanine

methyl salicylate   (1.112)

Methyl salicylate (MeSA) is widespread in the plant kingdom; a volatile molecule, many plants make use of it, often to send messages (pages 144–145). Many people encounter it in the embrocation that you put on your aching muscles.

Oxidative cleavage breaks up carotenoids like β-carotene, creating (1.113) strongly smelling β-ionone (1.109) and β-damascenone (1.110). They are present in tiny amounts compared to other compounds such as (Z)-3-hexenal, but because they have low aroma thresholds, they "punch above their weight" and make significant contributions to the aroma of a tomato.

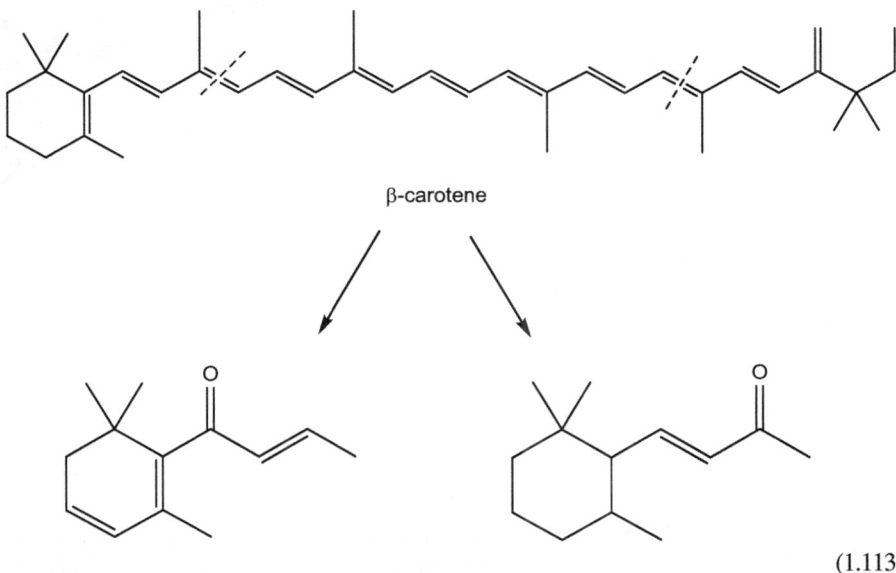

β-carotene

(1.113)

It was around a thousand years ago that farmers in Mexico or Central America were the first to domesticate tomatoes. We would not recognise the small, lumpy and ugly fruit. As time went on, farmers bred them, selecting characteristics that gave features like bigger fruit, which we see today in "heritage" tomatoes. More recently, breeders have increasingly selected traits like size, uniform ripening and high yield, which has come at a cost, such as less sugar formation. Practices like shipping tomatoes unripe and then ripening them with the plant hormone ethene just before sale have meant that the tomatoes do not get the extra input from 'ripening on the vine', so that, again, the tomatoes tend to lack taste.

And not just taste. It's been found that certain wild tomatoes (*Solanum habrochaites*) – unlike cultivated tomatoes (*Solanum lycopersicum*) – emit a molecule called epizingiberene (1.114), which makes them more resistant to herbivores, especially the sweet potato whitefly (*Bemisia tabaci*) pest. At some point over the ages, 'cultivated' tomatoes lost the gene that produces epizingiberene, presumably by concentrating breeding on producing tomatoes with larger, redder fruit. Epizingiberene is the epimer of zingiberene (1.115), a natural terpenoid that causes the taste of ginger (page 79). In contrast to epizingiberene, zingiberene has no effect on the whiteflies, indicating that somewhere in the whiteflies there is a receptor binding site with a specific shape that can dock epizingiberene molecules but not zingiberene. Nature really is *that* picky.

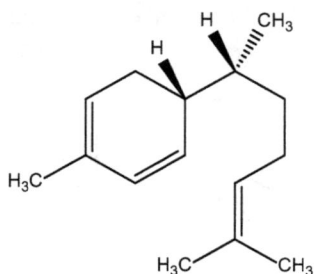

(1.114)

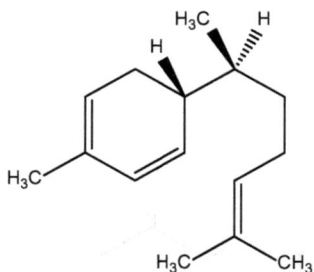

(1.115)

It seems that centuries ago, some tomatoes may have smelled like cucumbers. A few years ago, Japanese and Israeli scientists crossed a cultivated tomato with a wild one, finding that as well as the usual tomato volatiles, these tomatoes produced (1.116) large amounts of C_9 aldehydes, notably "cucumber aldehyde", (2E, 6Z)-nona-2,6-dienal in particular, as well as (2E)-nonenal (1.26), which also has a cucumber smell.

Tomato eaters in the past would have regarded the smell due to 'cucumber' aldehydes as an undesirable flavour, so it looks as if the gene making them may have been bred out of cultivated tomatoes to produce better flavours.

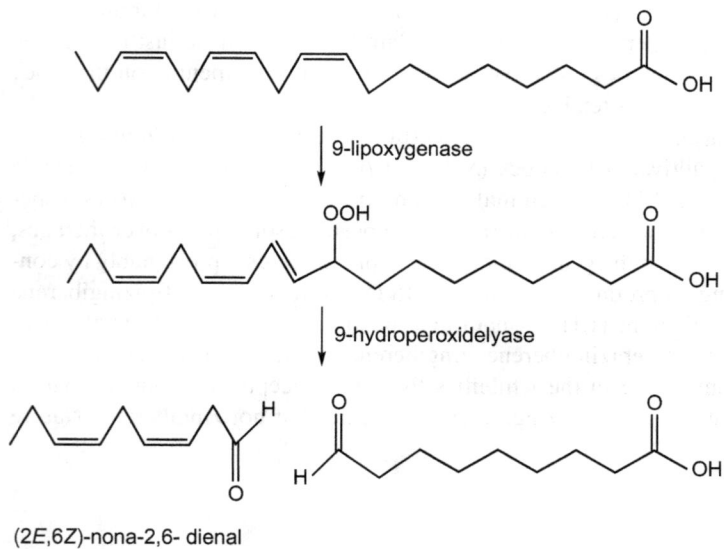

(2E,6Z)-nona-2,6- dienal

(1.116)

STRAWBERRIES

People have been eating strawberries for thousands of years. Then, of course, it would have meant what we now call wild strawberries. The Roman poet Ovid mentioned *Fragaria* twice and very likely meant *Fragaria vesca*. Today, they can be found growing wild in the forests of Central Europe, north into Scandinavia and east into Russia.

Cultivated strawberries are a hybrid *Fragaria × ananassa*, which resulted from a fortuitous cross between Chilean *Fragaria chiloensis* and North American *Fragaria virginiana* (in Brittany) 300 years ago.

The characteristic colour of strawberries is due to anthocyanins – particularly pelargonidin 3-glucoside and, to a lesser extent, cyanidin 3-glucoside (1.117) – which are biosynthesised from the amino acid L-phenylalanine.

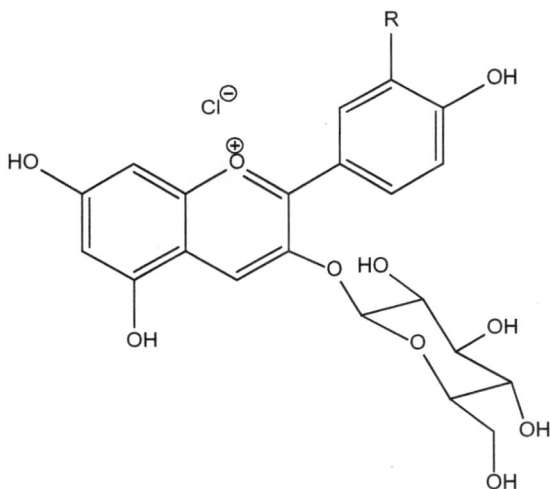

Cyanadin, R = OH; Pelargonidin, R = H   (1.117)

Strawberries are very high in Vitamin C content and rich in antioxidants. Around 90% of a ripe strawberry is water, but it also contains sugars, about 80% glucose and fructose, with less sucrose. Sugar content rises from around 5% in unripe fruit to 6%–9% upon ripening, stimulating the production of secondary metabolites including anthocyanins and furanones. The acid content is largely citric acid, with some malic acid and ellagic acid, the latter reportedly an anti-carcinogen. The acidity decreases as the fruit ripens, resulting in an increasing sugar/acid ratio that makes ripe strawberries taste sweeter. Additionally, with ripening, enzymatic degradation of cell walls, due to auxin, causes the plant tissue to soften.

Consumers look for the right combination of sweetness, acidity and flavour, the latter depending upon the balance between several molecules, notably esters. A focus on breeding larger strawberries with long shelf life and disease resistance has come at the expense of flavour, with terpenoids being the main type of molecules that have been depleted, along with a decrease in some esters that significantly impact flavour.

Through the research of scientists such as Peter Schieberle (Garching), over 350 different volatiles have been identified in strawberries. A study of the molecules extracted from fresh strawberry juice identified the 15 molecules having the most impact (1.118 – 1.132). The list was led by 4-hydroxy-2,5-dimethyl-3(2*H*)-furanone (with a caramel-like note, (1.118)) and several fruity esters, notably methyl butanoate (1.121), ethyl butanoate (1.122) and methyl 2-methylpropanoate. The 'green grass' smell from (*Z*)-3-hexenal (1.120) was also an important contributor.

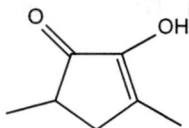

(1.118)

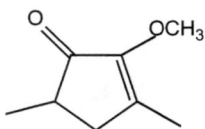

(1.119)

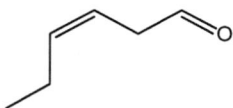

(1.120)

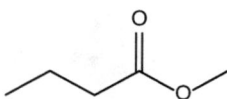

(1.121)

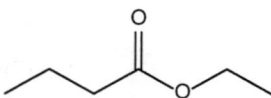

(1.122)

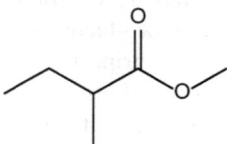

(1.123)

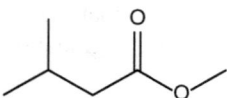

(1.124)

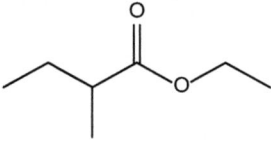

(1.125)

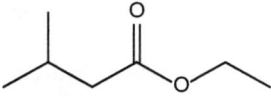

(1.126)

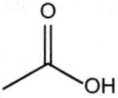

(1.127)

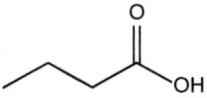

(1.128)

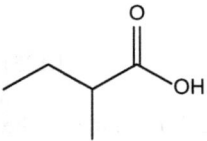

(1.129)

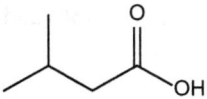

(1.130)

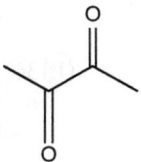

(1.131)

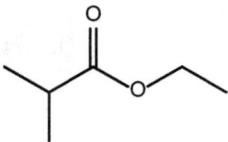

(1.132)

In the fruit, esters are formed from the reaction (1.133) between an acyl coenzyme A and an alcohol, catalysed by the alcohol acyltransferase enzyme (AAT); the amounts of esters and their identities depend greatly on the cultivar. The aroma thresholds (the concentration below which the molecule cannot be smelled) of esters vary enormously; thus, for butyl ethanoate, the aroma threshold is 5000 ppb, whilst for its isomer, ethyl butanoate, it is 0.13 ppb.

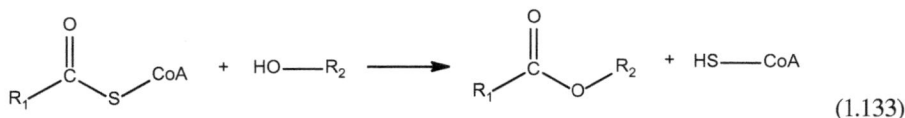

$$(1.133)$$

Cyclic esters (lactones), which are generated by cyclisation reactions of hydroxy acids, are significant contributors to strawberry aroma. The most important of these are γ-decalactone (1.134) and γ-dodecalactone (1.135). However, too much lactone can give the strawberry an unwanted peach note.

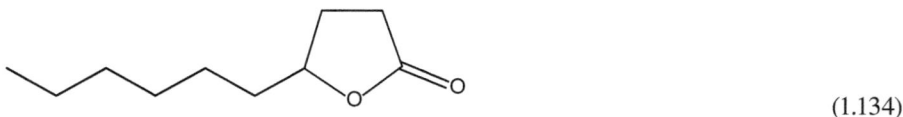

$$(1.134)$$

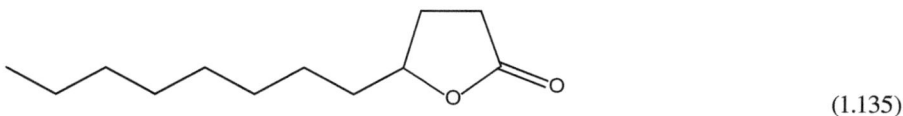

$$(1.135)$$

Butanoic, 2-methylbutanoic and hexanoic acids can also be important to the aroma of strawberries, whilst some sulfur compounds are present in strawberries at low concentrations; the most abundant of these are methyl thioacetate (1.136) and methyl thiobutyrate (1.137), which may affect the odour of some ripe cultivars.

$$(1.136)$$

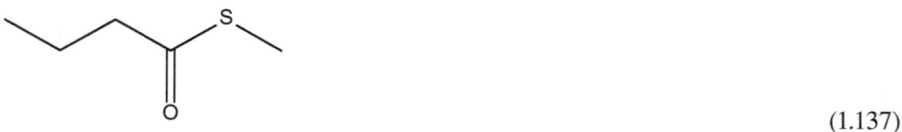

$$(1.137)$$

WILD STRAWBERRIES

Compared with cultivated strawberries, wild species have stronger aromas and richer flavours because they contain greater quantities of odorous volatiles, including some additional ones (1.138–1.146). Wild strawberries can usually be identified by the presence of the esters methyl anthranilate and methyl cinnamate; methyl anthranilate (1. 92) contributes an aromatic and flowery odour note, whilst methyl cinnamate (1. 144) provides a pleasant and spicy note.

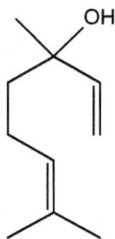

(1.138)

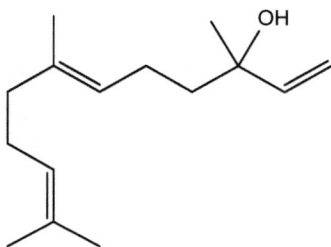

(1.139)

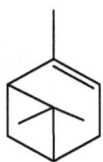

(1.140)

(1.141)

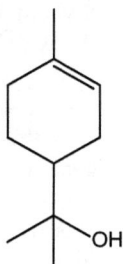

(1.142)

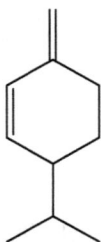

(1.143)

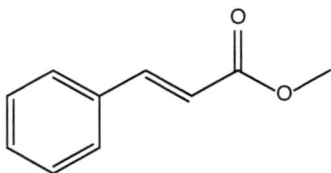

(1.144)

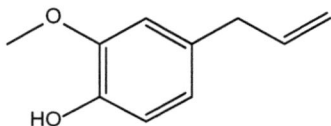

(1.145)

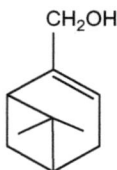

(1.146)

Examination of the vapour above ripe cultivated *Fragaria ananassa Elsanta* species showed that the only terpenoids present were linalool (1.138) and nerolidol (1.139), whilst wild *Fragaria vesca* emitted a wider selection of terpenoids, including α-pinene (1.140), β-myrcene (1.141), α-terpineol (1.142) and β-phellandrene (1.143) as well as myrtenyl acetate and myrtenol.

A comparison of wild Finnish strawberries showed much higher levels of esters and mesifurane (2), as well as the presence of methyl anthranilate (1.92), methyl cinnamate (1.144) and terpenoids like eugenol (1.145) and myrtenol (1.146) in the wild strawberries, which were not found in the cultivated variety.

Wild strawberries have a synthase gene to make α-pinene, which they synthesise from geranyl diphosphate. A cytochrome P450 enzyme uses α-pinene as the substrate in a hydroxylation reaction at the ring methyl group (1.147) to generate myrtenol, which can be esterified using an alcohol acyltransferase (AAT) enzyme to form myrtenyl acetate (1.148). Similarly, α-pinene acts as the source of other terpenoids, which ultimately derive structurally from isoprene. This α-pinene synthase gene is not operative in cultivated strawberries.

These molecules in strawberries are more than just a pleasant smell. For example, methyl anthranilate (1.92) and γ-decalactone have been found to inhibit strawberry pathogen growth. Terpenoids are part of the plant's defence against pathogens and parasites; it is possible that they also assist in resistance to diseases and herbivores.

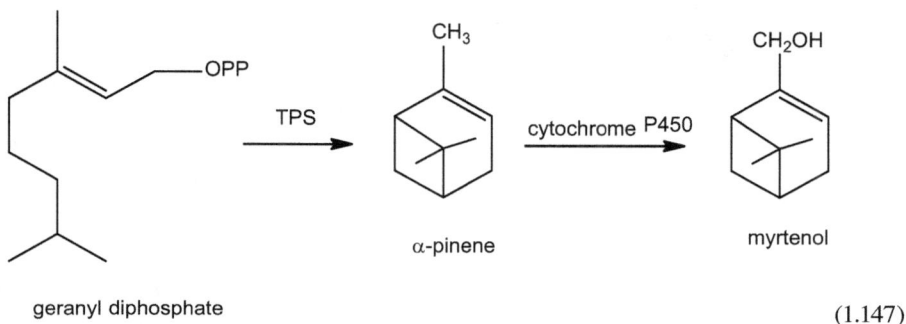

geranyl diphosphate   (1.147)

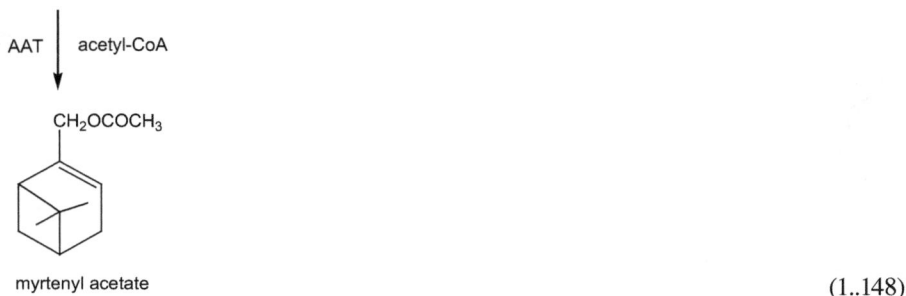

myrtenyl acetate   (1..148)

Modern strawberry breeding easily loses the gene for methyl anthranilate synthesis, which is why this ester is not found in modern cultivars of *Fragaria × ananassa*. It has been found that back-crossing *Fragaria ananassa* cultivars with either *F. vesca* or *F. virginiana* yields cultivars with small berries but with aromas more closely resembling those of wild strawberries.

ORANGES AND LEMONS

Just as in the nursery rhyme, oranges and lemons belong together in people' minds. Although they are both citrus fruits, they appeal to different tastes and different markets. Whilst the most popular fruit juice, orange juice, is widely consumed, lemons are associated with pancakes and also with domestic cleaning products. Hence, it is not totally surprising that they involve rather different molecules.

Lemon oil is a mixture of organic compounds, of which by far the most abundant (ca. 94%) is (R)-(+)-limonene (1.149). It does have a very weak smell (much of the smell of commercial limonene is caused by impurities), but the chemical primarily responsible for the odour of lemon oil (and lemon juice) is known as citral, which is a mixture of the two geometric isomers, geranial (1.150) and neral (1.151).

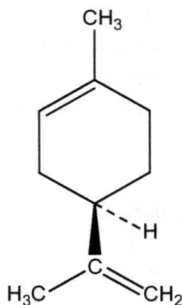

(R)-(+)-limonene

(1.149)

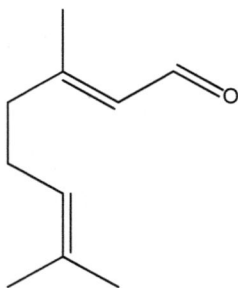

Geranial

(1.150)

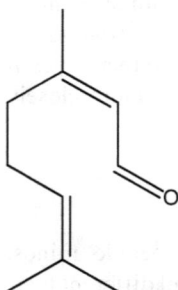

Neral

(1.151)

They have rather similar, though not identical, sharp, lemony smells, with geranial smelling rather fresher. Another lemony-smelling ingredient, though just present in trace amounts, is citronellal (1.152), used commercially in some insect repellents.

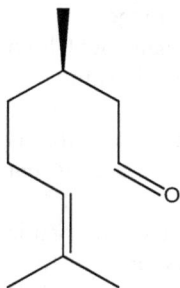

Citronellal (1.152)

Other ingredients include various sesquiterpenes and esters, principally geranyl acetate. Lemon peel oil and lemongrass oil are amongst the substances that contain citral, with lemongrass oil being the best source of it, often containing more than 70% citral. Lemon oil mainly comprises (R)-(+)-limonene, but its smell is largely due to the small amount of citral that it contains. There is less citral, just a trace, in orange oil.

Citral is an ingredient of Nasonov pheromone; it is released by (male) honeybee drones from their abdomen to recruit other males and guide swarms. Beekeepers use a man-made version of Nasonov pheromone, also containing citral, to recruit swarms.

ORANGE JUICE AND LIMONENE ISOMER SMELLS

Oranges contain a number of 'odour impact' substances, unlike lemons. Orange oil usually comprises at least 90% limonene. Limonene has two optical isomers, but the only one found in oranges is (R)-(+)-limonene (1.149), not (S)-(-)limonene (1.153).

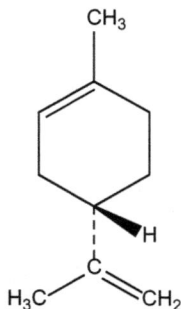

(S)-(-)-limonene (1.153)

For half a century, it has been widely believed that one isomer of limonene smells of lemons and the other of orange, with statements such as 'R-(+)-limonene smells of oranges' and 'S-(−)-limonene smells of lemons'. This belief seems to derive from a statement made in a paper published in 1971 and has since appeared in influential textbooks. It is *not true*. (*R*)-(+)-limonene has a citrus smell with an orange note, but *S*-(−)-limonene does not have a citrus smell. (*R*)-(+)-limonene is associated with both orange and lemon. The issue is complicated by the presence of impurities in commercial limonene samples.

Limonene is attractive as a substance produced from renewable sources that is biodegradeable; it has various commercial uses such as a cleaning agent and solvent and is also insecticidal.

The smell of orange juice and orange oil comes from a number of small organic molecules, notably various aldehydes, including the saturated C_8–C_{11} aldehydes from octanal (1.82) to undecanal (1.154), plus unsaturated aldehydes including neral (1.151) and the sinesals (1.155) and (1.156); various esters, notably ethyl butanoate (1.157), ethyl octanoate and ethyl decanoate, also contribute.

Freshly squeezed orange juice has a different smell from the packaged product; this is due to the presence of the aldehyde ethanal (1.158), which causes the fresh, pungent smell of the product. Ethanal's low boiling point of 20°C means that it easily escapes.

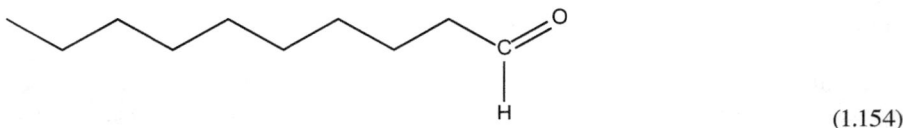

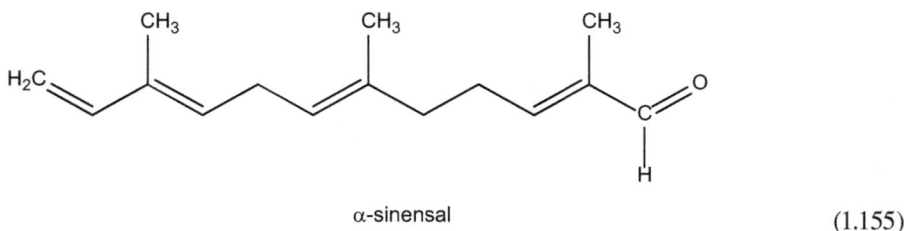

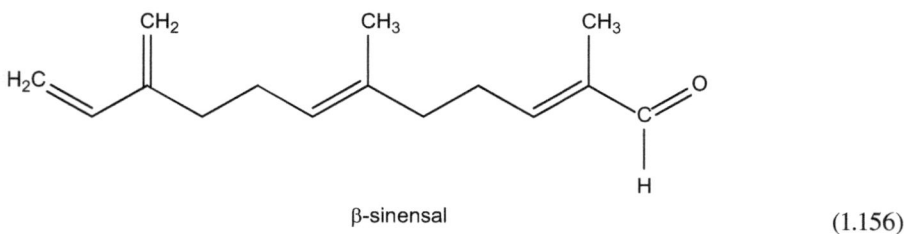

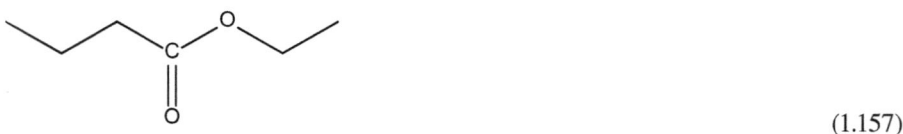

(1.154)

α-sinensal

(1.155)

β-sinensal

(1.156)

(1.157)

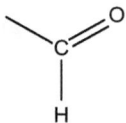

(1.158)

Geranial and neral, responsible for the smell of citral, are commercially in demand; however, these unsaturated aldehydes are easy to oxidise, and thus, there is a quest for more stable alternatives like geranylnitrile (1.159), which is more stable than citral and also has a lemon smell.

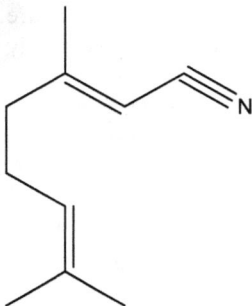

(1.159)

Lemon odours are also found with molecules of rather different shapes, such as two found in fried chicken: 3-butyl-2-thiophenecarbaldehyde (1.160) and 3-(3-methylbut-2-en-1-yl)-2-thiophenecarbaldehyde (1.161), both of which have strong citrus and citral-like smells.

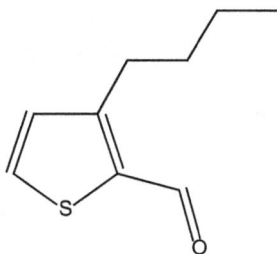

(1.160)

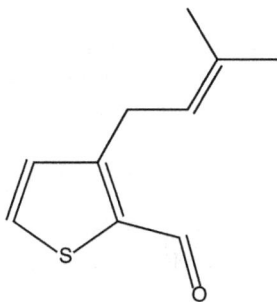

(1.161)

2 Vitamins

INTRODUCTION

Vitamins are hard to pigeonhole. As Coultate puts it, they 'are an untidy collection of complex organic nutrients that occur in the biological materials we consume as food (and in others we do not)'. They can be defined as organic substances that are essential, in tiny quantities, for an organism to maintain its health, which generally have to be obtained through the diet.

They can be divided into two groups: (1) fat-soluble vitamins A, D, E and K_1. (2) water-soluble vitamins, B_1 (thiamin), B_2 (riboflavin), B_3 (nicotinamide), B_5 (pantothenic acid), B_6 (pyridoxine), B_7 (biotin), B_9 (folic acid), B_{12} (cobalamin) and C (ascorbic acid).

Some ancient civilisations noted diseases linked to vitamin deficiency, without being clear about the cause. Thus scurvy was described c. 1550 BC by Egyptians in the Eber Papyrus. It came into prominence following Europeans' discovery of the Americas and the long ocean voyages linked with this, when thousands died from scurvy. In the mid-18th century, the Scottish doctor James Lind investigated preventing scurvy in a clinical trial, giving oranges and lemons to some sailors but not to others. Following the success of the experiment, citrus fruits became a dietary staple for sailors of the Royal Navy.

The term 'vitamin' was coined in 1912 by the Polish biochemist Casimir Funk, who believed that the substance he was investigating was a 'vital amine', hence the term vitamin. In 1916, the American biochemist Elmer McCollum further differentiated them, using letters to distinguish between them (initially A–D). Subsequently, it was found that vitamin B involved more than one factor, so it was sub-classified into B_1, B_2 and so on.

VITAMIN A

Vitamin A, as you might suspect from the 'A' in its name, was the first vitamin to be identified. In 1912, the English biochemist Frederick Gowland Hopkins showed that rats needed unknown substances, in addition to fats, proteins and carbohydrates, for their growth (he was awarded the Nobel Prize for Physiology or Medicine in 1929 for this). They were called vitamin A in 1920, and the structure of beta-carotene was described by Paul Karrer, a Swiss chemist, in 1931.

Either vitamin A or its precursors in the diet are necessary to maintain the immune system as well as good vision (its deficiency leads to 'night blindness', the loss of vision in dim light), not to mention skin and embryo development. Its main dietary sources are fish liver oil, animal livers, milk fat and egg yolk, whilst it can also be synthesised from the enzymatic breakdown of carotenoids like β-carotene (2.1).

DOI: 10.1201/9781003658115-2

Carotenoids are especially present in green, yellow and leafy vegetables, notably carrots, spinach, tomatoes, etc.

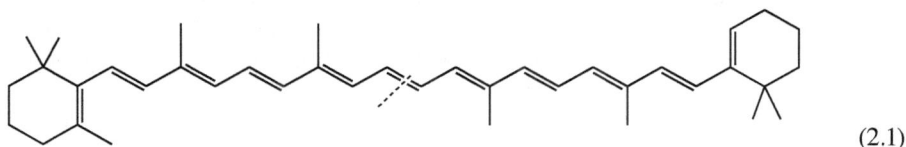

(2.1)

For example, cleavage of the central double bond in β-carotene affords retinol ((2.2), vitamin A), whose oxidation produces retinal, which is required for the process of vision. One should specify the isomer of retinal, as two isomers are involved.

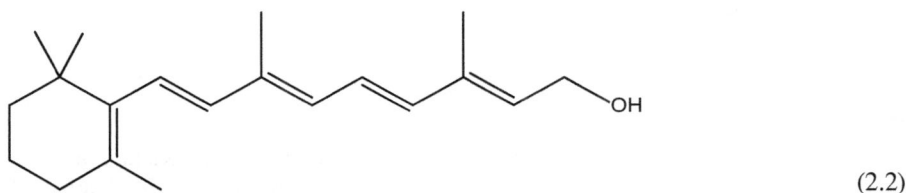

(2.2)

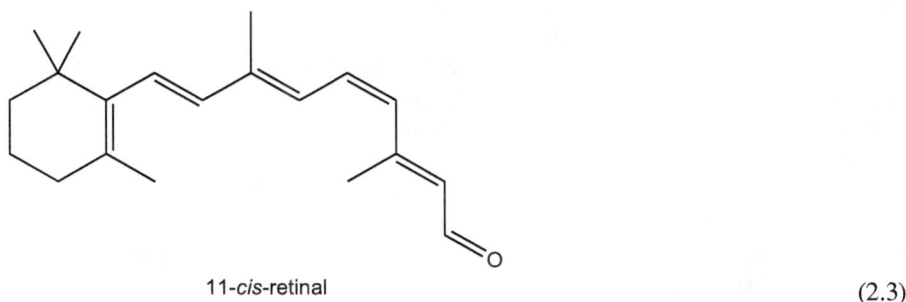

11-*cis*-retinal

(2.3)

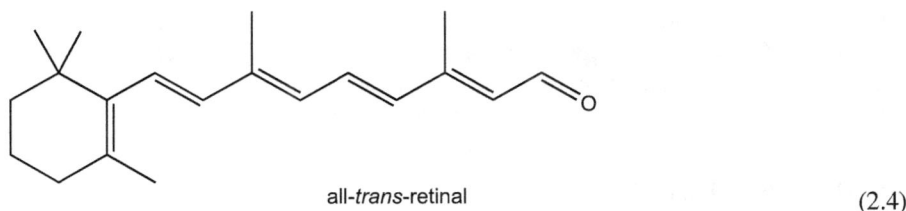

all-*trans*-retinal

(2.4)

11-cis-retinal (2.3) is found in the rods and cones of the eyes, where it is linked to proteins called opsins, forming the light-sensitive chromoproteins known as rhodopsins. There are three slightly different rhodopsins that absorb slightly different wavelengths of incident light – red, blue and green.

When light hits the rod or cone, it is absorbed by the *cis*-retinal molecule, causing the double bond at carbon-11 to change to a single bond; this means that it is no longer a rigid molecule. Rotation can occur about the 'new' single bond, changing to the *trans*-arrangement when the double bond re-forms. The change in molecular geometry from 11-*cis*-retinal to all-*trans*-retinal (2.4) has a knock-on effect upon the structure of the opsin protein. This conformational change causes a signal to be sent down the optic nerve to the brain, whilst the all-*trans*-retinal molecule, which no longer fits in the opsin, is released, subsequently being converted back to 11-cis-retinal. This can be reattached to another opsin before undergoing another cycle of the process.

In addition to the alcohol retinol, esters are important. Enzymes esterify retinol in the small intestine; these esters are transported in the circulatory system. Animal tissues store retinol mainly as esters like retinyl palmitate (2.5).

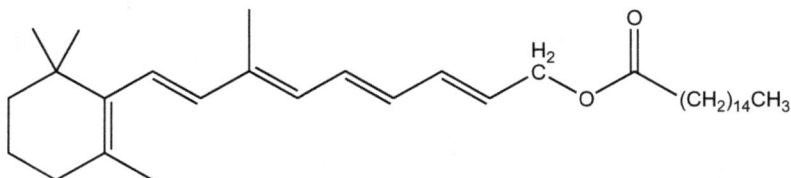

$$(2.5)$$

Retinyl palmitate is the common form of vitamin A used to supplement margarine in the UK, to give a vitamin A level comparable to that of butter. The milk 'derivatives' – butter and cheese – are much more concentrated in vitamin A than milk.

A shortage of vitamin A in the diet is not normally a problem in the developed world, but in developing countries, vitamin A deficiency is a real problem, affecting pregnant women and young children, the latter of whom can suffer from blindness or worse. Apart from vitamin A supplements, 'Golden rice' is a possible answer. This is rice that has had the genes for making carotene inserted into its DNA (leading it to be yellow in colour), making it a good source of vitamin A. This is not linked to the story that eating carrots helps you see in the dark. That was a fiction propagated during World War II to cover up the fact that British Beaufighter night fighters had airborne radar that made them very effective at intercepting Luftwaffe night bombers.

An excess of vitamin A can also be a health hazard. This has been demonstrated by some 'health fanatics' and, more famously, by some polar explorers who have eaten the flesh, especially the liver, of animals – most notably in 1913 when Xavier Mertz died after eating the liver of his huskies. It has been stated that a small portion of polar bear liver is equivalent to over 2 years' supply of retinol.

VITAMIN B$_1$, THIAMIN

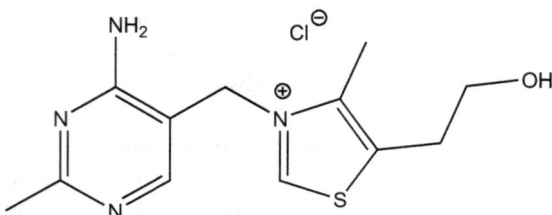

$$(2.6)$$

Vitamin B_1 has an important part in the history of vitamins. Just as scurvy was linked, especially in the Western world, with the discovery of vitamin C, so the B vitamins are linked with a disease called beriberi, which was particularly problematic in Southeast Asia. Beriberi was often associated with symptoms including numbness of the hands and feet, difficulty walking and weight loss. For a long time, it was thought to be caused by an infection. Takaki Kanehiro, a doctor in the late 19th century in the Japanese Navy, noticed that ordinary sailors, whose diet was largely white rice, were more likely to get beriberi than officers, who had a more balanced diet. Then, in 1897, a Dutch doctor, Christiaan Eijkman, working in the Dutch East Indies (now Indonesia), noted that white rice, which had been milled to remove the outer parts (mainly the husks), lacked an essential ingredient that prevented people from getting beriberi. In the early 20th century, the English biochemist Frederick G. Hopkins developed these findings, whilst in 1912 the Polish biochemist Casimir Funk isolated the substance responsible from brown rice, calling it a '*vitamine*' (he believed that these essential substances were all amines). It was not until the 1930s that the structure of thiamin was established by an American chemist, Robert R. Williams. So, it took a half-century of research on an international scale to establish what vitamin B_1 was. Amongst other roles, vitamin B_1 acts as a component of a coenzyme in carbohydrate metabolism (thus, in releasing energy), in nerve action and in muscle function.

Around 1920, the time that vitamin B_1 was discovered, there were only two vitamins: fat-soluble, which prevented night blindness, also known as vitamin A, and water-soluble, which prevented rickets, named vitamin B (subsequently, a substance that prevented scurvy became known as vitamin C). Over the next few years, more substances were discovered in 'vitamin B', leading to them being called B_1, B_2, B_3, etc.

Small amounts of B_1 are found in many foodstuffs, but only some are useful sources. It is abundant in pork, beef, fish and animal organs like liver, kidney, brain and heart. Whole grain bread and potatoes are good dietary sources, whilst (as we have seen) it is in the pericarp and germ of cereals. Human and cow's milk both contain B_1. As already noted, 'polishing' rice removes most of the vitamin B_1 content of the rice, as B_1 is in the outer part of cereal grain hulls. In practice, in most of the world, people get enough vitamin B_1 from their diet, whether from foods that naturally contain the vitamin or from foods fortified with B vitamins, such as breakfast cereals. Thus, diseases like beriberi are associated with malnutrition, notoriously in some World War II prison camps.

VITAMIN B$_2$, RIBOFLAVIN

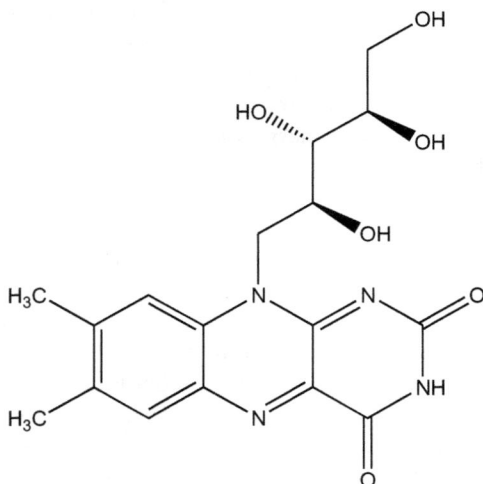

(2.7)

Interest in the B vitamins began with the discovery of a substance that helped prevent beriberi (see vitamin B$_1$). After some time, it was realised that what they called vitamin B lost its ability to prevent beriberi after being heated but retained the ability to stop stunted growth in rats, indicating that it was a mixture. Within a few years, the two compounds responsible were isolated and given the names thiamin (vitamin B$_1$) and riboflavin (vitamin B$_2$), respectively, with the isolation of riboflavin (2.7) from egg yolk being reported in 1933. The German chemist Richard Kuhn was awarded the 1938 Nobel Prize for Chemistry for his work on vitamins, including B$_2$. The name riboflavin reflects the involvement of the sugar ribose in its structure. Some more 'B vitamins' remained to be identified. Riboflavin had already been 'discovered' in 1879 by an English chemist, Alexander Wynter Blyth, who obtained it from cow's milk; he called it lactochrome, referring to its yellow–green colour when exposed to light. Riboflavin is synthesised in plants but not by animals; it is an essential human nutrient and is widely available in dietary sources (e.g. milk and milk products, eggs, leafy vegetables and meat products such as heart, liver, kidney and fish liver), and no important deficiency diseases have been described – deficiency (ariboflavinosis) is associated with symptoms such as cracking of the skin and inflammation of the tongue. In the body, it is converted into the 5'-phosphate derivative (2.8).

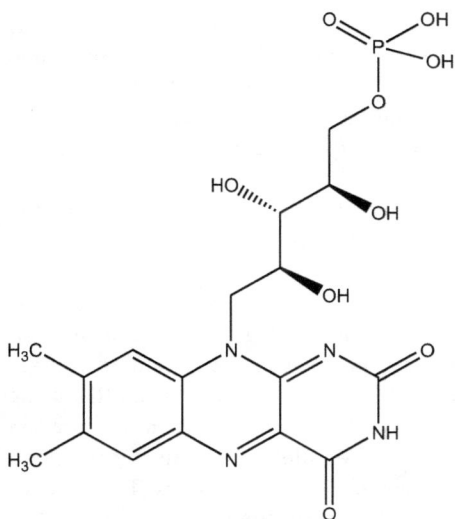

(2.8)

We now know that riboflavin is found in all plant and animal cells. It is an essential coenzyme in many metabolic pathways, such as protein metabolism and respiration. It is vital to the energy metabolism of fats, ketone bodies, carbohydrates and proteins, as well as the immune and nervous systems, the formation of red blood cells and cell reproduction.

Riboflavin is the precursor to the flavoenzymes flavin mononucleotide and flavin adenine dinucleotide; flavin adenine dinucleotide is especially important in catalysing a large number of redox reactions.

The yellow–orange colour of riboflavin means it is sometimes used as a food colouring; it is a rare case of a vitamin with an E number, E101.

VITAMIN B$_3$, NIACIN

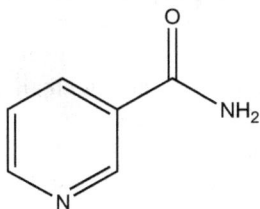

(2.9)

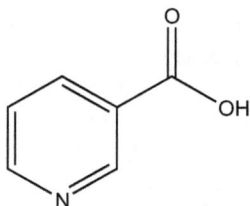

(2.10)

Vitamin B_3 is the historical name for niacin, a generic term for nicotinamide (2.9) and nicotinic acid (2.10), though it is scarcely used now. When choosing the name for the drug, it was pointed out that 'nicotinic acid' combined two undesirable words, 'nicotinic' having connotations with tobacco and 'acid' with a corrosive substance. Niacin is a name resulting from a combination of **ni**cotinic **ac**id + vitam**in**, making it less objectionable.

It is a reminder of how the first vitamins to be identified related to specific diseases, such as B_1 with beri beri, C with scurvy and D with rickets.

In the case of niacin, the disease in question was pellagra. It was associated with dermatitis (inflamed and peeling skin, mouth sores), diarrhoea and, in more extreme cases, dementia and death – it had a high fatality rate. Pellagra was unknown until the mid-18th century, but after that, it became very common in southern Europe and also in the USA from 1900 – between 1900 and 1940, it is estimated that some 100,000 Americans died from it. In the southern states of the USA, it appeared in the spring when diets shifted to a corn (maize) based diet, especially for the poor. It caused 7,000 deaths a year for several decades in 15 southern states. This novelty was a puzzle, as in South America and the Yucatan Peninsula the natives had eaten maize for hundreds of years, with no sign of pellagra. The mystery was partly solved by Joseph Goldberger, an (Hungarian-born) American doctor. He saw that it was not a contagious disease and carried out experiments changing people's diets, soon establishing that maize was indeed the cause. This was unacceptable to some politicians in the Deep South, who seemed to feel that it was more acceptable for pellagra to be infectious than for it to be linked with poverty. Finally, in the late 1930s, it was found that vitamin B_3 prevented pellagra, leading the USA (and Europe) to embark on programme of adding niacin to breakfast cereals. Pellagra soon became a thing of the past in most places, though it can still occur in times of famine and war.

The reason for that resurgence in pellagra is related to food processing. In cereals like maize, the nicotinic acid they contain is largely esterified with polysaccharides, polypeptides and glycopeptides, making it unavailable for absorption by consumers (the inhabitants of the USA and Europe). For hundreds of years, American natives had treated corn with a hot lime solution to soften it before milling (a process called nixtamalisation). In this process, the alkaline lime hydrolyses the esters, freeing the nicotinic acid, allowing Native Americans to obtain flour rich in nicotinic acid (*tortillas*, anyone?). Inhabitants of the USA used mechanical milling without alkali, resulting in milled corn with low levels of nicotinic acid. Humans can also convert the dietary amino acid tryptophan into nicotinamide, particularly in their liver, although the body preferentially uses tryptophan for protein synthesis. There are plenty of food sources for vitamin B_3 – lean meat, liver, cereals, yeast, mushrooms, whole grains and peanuts are good examples.

The body converts nicotinic acid and nicotinamide into nicotinamide adenine dinucleotide (NAD: (2.11)), one of two key related coenzymes. NAD has a nicotinamide linked to ribose and an adenine also bonded to ribose, with the two nucleosides joined by a phosphate link. NAD is essential for human life, and in turn, vitamin B_3 is required by the body to make NAD. NAD acts as a coenzyme in glycolysis and the Krebs cycle, which are processes in the body's energy production.

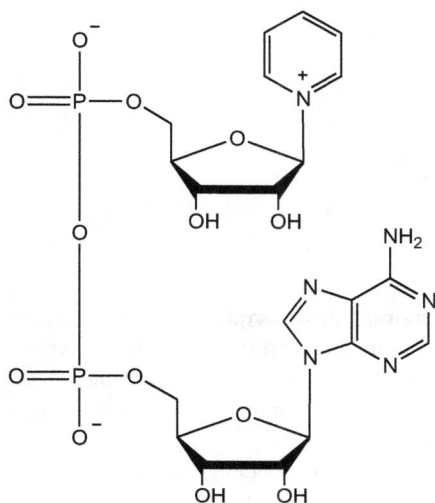

(2.11)

The second of these coenzymes, NADP (2.12), has a phosphate added to one of the riboses of NAD. NADP is involved in the synthesis of large molecules like fatty acids and steroids such as cholesterol.

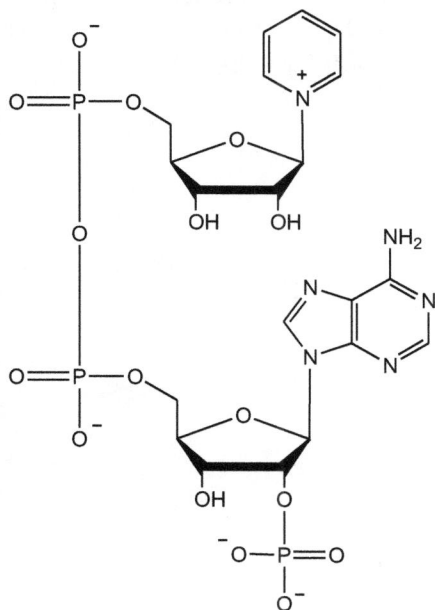

(2.12)

VITAMIN B$_5$, PANTOTHENIC ACID

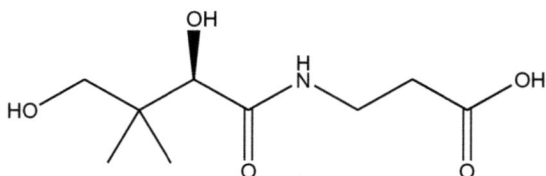

(2.13)

Pantothenic acid (2.13) was one of the B vitamins discovered in the burst of activity between the two world wars. In 1933, it was found to promote the growth of yeast; subsequently, it was discovered to support growth in animals such as rats and chickens. Its structure was reported in 1940. Pantothenic acid got its name from the Greek πάντοθεν because of its universality in foods.

It is vital for animals, as it is needed for the synthesis of coenzyme A (CoA). The metabolic breakdown (catabolism) of proteins, fats and carbohydrates in food generates pyruvate ion, which combines with CoA to form acetylcoenzyme A, the starting point for the citric acid cycle (Krebs cycle), in which, overall, acetate ions are converted into carbon dioxide and water, releasing energy. Thus, CoA is vital for cells to generate energy. It is also used in acetylation reactions.

Potatoes, tomatoes, many cereals, dairy food and eggs are regarded as good sources of pantothenic acid, but because of its wide distribution in foods, pantothenic acid deficiency is exceptionally rare (e.g. amongst starving prisoners of war). The calcium salt is regarded as a good supplement.

VITAMIN B$_6$, PYRIDOXINE

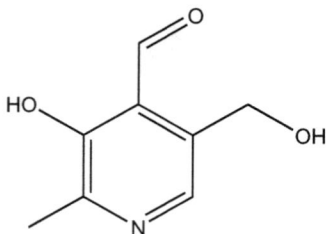

(2.14)

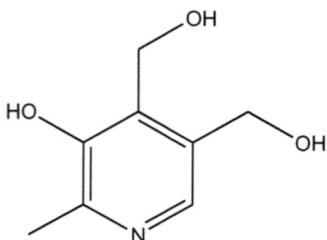

(2.15)

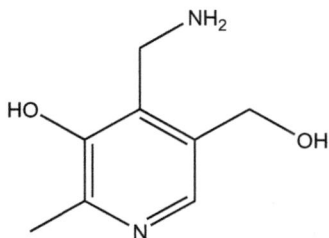

(2.16)

In 1934, the Hungarian Paul György identified a substance that he called vitamin B_6, which four years later was isolated from sources including rice bran and yeast by several groups, the year in which Richard Kuhn was awarded the Nobel Prize for Medicine, partly for his work on B_6. Synthesis of pyridoxine was reported in 1939.

Vitamin B_6 is composed of the substances pyridoxal (PL: (2.14)), pyridoxine (PN: (2.15)) and pyridoxamine (PM: (2.16)), plus their phosphate derivatives, which are all interconvertible. The most important of these is pyridoxal 5′-phosphate (PLP: 2.17), due to its biological activity as a cofactor for many important reactions involving oxidoreductase, transferase, hydrolase, lyase and isomerase enzymes, such as the conversion of α-keto acids into amino acids and L-amino acids into D-amino acids, as well as the desaturation of fatty acids to polyunsaturated fatty acids. PLP is a cofactor in syntheses including that of haemoglobin, as well as neurotransmitters like serotonin, dopamine, epinephrine, norepinephrine and GABA.

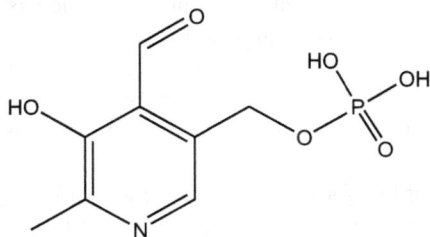

(2.17)

Beyond that, vitamin B_6 is involved in processes like cellular metabolism.

Humans need to obtain vitamin B_6 from their diet, as they cannot synthesise it. There are high levels of it in meat such as beef and pork, as well as poultry and fish, whilst plant sources provide sufficient amounts for vegetarians and vegans to avoid serious risk of deficiency. Vitamin B_6 deficiency leads to a number of symptoms, including dermatitis. The most common B_6 dietary supplement is the hydrochloride of pyridoxine.

VITAMIN B$_7$, BIOTIN

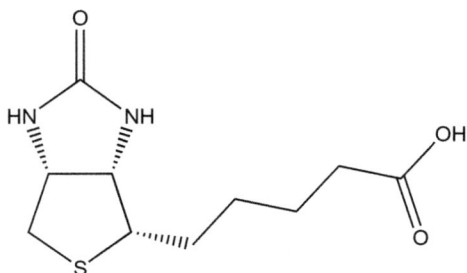

(2.18)

Biotin (2.18) has gone by two other names: it was formerly called vitamin B$_7$, and before that, Paul György referred to it as vitamin H, as he did not realise that it was one of the group of B vitamins.

Biotin has three stereogenic (chiral) centres. The isomer shown is the sole isomer with the maximum biological activity.

It has an important role as a coenzyme for several carboxylase enzymes; these are involved in the catabolism of amino acids (particularly the branched-chain amino acids (leucine, isoleucine and valine) and fatty acids, as well as the synthesis of fatty acids and the metabolism of pyruvate and lactate.

Biotin is found in most foodstuffs – good sources include liver, eggs, bananas, yeast and avocado – but 'normal' diets contain an ample amount. Deficiency is unknown, except in the (unlikely) case of consuming a large amount of raw egg white, which contains a protein called avidin that complexes biotin very strongly; cooking eggs destroys (denatures) avidin, but biotin is resistant to heating.

In recent years, the bacterial protein streptavidin has been shown to bind even more strongly to biotin.

Biotin has gained commercial popularity of late for its claimed benefits in giving its users longer, healthier hair and nails. There is some evidence for benefit in the case of people with either biotin deficiency or conditions like brittle nail syndrome, but evidence of benefit appears lacking in healthy individuals.

VITAMIN B$_9$, FOLIC ACID

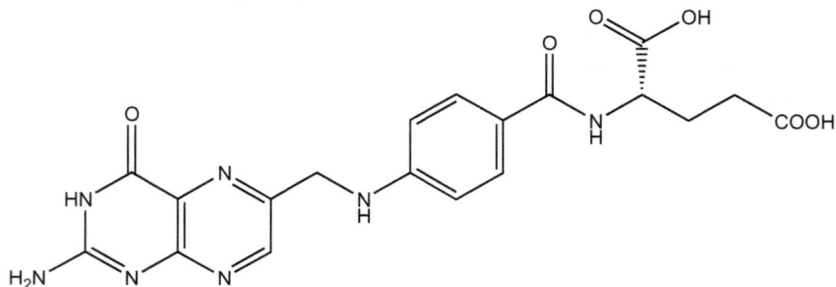

(2.19)

In 1930, the British scientist Lucy Wills was studying pregnant textile workers in India, amongst whom severe anaemia was endemic; she established that brewer's yeast treated the condition successfully. Within a few years, folate was found to be the chemical responsible, and folic acid was isolated from spinach leaves in 1941 (folic is derived from the Latin noun *folium* (leaf)).

Folate is a component of coenzymes involved in making DNA and is required for the metabolism of amino acids, as well as for cell division and maturation of red blood cells. The human body cannot synthesise folate, so it must be obtained through the diet. Folate occurs naturally in a wide range of dietary sources such as green leafy vegetables (like lettuce and spinach), asparagus, broccoli, Brussels sprouts, legumes, cereals, whole grains, yeast, lima beans and liver.

Neural tube defects (NTDs) are a serious problem caused by inadequate levels of folic acid, affecting pregnant women and their unborn children; this means that the neural tube fails to close early enough in the development of the embryo, resulting in congenital abnormalities, particularly spina bifida, which can even be life-threatening. The U.S. Public Health Service mandated the fortification of enriched grain products with folic acid by the end of 1997. Such mandatory fortification of foods has spread to over 60 other countries. Several countries have reported significant decreases in NTDs (19%–55%) since the introduction of this folic acid fortification. An estimated 300,000 NTDs occur worldwide, so this remains a serious health problem susceptible to further improvement. Folic acid is the synthetic form of folate used commercially in supplements and fortified foods; it is converted into folate in the body.

VITAMIN B$_{12}$, COBALAMIN

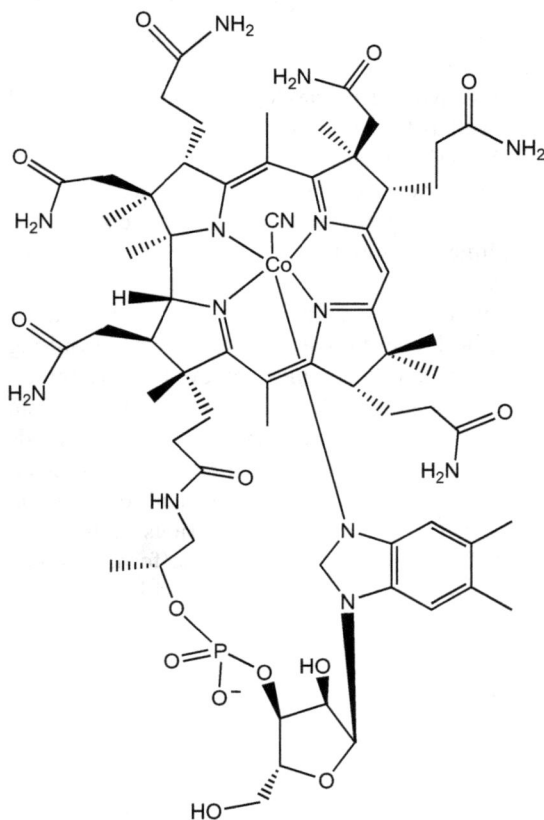

(2.20)

Although the disease pernicious anaemia – resulting in a shortage of red blood cells and all that implies – was first noted in the mid-19th century, it was not until the 1920s that George Whipple suggested that eating liver might be a suitable treatment; in 1934, George Whipple, William P. Murphy and George Minot shared the Nobel Prize for Physiology or Medicine for a liver-based treatment for pernicious anaemia. The compound responsible, which became known as vitamin B$_{12}$, was first isolated in 1948 as little red crystals of cyanocobalamin (from *Lactobacillus lactis*).

As an extremely large molecule (and the largest vitamin known, as well as the only one to contain a metal), it posed a challenging problem for the great crystallographer Dorothy Hodgkin, but solve it she did; published in 1956, it was the first structure of a metalloenzyme to be reported. It features a cobalt atom at the centre of a corrin ring system. Along with her work on the structure of penicillin, this was a major factor in her being awarded the Nobel Prize for Chemistry in 1964 (she is still the only British woman scientist to be awarded a Nobel). Vitamin B$_{12}$ was also a factor in the awarding of Nobel Prizes in Chemistry to Alexander Todd in 1957 and Robert Burns Woodward in 1965.

Cyanocobalamin, as shown in the structural diagram (2.20), is the most stable form, the form in which vitamin B_{12} is usually administered. The cyanide (CN) group is replaced by methyl or adenosyl in the biologically active forms, in which cobalt forms σ-alkyl linkages. The methyl derivative is involved in the activity of B_{12} in methyltransferase enzymes, whilst adenosylcobalamin is implicated in isomerase enzymes. The presence of the low-spin Co^{3+} ion in cyanocobalamin and other derivatives, with its $3d^6$ electron configuration, is responsible for the stability of cyanocobalamin. As a transition metal, cobalt can adopt other oxidation states (+2 and +1) that enable it to supply deoxyadenosyl radicals or CH_3^+ cations in its reactions.

In the human body, vitamin B_{12} has roles including DNA synthesis (together with folic acid), where its deficiency leads to the red blood cell problem already mentioned. It is also involved in carboxylic acid synthesis. In cows and sheep, vitamin B_{12} is synthesised in the rumen by microorganisms. The human body cannot make vitamin B_{12}; it has to get it from the diet, which includes meat, poultry, eggs and dairy products; thus, vegetarians and vegans are at risk of B_{12} deficiency (as are pregnant women and the elderly). However, supplements are readily available, and breakfast cereals are often fortified with it. It is also a growth factor in many animals, leading to the fortification of some animal diets. B_{12} deficiency in humans leads to symptoms including a smooth tongue, gastrointestinal disturbances and nervous symptoms. It is also suggested that B_{12} can combat cognitive impairment, especially in the elderly, such as Alzheimer's disease and dementia.

VITAMIN C, L-ASCORBIC ACID

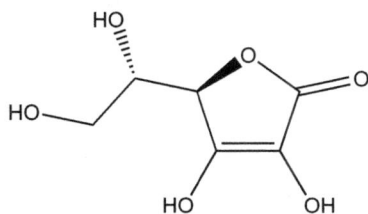

(2.21)

Scurvy is a disease with a long history. Although it is associated with sailors in the age of exploration, its symptoms have been known for thousands of years, described in the Egyptian Eber papyrus around 1500 BC and by the Greek writer Hippocrates a thousand years later. People knew ways of treating it. In 1536, the local Iroquois Indians told the French explorer Jacques Cartier how to make tea from the eastern white cedar tree to cure his sick sailors. Famously, in 1747, the Scottish naval surgeon James Lind (1717–1794) carried out clinical trials on board HMS *Salisbury* to show that oranges and lemons were the most effective remedy for scurvy. However, it was a good half-century before this practice was universally adopted in the Royal Navy.

It was not until 1928 that the Hungarian biochemist Albert Szent-Györgyi isolated from paprika a compound $C_6H_8O_6$, which he called hexuronic acid (Szent-Györgyi earlier coined the name *ignose* on the grounds that he did not know what it was); it turned out to be L-ascorbic acid (in Greek, that means 'no scurvy'). In 1932, Norman Haworth's team at the University of Birmingham synthesised it and deduced its molecular structure (2.21). All those hydrophilic –OH groups explain why it is a water-soluble vitamin, and the C=C bond suggests why it is an antioxidant (and also easy to oxidise). It was called vitamin C, as it was the third vitamin to be recognised. Their work on vitamin C was acknowledged when they received Nobel Prizes in 1937, when Szent-Györgyi was awarded the prize for Physiology or Medicine, and Haworth shared the prize for Chemistry. Many animals can synthesise their own vitamin C, but humans and many other primates – and guinea pigs – cannot; most vegetables and fruits are a good source.

So, what does vitamin C do? The answer is a list as long as your arm (which, in fact, vitamin C helped make). It is vital in the formation of collagen, the most abundant protein in vertebrates, which is involved in connective tissue, without which the typical symptoms of scurvy arise – wounds that fail to heal; painful joints vulnerable to fractures; bleeding gums; and internal bleeding. It is needed for the synthesis of cartilage, dentine, bone and teeth, as well as the formation of new tissue which is vital for healing wounds. As a reducing agent, it may assist in the uptake of iron, reducing it to the Fe^{2+} state, the state in haemoglobin and myoglobin. It is active against free radicals, and it is essential for the synthesis of the amino acids tyrosine, phenylalanine and tryptophan. Linus Pauling advocated taking it for the common cold, though this advocacy did not gain him a third Nobel Prize.

So that is vitamin C, just the stuff for a scurvy knave like Captain Jack Sparrow.

VITAMIN D, CALCIFEROL

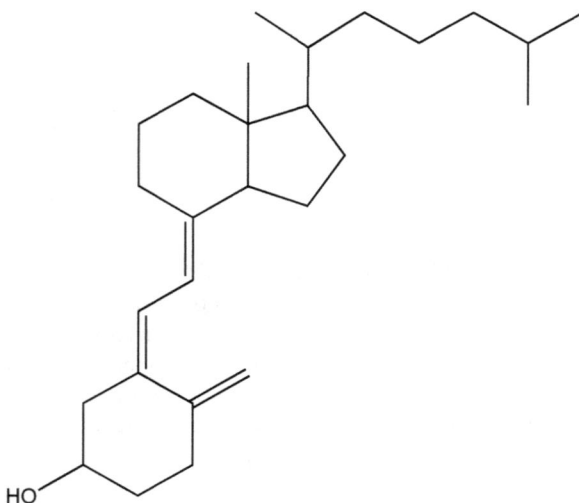

(2.22)

Like vitamin B, for example, vitamin D is a group of substances rather than just one, with vitamin D_3 (cholecalciferol) being the best known. The human body gets vitamin D_3 from its diet, and it is also biosynthesised from cholesterol via 7-dehydrocholesterol, which is converted into cholecalciferol in the epidermis of the skin through the action of sunlight (2.23).

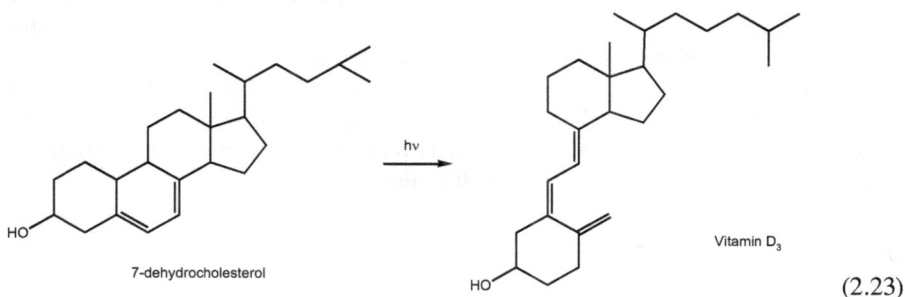

7-dehydrocholesterol

hv

Vitamin D₃

(2.23)

In the liver, cholecalciferol is hydroxylated to convert it into calcifediol; this, in turn, undergoes a further hydroxylation by the kidneys, turning it into calcitriol (2.24), a hormone that controls calcium metabolism and promotes the synthesis of the proteins that transport calcium and phosphate ions in the body. Calcitriol may be termed the physiologically active form of vitamin D.

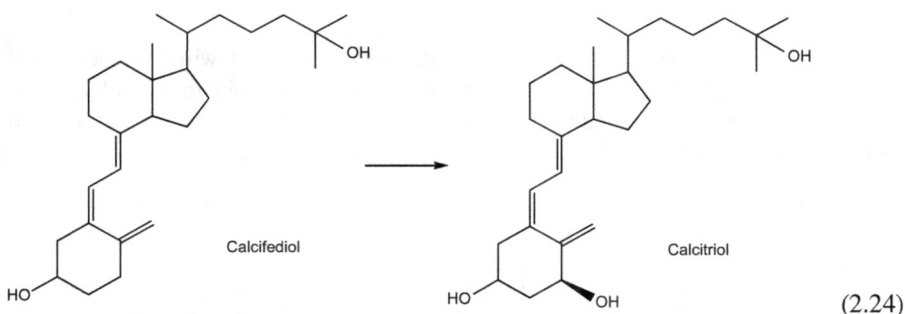

Calcifediol

Calcitriol

(2.24)

When human skin receives adequate exposure to sunlight, the body can make enough vitamin D_3 for its daily needs, around 2.5–10 μg (micrograms). Dietary sources of vitamin D_3 are limited. There are high levels in fish liver oils, especially halibut liver oil, whilst some food sources are fortified with D_3 in certain countries (like the USA), particularly milk products. Many breakfast cereals have vitamin D_3 added. Generations of children (like the present writer) were given dietary cod liver oil to boost their vitamin D levels, and even in recent years, the UK Government has suggested that the population should take vitamin D supplements in the relatively sunless winter months.

Historically, vitamin D deficiency has been linked with rickets, a childhood disease caused by a lack of proper bone development. It was first described in the 17th century and was linked with urban poverty; in late Victorian times, around 80% of poor London children showed symptoms – typically bowed legs – associated by some, correctly, with a shortage of sunlight. In the early 20th century USA, it was a major public health concern. Changes in the typical American diet – higher protein and milk levels, plus vitamin D fortification – led to marked improvements in a short period of time. Rickets is nowadays mainly confined to low-income parts of Africa, Asia and India, though in the UK, some people in immigrant communities who have darker skin (whose high melanin levels hinder vitamin D synthesis) and consequently lower exposure to sunlight (especially in the winter months), together with dietary factors (low levels of meat, fish and eggs), can be vulnerable.

The discovery of vitamin A in cod liver oil early in the 20th century led to tests on dogs, which revealed that the dogs did not develop rickets. Next, it was found that the same outcome occurred when the cod liver oil had its vitamin A content removed, leading to the discovery of vitamin D as the cure.

VITAMIN E

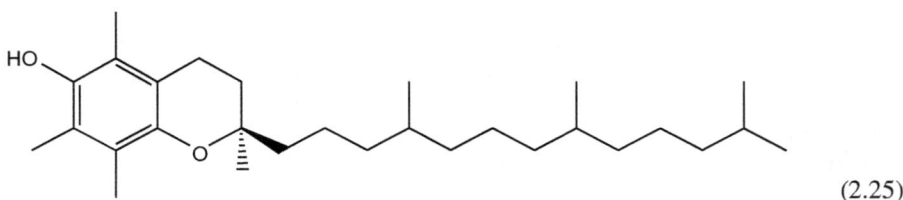

(2.25)

'Vitamin E' represents a family of similar compounds, of which α-tocopherol (2.25) has the highest antioxidant power. The tocopherols differ in the substituents around the benzene ring; additionally, the tocotrienols (2.26) have an unsaturated hydrocarbon side chain (tail).

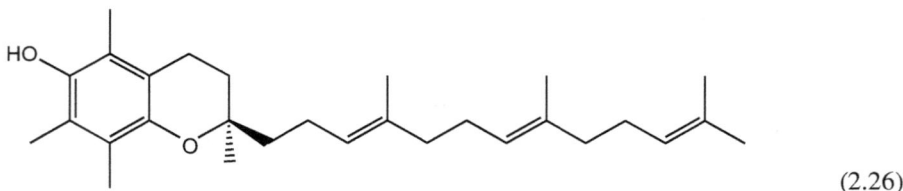

(2.26)

Its most important role is in stopping lipids from being oxidised, thus helping protect membranes, notably by countering the effect of peroxides. It also protects other substances, including vitamin A. It is found in vegetable oils, notably plant seed oils like wheat germ oil, palm oil and cottonseed oil. Other sources include cereals and bread, as well as some vegetables, including tomatoes, asparagus and almonds. In practice, people eating a balanced diet can get the vitamin E they need from these sources.

Vitamin E was discovered in the 1920s following research by Herbert Evans and Katherine Bishop of the University of California in Berkeley, who discovered that an extract of wheat germ cured infertility in some laboratory rats. The compound responsible was later isolated and subsequently named vitamin E. This has given vitamin E a reputation (yet to be authenticated) for enhancing sexual performance, which has been exploited in parts of the supplements industry.

Vitamin E acetate – the acetate of tocopherol (2.27) – gained a bad reputation a few years ago when it was found to be added to vaping liquids as a thickening agent, and it was linked to cases of lung disease. Linked with this was the discovery that the carcinogen benzene, as well as alkenes and the highly toxic ketene, were formed when vaporising vitamin E acetate (see chapter 9, p. 204).

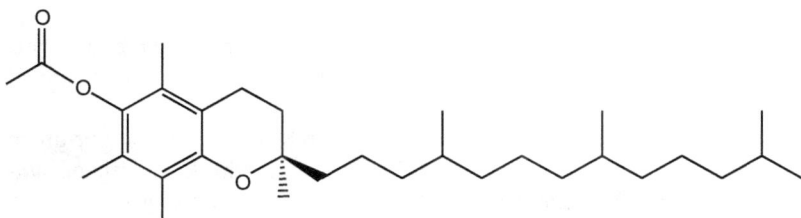

(2.27)

VITAMIN K

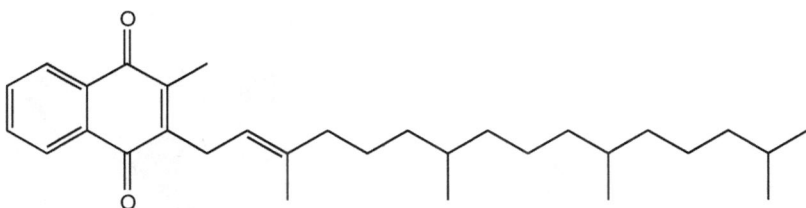

Vitamin K$_1$ (2.28)

The term 'vitamin K' represents a family of molecules with similar structures, used in a way similar to how 'vitamin B' represents a group of similar vitamins. They are important in helping blood to clot and thus in the healing of wounds. It is also believed that vitamin K helps to strengthen bones.

Plants, especially leafy green vegetables like spinach, broccoli and cabbage, synthesise vitamin K$_1$ (2.28) and use it in photosynthesis. It was discovered in 1929 when the Danish scientist Henrik Dam found that feeding chickens a cholesterol-depleted diet resulted in bleeding. This could not be stopped by feeding the chickens pure cholesterol, so evidently, the depletion process had removed another chemical too, which turned out to be what we now know as vitamin K. It received this designation because the initial publications in a German journal called it the 'Koagulationsvitamin'. Along with American Edward Doisy, who also conducted significant early work on vitamin K, Dam was awarded the 1943 Nobel Prize for Medicine.

Most of the vitamin K$_1$ that we consume gets converted in the body to vitamin K$_2$ (2.29), which has a similar structure and is the molecule responsible for making the proteins that cause clotting. K$_2$ is also found in liver, milk, cheese and fermented soy products.

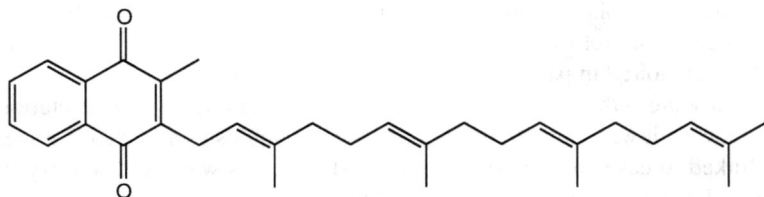

Vitamin K$_2$

(2.29)

Three more vitamins, K$_3$ (2.30), K$_4$ (2.31) and K$_5$ (2.32), are synthetic and lack the extended side chain found in K$_1$ and K$_2$. They are sometimes used in nutritional supplements for animals, notably K$_3$.

Normal adults with a balanced diet get all the vitamin K they need from the green vegetables in their diet. However, new-born babies are born with low levels of vitamin K and are not yet able to make it efficiently, so they are sometimes given vitamin K supplements to avoid internal bleeding.

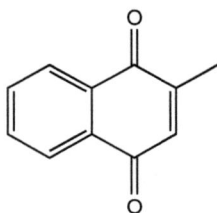

Vitamin K$_3$

(2.30)

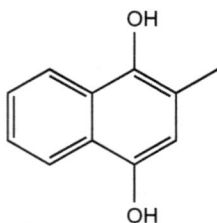

Vitamin K$_4$

(2.31)

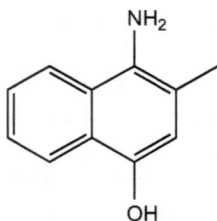

Vitamin K$_5$

(2.32)

3 'Hot' and 'Cold'

INTRODUCTION

No one knows exactly when humans first started adding flavouring molecules to their otherwise bland meals. We take these spices for granted now. Besides their taste, they often provide other sensations.

SPICES AND 'HOT'

Spices have been around for a long time; most have an Asian origin, with chillis being the principal exception. Thousands of years ago, a wide range of spices was available in the East – pepper (India), ginger (SE Asia), nutmeg (the Banda Islands), cloves (the Moluccas), cinnamon (Sri Lanka), turmeric (SE Asia) and cardamom (India). Over 2000 years ago, these spices were traded across Asia to China, Arabia and the Mediterranean world. China connected to the Mediterranean via the Silk Road and from the southeast via Arabia and Africa. Vietnamese archaeologists recently studied footed grinding slabs, traditionally associated with India and Sri Lanka since around 500 BC, discovered in the Mekong area of southern Vietnam. One of these slabs was dated to 207–326 AD by ^{14}C dating of associated material. Plant micro-remains found on the surfaces of the slabs were identified as ginger, turmeric, clove, nutmeg and cinnamon (amongst others) – all indispensable today for the preparation of curries. It is now believed that South Asians introduced their culinary traditions into Southeast Asia around 2,000 years ago. In 1271, Marco Polo witnessed large amounts of pepper being imported into China; even then, this trade dated back over a millennium.

Ancient Egypt used spices – peppercorns were found on the mummified body of the Pharaoh Ramesses II (Ramesses the Great, d. 1213 BC). Thus, there was a thriving trade in pepper long before the Western world became aware of it. Cinnamon is mentioned three times in the Old Testament (Exodus 30:23; Proverbs 7:17 and Song of Solomon 4:14) as well as in Revelation (18:13), indicating that this spice (and presumably others) was well known in Palestine over 2000 years ago. According to Herodotus, the Greeks knew of pepper in the 4th and 5th centuries BC, whilst the ancient Romans used and esteemed it as their spice of choice for cooking, especially in spiced meat dishes. The Romans traded with India, as coins from the reigns of Augustus and Tiberius have been discovered in that region. Medieval England (and, of course, Europe) was familiar with and used pepper, cloves, cinnamon, ginger and saffron. By the 15th century, Venice dominated the spice trade, which is why Portugal sought to find a route to the Indies. Around 1498, Vasco da Gama arrived on the Malabar coast of western India, opening up trading routes to the East; subsequent Portuguese expeditions were able to obtain spices from India, particularly pepper.

DOI: 10.1201/9781003658115-3

The Portuguese became European leaders in searching for spices, such as mace and nutmeg from Banda, cinnamon from Ceylon, pepper from India and Malabar, and cloves from the Moluccas.

At this point, Columbus was trying to reach Asia by sailing west from Europe, but America got in the way. Early European settlers there encountered a plant whose fruit mimicked the pungency of black pepper (*Piper nigrum*). The pods were red, so they called it red pepper, now classified as Capsicum (derived from the Greek kapto, meaning to bite). Chillis had been eaten in countries like Mexico since around 7000 BC, but did not appear elsewhere in the world until after the new settlers took it back to Europe, whence it spread into India, China and Japan. Unlike most new foods from the New World, it was rapidly incorporated into the cuisines. Now, some three-quarters of the world use chilli regularly in their diet.

As the 17th century progressed, the Portuguese lost most of their bases in Asia to the Dutch (and English). The English had a trade treaty with the headmen of Run, the remotest Banda Island. The Dutch invaded, and in the end, the Treaty of Breda (1667) saw the English give up their rights in exchange for rights to the island of Manhattan (not a bad bargain in the long term). The British transferred nutmeg production to the West Indies island of Grenada, today its leading producer.

Spices generally contain at least one significant constituent, though none have such a background as capsaicin in chillis.

CAPSAICIN

There has been a revolution in British cuisine since World War II, typified by the spread of Asian food. If one molecule can symbolise this, it is capsaicin (3.1). Capsaicin is the "hottest" of a family of structurally related molecules known as capsaicinoids, made by pepper plants of the genus Capsicum; they are secondary metabolites produced by the peppers as a defence against fungi and herbivores. The 'heat' of a chilli pepper comes largely from capsaicin, along with the very similar dihydrocapsaicin (3.2). Capsaicin accounts for some 70% of the capsaicinoid content of a chilli, whilst dihydrocapsaicin accounts for another 20%. Capsaicin itself comprises just over 0.1% of the mass of a chilli.

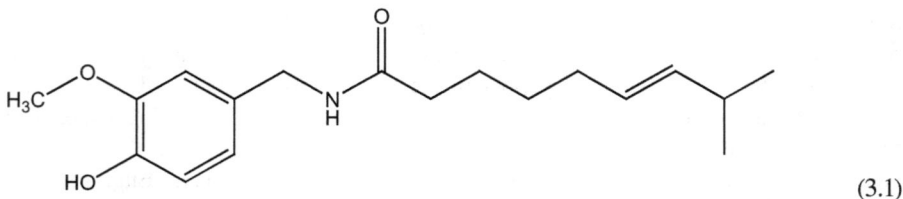

(3.1)

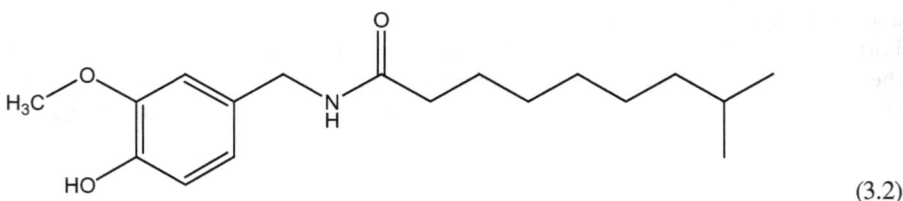

(3.2)

Other capsaicinoids include nordihydrocapsaicin (3.3) and homocapsaicin (3.4).

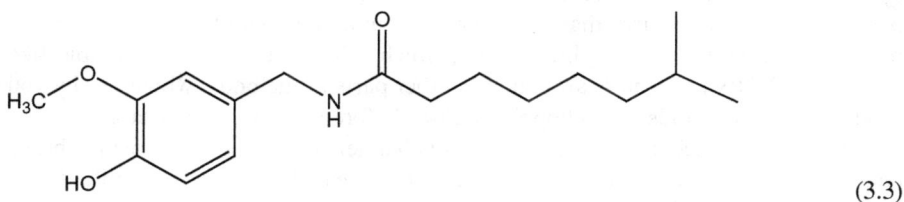

(3.3)

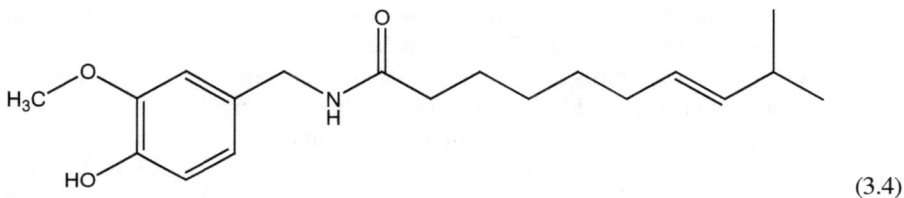

(3.4)

As already noted, chilli peppers have been part of the diet in Central and Southern America from Peru to Mexico for 6,000 years and more, before being brought back to Europe in the 16th century, and thence to Asia and the Indian subcontinent. Plants use an enzyme, *capsaicin synthase,* to make capsaicin from vanillylamine and 8-methyl-6-nonenoic acid (3.5).

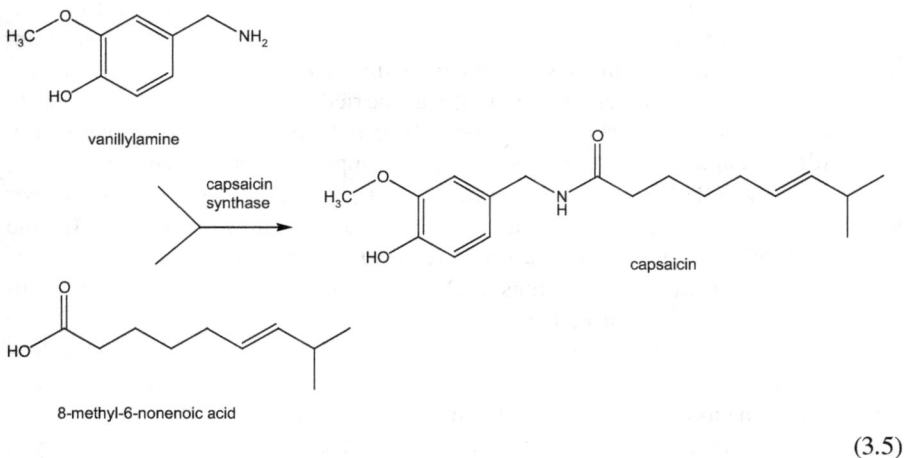

(3.5)

THE TRPV1 RECEPTOR

Just how capsaicin produces a "hot" response in the body has only been discovered relatively recently (1997). The receptor involved is referred to as TRPV1 (**t**ransient **r**eceptor **p**otential cation channel, **v**anilloid subfamily member **1** (capsaicin belongs to the vanilloid family of molecules)). When this is activated, a burning sensation results; receptors are found particularly in the mouth, lips, throat and tongue, accounting for the "burning" effects of curries. Much of a capsaicinoid molecule is hydrophobic and not water-soluble, which is why reaching for the water jug or a beer is not the answer to a curry that is too hot for you; it is thought that milk is the best option, as it contains the lipophilic casein, which is better at removing the lipid-like capsaicins. TRPV1 receptors are found in other parts of the body, which is why you should wash your hands after chopping chillis, before you go to the lavatory.

The capsaicin receptor is a protein cation channel containing six transmembrane helices, found in the outer membrane of certain nerve cells, the active receptor being made up of four of these subunits; when capsaicin binds to the receptor, the channel opens and Ca^{2+} ions enter the cells, causing a signal to be sent to the pain-processing centres in the brain. The receptor is activated by capsaicin and other capsaicinoids, as well as by temperatures above $\sim 42°C$ ("noxious heat"). The receptor integrates these responses, which is why chillis (and curries) taste "hot"; another way of looking at it is to think of noxious heat and capsaicin both pressing the same 'switch' when they activate the same receptor, so that the brain gets the same signal, which it interprets as 'hot'. There are other ingredients in spices that also act as agonists to TRPV1. Capsaicin binding to TRPV1 appears to involve both hydrogen bonds and van der Waals interactions.

Other TRP receptors mediate the feelings of heat between extreme cold and extreme heat. TRPV1 itself has been suggested to be involved in body temperature regulation. The "hot" response to capsaicin is thought to release endorphins in the brain, which gives a sense of well-being, a possible reason for addiction to curry! Upon exposure to noxious heat, linoleic acid in cell membranes in mouse and rat skin forms metabolites including 9- and 13-hydroxyoctadecadienoic acid (9- and 13-HODE), which activate TRPV1 and can thus produce inflammation and pain. Much of the research on the TRPV1 receptor that detects capsaicin (and the corresponding TRPM8 receptor for menthol) was carried out by the research group of David Julius, who in 2021 shared the Nobel Prize in Physiology or Medicine jointly with Ardem Patapoutian for their discoveries of temperature and touch receptors.

In birds, the TRPV1 channel is not activated by capsaicin, so that chilli pepper seeds consumed by birds pass through the digestive tract and get dispersed. On the other hand, other mammals like squirrels respond to capsaicin in the same way as humans, so coating bird food like nuts with chilli powder dust (or Tabasco sauce) will keep the squirrels off the bird table.

In high concentrations, capsaicin is unpleasant, so it is used in pepper sprays to subdue criminals or fight off muggers (or grizzly bears, for that matter). Perhaps surprisingly, capsaicin has been used for pain relief for hundreds of years; the Mayan people used chillis to treat asthma, coughs and sore throats, whilst the Aztecs employed them for toothache. This application depends on the fact that applying capsaicin to the skin initially causes pain, but the nociceptive neurons desensitise, and thus the skin becomes desensitised after a while. Capsaicin is an ingredient in pain-relieving creams and can alleviate the pain associated with arthritis and shingles as well as post-operative pain. A mixture of capsaicin and a lidocaine derivative (QX-314) has been suggested as a local anaesthetic; capsaicin binds to its receptor and keeps pores open for the QX-314 to enter, allowing it to block Na^+ channels from the inside.

The hotness of a chilli pepper is conventionally assessed using a panel of five tasters, with the pepper source being diluted until the tasters can no longer detect it. The dilution factor is the "hotness", quoted in Scoville units. High-performance liquid chromatography has been used as a more reliable quantitative technique, but the search is on for a cheap, reliable method. Scientists have found that they can measure capsaicinoid content by adsorbing capsaicin onto multi-walled carbon nanotubes and measuring the current when the capsaicin undergoes electrochemical oxidation. Neat capsaicin has a Scoville value of 16 million, whilst dihydrocapsaicin is slightly less at 15 million, with the other capsaicinoids clocking in between 8.6 and 9.1 million. Police pepper sprays are unsurprisingly the hottest application of capsaicin, at 5 million Scoville units. Among peppers, the Dorset Naga is the hottest yet bred, at 1.6 million and the hottest habanero around 600,000, mild peppers like jalapeno average around 5,000 and the capasaicinoid-free green bell peppers are zero.

There is some evidence that capsaicin treatment can improve cerebrovascular function and cognition, potentially helping to stave off dementia.

BLACK PEPPER

Black pepper (*Piper nigrum*) is the most widely consumed spice, with Kerala supplying the best varieties. It contains over 59 volatiles, with the most abundant alkaloid piperine (3.6), which is responsible for its pungency, first identified in 1820 by Hans Christian Orsted.

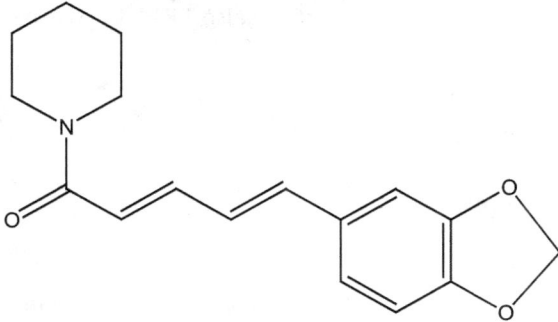

(3.6)

Like capsaicin, it is an agonist at the human TRPV1 receptor.

GINGER

Ginger (*Zingiber officinale*), which originated in Asia (though Jamaica today supplies ginger of very high quality), is more complex. Fresh ginger contains gingerol (3.7) ([6]-gingerol), which activates the TRPV1 receptor.

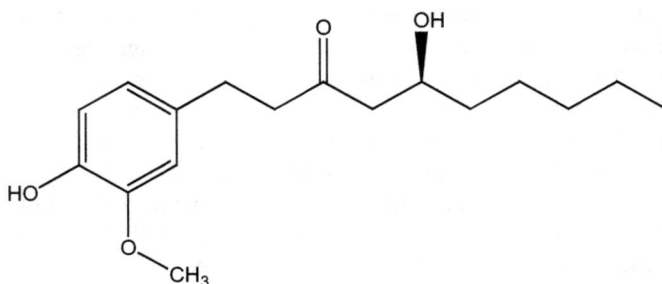

(3.7)

Upon cooking, gingerol transforms into zingerone (3.8), which is still a TRPV1 agonist and has a smell described as 'spicy-sweet'. It is only about one-thousandth as hot as the capsaicin in chilli peppers.

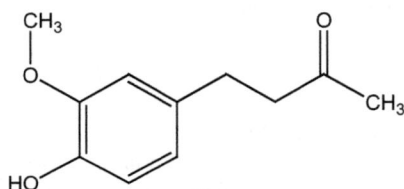

(3.8)

Drying or gentle heat converts gingerol by dehydration into shogaols, such as 6-shogaol (3.9). These are more pungent than ginger (and are also TRPV1 agonists).

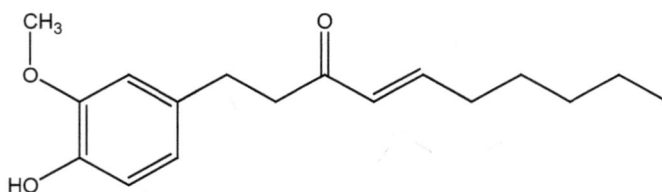

(3.9)

Evidence suggests that gingerol, zingerone and the shogaols bind to TRPV1 in a similar manner to capsaicin.

CLOVE

Clove (*Syzygium aromaticum*) gets its name from the Latin *clavus* (nail) (or the French *clou*), just think of the shape of cloves. The active molecule is eugenol (3.10), whose analgesic properties have long been utilised in dentistry, where it inhibits TRPC5 channels, which sense cold. It is also used in mouthwash and gargles because of its antimicrobial properties. On a less healthy note, it is the flavouring of the Indonesian kretek cigarettes.

Eugenol makes up as much as 85% of clove oil. It has been shown to activate other nerve channels, not just TRPV1 but also the TRPV3 and TRPA1 receptors.

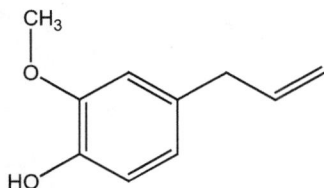

(3.10)

A possible application of eugenol resembles one already described for capsaicin (see p. 78); in combination with the lidocaine derivative QX-314, eugenol opens TRPV1 channels for QX-314 to block them, thus acting as a long-lasting local anaesthetic.

NUTMEG

Nutmeg (*Myristica fragrans*)
Various tropical sources exist for nutmeg, with Grenada making up some 40% of the supply.

Although it comprises only about 2% of nutmeg oil, the slightly peppery myristicin (3.11) is possibly the best-known ingredient of nutmeg.

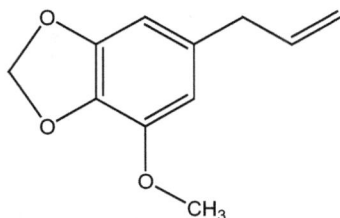

(3.11)

Much of the smell of nutmeg comes from compounds like **methyleugenol** (3.12), which belongs to the same family as eugenol (3.10).

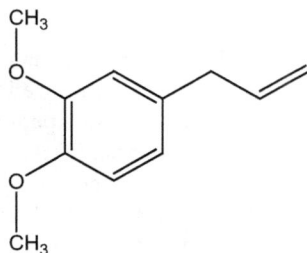

(3.12)

A compound found in nutmeg named Δ8'-7-ethoxy-4-hydroxy-3,3',5'-trimethoxy-8-O-4'-neolignan (3.13) has been found to activate the TRPM8 receptor in a manner similar to menthol, suggesting that this compound may have cooling properties.

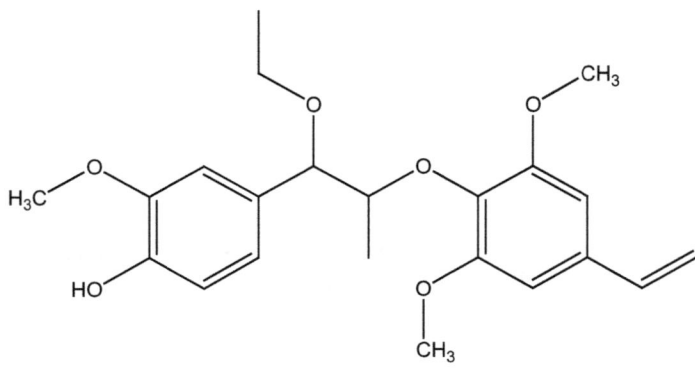

(3.13)

CINNAMON

Sri Lanka is the source of the best cinnamon (*Cinnamomum verum* or *Cinnamomum zeylandicum*), producing around 60% of the world supply. Cinnamaldehyde (3.14) contributes to its aroma as well as its antimicrobial properties. It is the main constituent of cinnamon bark oil (eugenol (3.10) is another component).

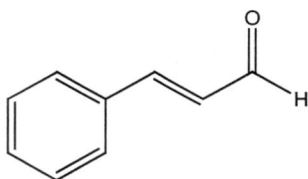

(3.14)

Like compounds in Szechuan peppers (page 82) and mustard (page 89), cinnamaldehyde produces a response from TRPA1 receptors, which are activated by noxious cold, producing burning or tingling sensations.

SZECHUAN PEPPERS

If you want a pepper offering something different from the others, then Szechuan (Sichuan) peppers are your spice. They are nothing like chillis to start with – the dried berries of *Zanthoxylum simulans*, known as Chinese prickly ash, look like little peppercorns; there is also the very similar *Zanthoxylum piperitum*, or Japanese Prickly ash. They taste different from chilli peppers. Whilst not as hot, they have other sensations too. To quote Harold McGee in *Of Food and Cooking*, "they produce a strange, tingling, buzzing, numbing sensation that is something like the effect of carbonated drinks or of a mild electric current (touching the terminals of a nine-volt battery to the tongue)", as well as a weak lemon taste. They are often used mixed with cloves, cinnamon, star anise and fennel to create "Chinese five-spice" powder.

Szechuan peppers contain a family of related molecules, the sanshools, which differ from each other much like the various natural capsaicinoids that have closely related structures to capsaicin. Like the family of capsaicinoids, the sanshools are all

amides with a long carbon chain attached, but the sanshools do not contain the phenolic group found in all the capsaicinoids. The various sanshools (3.15–3.18) produce
different sensations from each other, with α-sanshool associated with tingling and
burning and hydroxy-α-sanshool with numbing and tingling; those apart from the
α-sanshools are bitter.

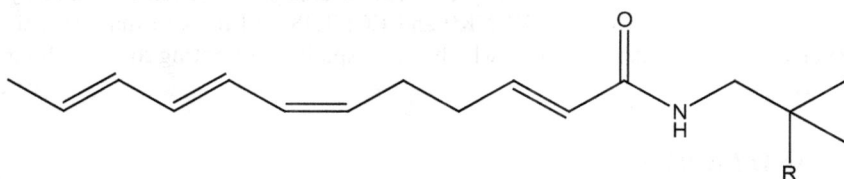

a-Sanshool, R = H; hydroxy-a-sanshool, R = OH

(3.15)

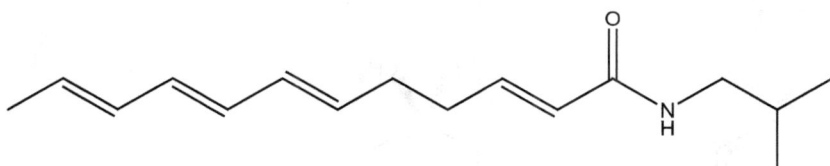

β-Sanshool   (3.16)

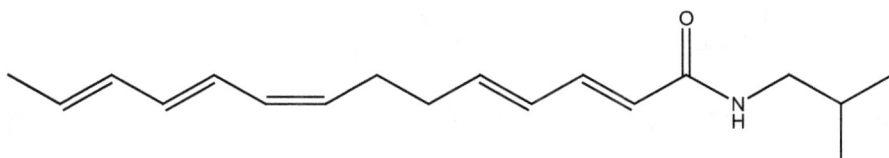

γ-Sanshool

(3.17)

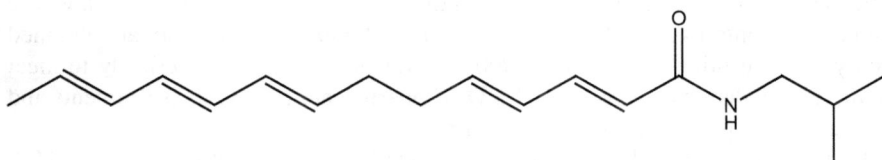

δ-Sanshool

(3.18)

Scientists are still evaluating how the sanshools work. Hydroxy-alpha sanshool is believed to be the most potent active ingredient in the Chinese peppers;

it has been reported as an agonist at the TRPV1 channel and the TRPA1 channel. Hydroxy-α-sanshool and synthetic sanshools with different carbon backbones were prepared by a team at the Nestlé Research Centre in Switzerland, who found that changes in the carbon chain, including its degree of saturation, have different effects upon the TRPV1 and TRPA1 channels. The Julius group at the University of California suggests that that hydroxy-α-sanshool creates its action potential by inhibiting K^+ channels – KCNK3, KCNK9 and KCNK18 – of the sort that are usually involved with senses like touch, which may explain the tingling and numbing sensations.

COOL AND MENTHOL

There aren't just 'hot' molecules; there are 'cool' molecules as well, substances that give you a cold sensation. The best known of these is menthol, 5-Methyl-2-(propan-2-yl) cyclohexan-1-ol. It contains three chiral carbons; nature makes just (−) menthol of the two optical isomers (3.19).

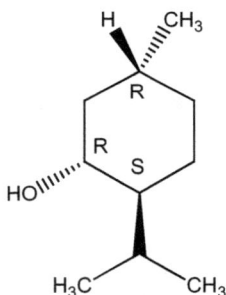

1R, 3R, 4S-(-)-menthol   1S, 3S, 4R-(+)-menthol   (3.19)

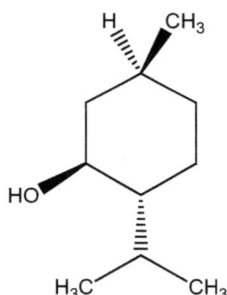

Mint makes menthol as a secondary metabolite, probably because of its antifungal or antibacterial properties. The best source is *Mentha arvensis*, corn mint, mainly grown in India; extraction by steam distillation followed by drying yields an oil that contains up to 85% (−)-menthol and that gives crystalline menthol when chilled. Other sources include peppermint oil from the hybrid mint *Mentha x piperita*, whose essential oil contains 30%–60% menthol. Some 10,000 tons of menthol are obtained each year from mint plants, with 3,000 tons (or more) made synthetically to meet the increasing demand – especially for OTC pharmaceuticals, such as liniments and decongestants, mouthwashes and toothpaste.

Menthol is not the only minty molecule found in peppermint oil – others (3.20) include the ketone menthone and the ester menthyl acetate; other esters like menthyl lactate also smell 'minty'.

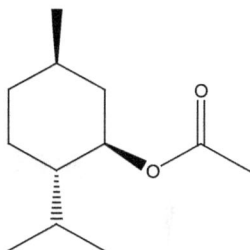

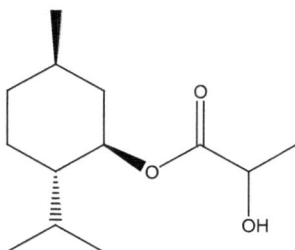

menthyl acetate   menthyl lactate   menthone

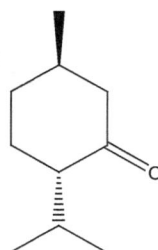

(3.20)

Both enantiomers of menthol can be made in the laboratory, but natural menthol contains only the (−) form. Compared with (−)-menthol, the (+) isomer has weaker cooling properties and a less pleasant smell, described as having "musty, bitter, phenolic and herbaceous notes". Menthol has three asymmetric (chiral) carbon atoms in its cyclohexane ring; the molecule therefore exists as four pairs of optical isomers. The other isomers are known as isomenthol, neomenthol and neoisomenthol, but menthol is the most stable molecule as the three bulkier groups (OH, CH_3 and $CH(CH_3)_2$) all occupy equatorial positions (3.21).

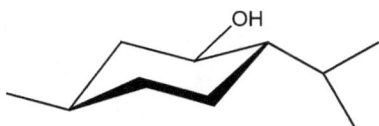

(-)-menthol   (3.21)

The biosynthesis of menthol involves nine enzyme-controlled reactions. It begins with isopentenyl diphosphate and dimethylallyl diphosphate linking to form geranyl diphosphate, which upon cyclisation affords limonene. Menthol is formed after six more stages. Expanding demand for menthol means that synthetic pathways have been increasingly sought. The best known of these is the Noyori synthesis of (−) menthol from myrcene (3.22). A key step involves the use of a chiral rhodium complex in the asymmetric isomerisation of geranyldiethylamine to (R)-citronellal enamine. Upon hydrolysis of the enamine, (R)-citronellal is formed in 96%–99% enantiomeric excess (even better than the 80% ee in naturally occurring (R)-citronellal from rose oil). Zinc bromide catalyses the cyclisation of (R)-citronellal to isopulegol, which is easily reduced to (−)-menthol. Ryoji Noyori shared in the 2001 Nobel Prize for Chemistry for this research.

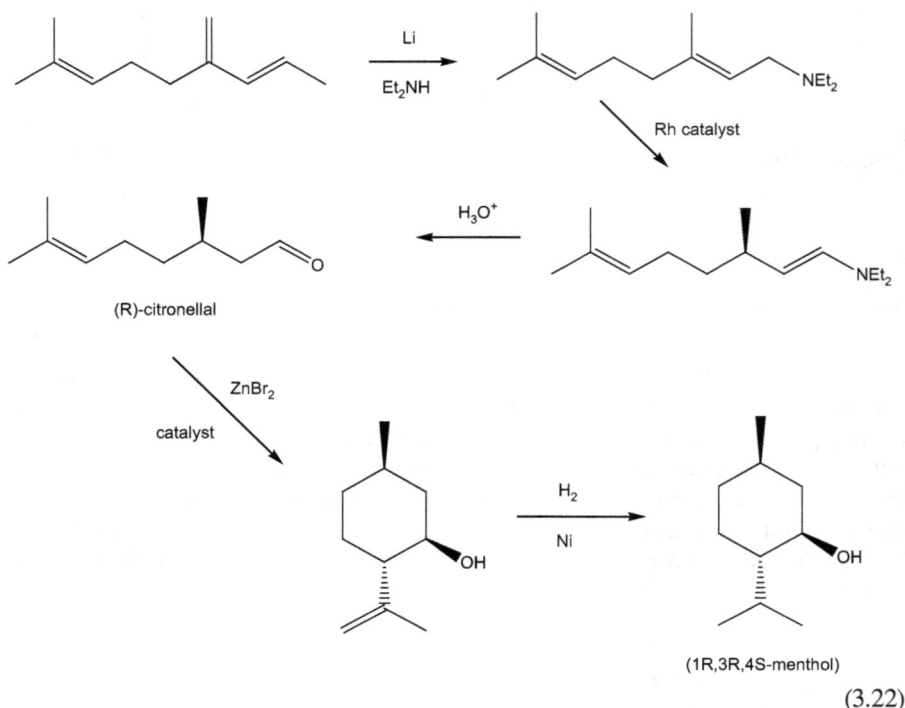

(R)-citronellal

(1R,3R,4S-menthol)

$$(3.22)$$

Menthol is generally regarded as a very safe chemical. One case has been reported of fatal menthol intoxication, caused by accidental exposure of a worker in a peppermint factory, whilst exposure to high doses of menthol in peppermint oil has caused near-fatality.

However, there is a bigger downside to menthol. Although it is used as a flavouring in chewing gum, as well as in sweets and drinks such as **Crème de menthe**, the American Mint Julep and peppermint tea, around a century ago it was introduced as a flavouring in cigarettes, specifically the American "Spud" brand. Other brands have followed. Mentholated cigarettes can contain a barely perceptible amount (ca. 0. 03%) of menthol to up to 1.0%. Menthol has cooling and slight anaesthetic effects, allowing smokers to take a bigger 'drag' and hold the smoke in their lungs for longer, resulting in greater exposure to both carcinogens and addictive nicotine. Mentholated cigarettes have been particularly advertised among African-Americans (who make up 12% of the US population), and one analysis suggested that their use is responsible for a disproportionately high number of premature deaths.

Studies using peppermint oil show that it improves exercise performance, and more specifically, basketball players run faster and perform more push-ups after inhaling peppermint. Inhaling peppermint odour raises oxygen uptake, enhances mood, and improves memory accuracy. One suggestion is that using peppermint as an in-car fragrance would help maintain drivers' alertness.

How It Works?

Menthol binds to the transient receptor potential cation channel subfamily M member 8 (TRPM8) receptor. This activates a pore, opening a channel, with Ca^{2+} ions flowing into the cell; the resulting change in electric charge sends a signal to the brain. This message is the same as the one sent when the channel is opened by cool temperatures (in the range of ca. 8°C–28°C), so that when menthol binds to TRPM8, the brain perceives 'cold'. Molecular modelling studies suggest that the menthol molecule forms a hydrogen bond between its –OH group and one particular amino acid in TRPM8 (as a 'hand'), whilst using the isopropyl group as 'legs' to 'stand' on two others. With its three chiral carbons, menthol has eight stereoisomers; (–)-menthol exhibits the greatest cooling effects. A comparison of (–)-menthol (3.23) with the four other available stereoisomers ((+)-menthol, (+)-neomenthol, (+)-isomenthol and (+)-neoisomenthol) shows that different spatial orientations affect the docking, as might be expected.

Menthol is the Arthur Herbert Fonzarelli of molecules. So cool.

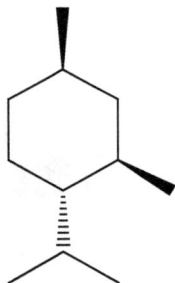

(3.23)

Carvone

Menthol isn't the only kind of minty molecule around, although it is unique in its 'cooling' effect. You'll know another one, though possibly not by its name, (R)-(–)-carvone (3.24). Used as a flavouring in mouthwashes and toothpaste, it is more familiar from its use in chewing gum. It comes from the spearmint plant, *Mentha spicata* (native spearmint), *Mentha gracilis* (Scotch spearmint) and others like *Mentha cardiaca* and *Mentha viridis*. The other optical isomer, (S)-(+)-carvone (3.25), also occurs naturally, as the scent of caraway.

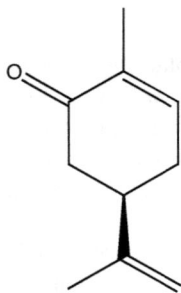

(R)-(-)-carvone   (3.24)

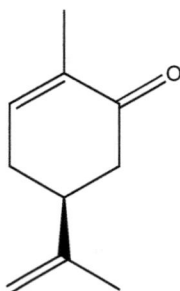

(S)-(+)-carvone   (3.25)

A New Minty Molecule

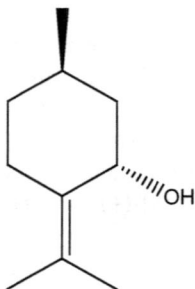

trans-pulegol   (3.26)

Along with his friend Jerrold Meinwald (1927–2018), Thomas Eisner (1929–2011) was a founding father of chemical ecology, the science of how living systems use chemicals to communicate with each other. In the late 1980s, he discovered a new 'minty' molecule while walking through a pine scrub in central Florida; when he brushed against a plant, the air almost immediately took on an intense peppermint smell. As a professor of biology at Cornell University in New York, Eisner knew his plants and identified it as a rare species, *Dicerandra frutescens*, a member of the mint family called scrub balm, which only occurs in a small area. These molecules were part of an oil contained in tiny capsules in the leaf; whenever a leaf was damaged by an insect, they acted as a powerful repellent. The molecule was the hitherto unknown (+)-*trans*-pulegol (3.26). Thus, conserving endangered species may lead not only to unknown molecules but also to new 'cool' molecules out there.

MUSTARD

Mustard has been used in cookery for thousands of years. It appears to have been enjoyed in ancient China; black mustard (*Brassica nigra*) seeds dating from around 4800 BC have been found by archaeologists there. Certainly, the Romans and Greeks used it 2,000 years ago, and the Romans brought it to Gaul, also known as France, where it was widely produced in the Middle Ages, not least because it was cheaper than pepper. The word *moutarde* was introduced into the French language in the 13th century, by which time mustard production was established at Dijon. In Norwich, Jeremiah Colman marketed Colman's Mustard from 1814 – mustard is the only spice grown in appreciable amounts in the UK. Brown mustard (*Brassica juncea*) is used to make French mustard, whilst yellow mustard (*Sinapis alba*) is used together with brown mustard to make English mustard.

Unlike capsaicin in chillis and menthol in mint, the active agent of mustard is not contained as an uncombined compound. Instead, mustard contains a compound called 2-propenyl glucosinolate, sinigrin for short (3.27).

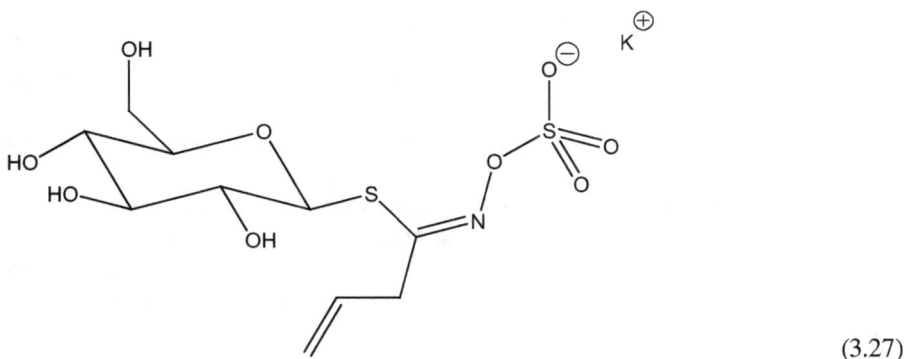

(3.27)

Sinigrin is a plant secondary metabolite. This means that it is a substance that is not essential to the life, reproduction or growth of the plant but has a 'secondary' purpose. Sinigrin's role is to be part of the plant's defence against predators like herbivores. It is stored in mustard seeds, and when the seed is chewed by an insect or animal, sinigrin is released and comes into contact with an enzyme, myrosinase. This catalyses the reaction in which sinigrin is broken down into smaller molecules, including allyl isothiocyanate (3-isothiocyanato-1-propene) (3.28).

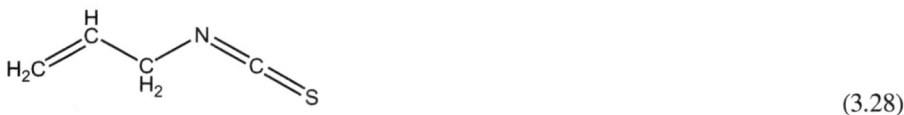

(3.28)

This is the molecule responsible for the repellent properties of mustard. It is one of several molecules that plants produce to ward off insect pests, as nicotine, cocaine and caffeine also do, by irritating nerve fibres in the insect.

And, of course, it has its effects on us too. Our bodies contain nerve cells that act as receptors; these respond to stimuli like temperature and pain, as well as to certain molecules. One important family of these is the Transient Receptor Potential (TRP) cation channels, such as TRPV1, which is activated by certain peppers like capsaicin (p. 76). Allyl isothiocyanate binds to another receptor, TRPA1, sensitive to 'cool' and also to 'pain'. Normally, when molecules activate it, they also produce the cool response, but allyl isothiocyanate binds covalently to it (3.29) and triggers the 'pain' response, explaining why mustard 'hurts'.

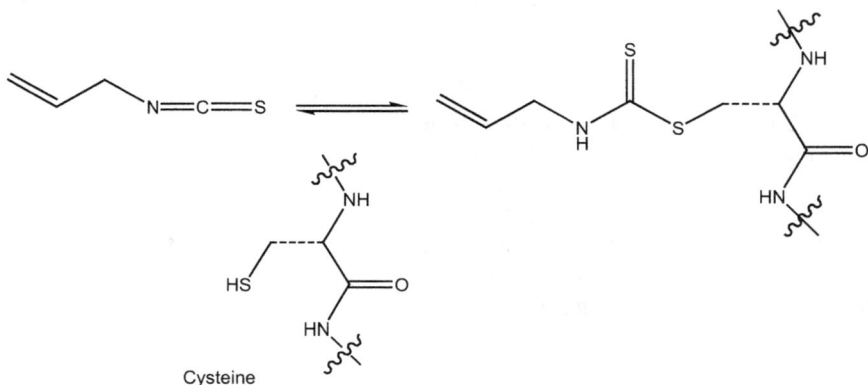

Cysteine

(3.29)

It seems that allyl isothiocyanate also activates the TRPV1 receptor, which is why mustard is 'hot' as well.

Mustard isn't the only source of allyl isothiocyanate; it can also to be found in horseradish and in the so-called 'Japanese horseradish', wasabi, an indispensable accompaniment to sushi. It is not the only isothiocyanate in mustards. Sinalbin (3.30) is a glucosinolate found in white mustard. This can act as a source of 4-hydroxy-benzyl isothiocyanate (3.31).

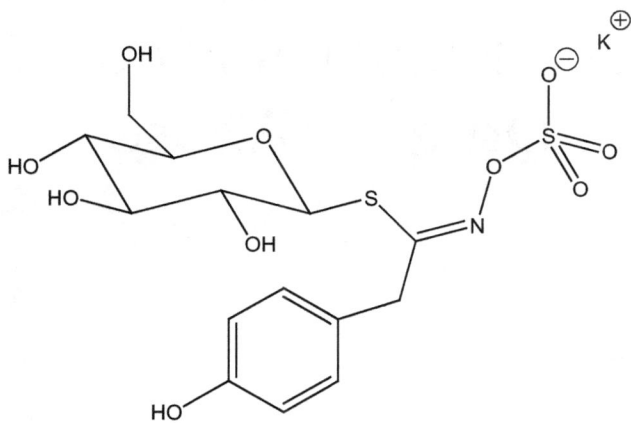

(3.30)

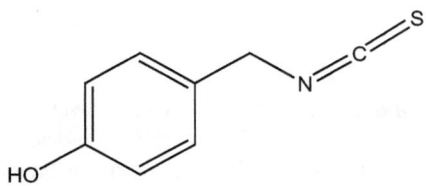

(3.31)

The Eastern Hemisphere mustard plant *Alliaria petiolata* is a source of benzyl isothiocyanate (3.32).

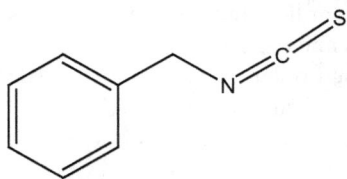

(3.32)

4 Abused Painkillers and Other Drugs of Abuse

OPIUM

In Xanadu did Kubla Khan
A stately pleasure-dome decree:
Where Alph, the sacred river, ran
Through caverns measureless to man
Down to a sunless sea.

Samuel Taylor Coleridge

Opium is obtained from the opium poppy (*Papaver somniferum*). After flowering, once the unripe seed capsules are cut, they exude a milky liquid, which, when dried, changes to a yellow–brown paste that we know as opium. Opium seems to have been used by humans since prehistoric times. Poppy seeds from the fourth millennium BC have been detected in Swiss villages, and in the same period, opium was being used by Sumerians in Mesopotamia. The opium poppy is certainly native to the Eastern Mediterranean, and opium from the second millennium BC has been found in Egypt. It was used medicinally from the first millennium BC, where writers as varied as Homer and Hippocrates wrote of it. Somnus, the Roman god of sleep, was associated with poppies, and the Romans viewed opium as a painkiller, drug and poison. In the second century AD, the Emperor Marcus Aurelius took it to help him sleep. In the succeeding centuries, it was widely traded by nations in the Near and Middle East, and at that time, it was primarily used as a medicine. Primitive versions of what we know as laudanum may have existed then. In the early 16th century, Phillipus Aureolus Theophrastus Bombastus von Hohenheim, more familiarly known as Paracelsus, developed an opium-based drug called laudanum. Ingredients may have included opium, saffron, nutmeg and musk, as well as crushed pearl. In the mid-17th century, a London physician named Thomas Sydenham concocted his version of laudanum, a liquid that involved opium, cinnamon, cloves and saffron soaked in sherry. Laudanum was promoted as a general cure-all (that included the plague). It could be used to treat cholera and dysentery, and drinking it was much safer than consuming much of the available 'drinking water'.

Until morphine could be isolated and purified in the early 19th century, laudanum was the painkiller of choice. It was used by both rich and poor. George Washington used it for his painful teeth, whilst Dr Samuel Johnson (1709–1786) took opium for the painful last year of his life. Medical practice relied on opium at that time. The great slavery reformer, William Wilberforce (1759–1833), was prescribed opium in

DOI: 10.1201/9781003658115-4

1788 for diarrhoea and fever by his doctor. In 1818, he was still taking the same dose (4 grains, three times a day) and refused to drop it. It was only with the Pharmacy Acts of 1868 and 1908 that opium and related products were regulated, so they could no longer be obtained from shops like grocers (or indeed street vendors) but were regulated by pharmacists.

Many great writers of the 18th and early 19th centuries took it. Samuel Taylor Coleridge (1772–1834) probably wrote his great poem the *Rubiyat of Kubla Khan* in the autumn of 1797. He said that it was the result of an opium-induced dream, though it was only some years later that he appears to have become an opium user, so this seems to have been under the influence of laudanum, which he started taking around 1796 for rheumatic pain in his knees (he was fully addicted by 1802 and from 1808 estranged from his family). Coleridge's contemporary, Thomas De Quincey (1785–1859), started taking opium for his neuralgia in 1804, but more regularly from 1813, at one point taking some 20 g of opium a day, about 10% of which was morphine. No wonder that he wrote an autobiography entitled *Confessions of an English Opium-Eater* (1821). Wilkie Collins wrote *The Moonstone* whilst under the influence of laudanum, and numerous other writers of the period took it, such as Byron, Keats, Shelley and Southey, as well as clergy including Lewis Carroll and George Crabbe. Sir Walter Scott was one of many writers who did not recognise what they had written under its influence. Elizabeth Siddall (1829–1862), who was the original "supermodel" for the Pre-Raphaelites, died of a laudanum overdose. 'Respectable' female users included Louisa May Alcott and Florence Nightingale.

Some people had recourse to opium in larger amounts. In the 1850s and 1860s, a large number of Chinese labourers went to the USA to build railways in the western states and also to work in mines. With them, they brought their habit of opium smoking. Rather fewer came to England, initially through their work as sailors with the East India Company. Chinese settlers brought opium shops (or 'dens') to built-up areas in the USA and Europe. The classical example of these in fiction occurs in the Sherlock Holmes story '*The Man with the Twisted Lip*' (1891), set in London's East End; another mention of such dens occurs in Charles Dickens' '*The Mystery of Edwin Drood*' (1870). Charles Dickens had been taking laudanum since he was injured in the 1865 Chislehurst train crash.

In the English Fenland – Cambridgeshire, parts of Huntingdonshire, southeast Lincolnshire and west Norfolk – the use of laudanum and opium tablets is well attested. At that time, before Fen drainage, malarial fever, known as the 'ague', was rife, and laudanum was a *much* cheaper treatment than quinine. Apart from its use as a febrifuge, laudanum helped relieve diarrhoea and pain and also made the drudgery of people's everyday lives of manual labour bearable. Chemists' shops in the Fenland sold it in immense quantities (it was estimated that half of Britain's opium imports ended up in the Fenland). As an over-the-counter cure-all, it was the equivalent of aspirin a century later for many people – but better. It was a painkiller, a sedative and a specific treatment for diarrhoea. Laudanum (and opium) was cheap and readily available. As well as in rural areas, life in the congested and unhygienic cities of England, following the Industrial Revolution, was often both unpleasant and unhealthy. The city air was frequently filthy (until the Clean Air Acts after World

War II), and the water supply was not chlorinated; cholera and dysentery were wide-spread. Diarrhoea was often a killer, especially for children.

Proprietary medicines (also known as 'patent medicines') flourished in the 19th century, many of them opium-based. The best-known of these was Dr Collis Browne's Chlorodyne, first marketed in 1856. It was devised by Dr Collis Browne (1819–1884), who had served with the British Army in Bengal and on the Northwest Frontier. He left the Army in 1856 and went into partnership with a chemist, John Davenport. Chlorodyne contained morphine and tincture of cannabis, plus other ingredients including chloroform, ether, liquorice and peppermint oil, and was recommended for a wide range of complaints, including diarrhoea, cholera, dysentery, toothache, meningitis, etc. Following the 1868 Poisons and Pharmacy Act, restrictions were introduced regarding the outlets for opiates, which were strengthened in 1892, leading to the removal of opium from these 'patent medicines'. Dr Collis Browne's Chlorodyne is still available today, but in a much modified form; in 1943, Chlorodyne was noted as containing small amounts of extract of opium and codeine; within another 25 years, the codeine was gone.

MORPHINE

Opium is a mixture that involves over 20 different alkaloid molecules. The main alkaloid present in opium, around 10%, is morphine; others include codeine and thebaine (whose name derives from poppy cultivation around Thebes in Egypt). Morphine and codeine are the only alkaloids that are painkillers. Morphine (4.1) is the better painkiller and was first isolated in the early 19th century by Friedrich Wilhelm Sertürner (1804), along with others. Named after Morpheus, the Roman god of dreams, it was found to be the best agent for dulling chronic pain and soon entered medicinal use. The emergence of the hypodermic syringe in the 1850s was speedily followed by the American Civil War (1861–1865); it is said that many Civil War battle victims in the 1860s became dependent upon morphine, as it was used 'promiscuously' on battlefield casualties. At the time of the Civil War, poppies were cultivated in states including Virginia, Tennessee, Georgia and South Carolina, providing the opium used to treat casualties who had diarrhoea, as well as being a source of morphine.

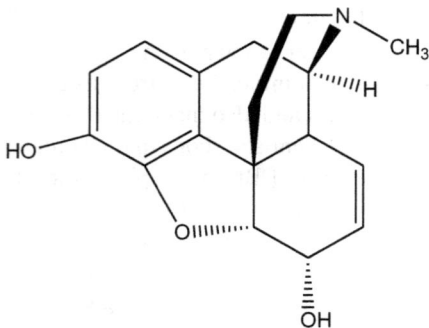

(4.1)

Shortly after the Civil War, in 1874, a London-based chemist, Charles Romley Alder Wright, who worked at St Mary's Hospital, utilised one of the newly fashionable organic synthetic routes. He acetylated morphine by heating it with ethanoic anhydride – a similar reaction was used 20 years earlier to make acetylsalicylic acid (aspirin) – and isolated diacetylmorphine (4.2), which is used medicinally in the UK under the name diamorphine. Alder Wright gave some diacetylmorphine to a colleague for animal testing. Striking effects were observed on dogs: – 'Great prostration, fear, sleepiness speedily following administration.' Some years later, in 1897, Heinrich Dreser at Bayer, the German pharmaceutical firm, tested it on animals and humans. Within a year, it was marketed as a cough medicine; it was viewed as a non-addictive yet faster-acting alternative to morphine, a wonder drug that was 'heroisch', leading it to be known as heroin. By the early 20th century, some individuals, particularly working-class men, were taking heroin recreationally.

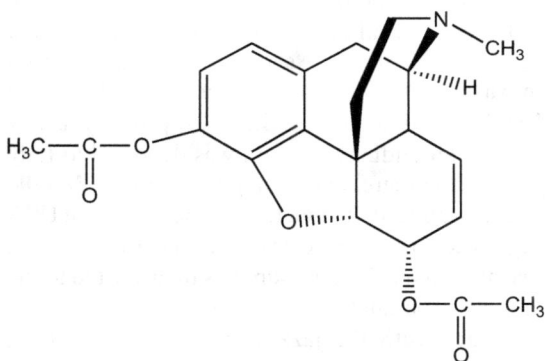

(4.2)

HEROIN

Heroin acts as a prodrug, as it is metabolised in the body to morphine, which is the active molecule in the body of a heroin user. However, when morphine is converted to heroin, two –OH groups are converted into ester linkages, making the molecule less hydrophilic and more lipid-soluble. Heroin is better at crossing the blood–brain barrier, causing its faster action compared to morphine.

After diacetylmorphine is taken into the body, it is rapidly converted into morphine by hydrolysis of the two ester linkages. The intermediate compound in which only one ester linkage has been hydrolysed, 6-monoacetylmorphine (6-MAM), may be an important molecule (4.3). Tests on rats have shown that it is the predominant opioid for most of the first half-hour after injection in both the blood and the brain, so it may cause many of the effects of heroin intake.

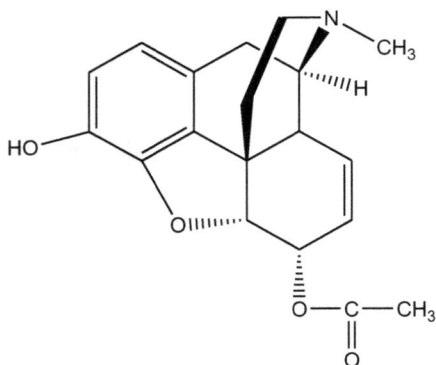

$$(4.3)$$

From the 1930s, after the end of Prohibition, heroin sales in the US started to be controlled by Mafia gangs under Salvatore Lucania (aka 'Lucky Luciano'). At that time, the opium supply to the USA was from Southeast Asia. World War II disrupted smuggling, and additionally, opium was used for medicinal purposes, so that at the end of the war, there were under 50,000 heroin addicts in the USA, possibly as few as 20,000 – so the war was a good thing for addicts. Luciano was deported to Italy in 1946; he set up a heroin supply operation there, using supplies from the Middle East. Thus, in the 1950s and 1960s, until around 1970, much of the heroin in the USA was brought in from Turkey through Marseille (hence *'The French Connection'*; think 'Popeye' Doyle), with the Corsican Mafia behind it. Supplies then shifted to the so-called Golden Triangle – Burma–Laos–Thailand.

After WW2, drugs became associated with the jazz scene – Billie Holiday, Chet Baker and Charlie Parker are names of heroin addicts that come to mind. Subsequently, heroin has continued to be associated with the world of entertainment, with victims including William Burroughs, Jerry Garcia of The Grateful Dead, Sid Vicious, Kurt Cobain, Janis Joplin, Jim Morrison and Peaches Geldof. Not to mention John Belushi, River Phoenix and Philip Seymour Hoffman, who succumbed to heroin-cocaine 'speedballs'. By the late 1950s, the number of American addicts had risen to around 100,000. The counterculture of the '60s saw increased drug use, and added to this, the Vietnam War saw many GIs serving there get hooked on heroin. By 1985, it was estimated there were 600,000–700,000 heroin addicts in the USA, mainly in urban areas; by which time Afghanistan became a major producer of heroin, with a new 'Golden Crescent', and subsequently, Colombia became an important supplier to the USA, with Mexico becoming a more important source. In the 2000s, most US heroin was sourced from Colombia and Mexico, with Mexico by itself having a 90% market share by 2016. Since 2000, the USA has had four 'waves' of mortality due to drugs; firstly, overdose deaths due to natural and semisynthetic alkaloids started to rise around the year 2000 and have risen steadily since; heroin-related deaths started to rise sharply around 2010, flattening around 2017; deaths due to fentanyl and other synthetic opioids rose very sharply from around 2013. Subsequently, a fourth, but smaller wave of overdose deaths has occurred due to methamphetamine and cocaine use along with fentanyls, in particular (p.97). Between 1999 and the end of 2015, over 183,000 deaths from prescription opioids were reported in the USA.

It is believed that there are around a quarter of a million heroin addicts in the UK.

FENTANYL

Paul Janssen (1926–2003), the son of a Belgian doctor, was a brilliant synthetic chemist. Additionally, he was an entrepreneur who set up the pharmaceutical firm Janssen Pharmaceutica, which developed a large number of new medicinal molecules, of which the fentanyl family is the best known. At the start of his career, Janssen knew that people were looking for a better painkiller than morphine, which had long been the standard. Pethidine ((4.4), also known as meperidine), had come into use but was less effective, so Janssen decided to develop molecules based on elements of the pethidine structure.

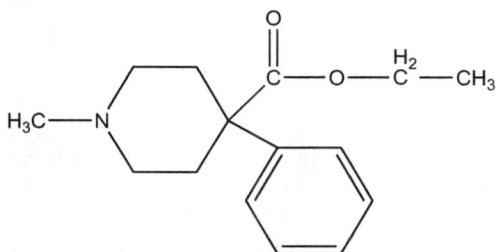

(4.4)

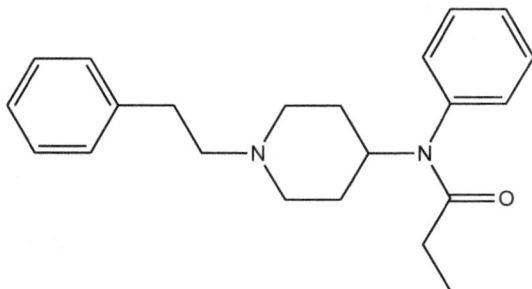

(4.5)

He was looking for molecules that were more lipid-soluble so they would enter the central nervous system faster and provide more rapid analgesic action. Janssen synthesised fentanyl (4.5) in 1960, and it was in use as a fast-acting painkiller in much of Europe within 3 years and in the USA in 1972 as the intravenous anaesthetic Sublimaze. A decade later, fentanyl transdermal patches (Duragesic) were introduced; these provide slow release into the bloodstream over 48 hours for palliative use.

Other fentanyl derivatives followed in the 1960s and 1970s, with sufentanil and carfentanil being the most successful. Fentanyl is about 100 times more potent than morphine; other derivatives are even stronger. Like morphine, they all target the mu-opioid receptor.

Fentanyl came off patent in 1980, and clinical sales rocketed, but it had already found its way onto the illicit drugs market in the USA. The first big outbreak occurred mainly in California between 1979 and 1988, with 112 deaths linked to the use of alpha-methylfentanyl. This was marketed as a 'street drug' called 'China White', with many deaths arising from people underestimating its potency (α-methylfentanyl (4.6) is said to be some 5,000 times stronger than morphine). Users taking it instead of heroin are prone to overdose on fentanyl (let alone the stronger α-methylfentanyl), often fatally, even in the hands of addicts who are medical professionals.

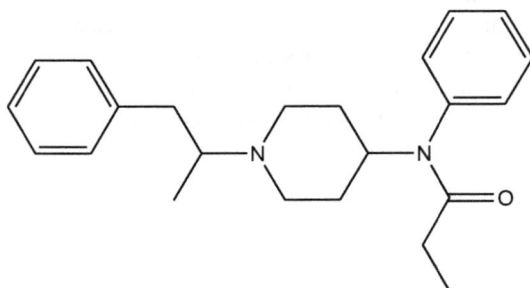

(4.6)

3-methylfentanyl, which exists in isomeric forms, was particularly associated with American fatalities in the 1980s and 1990s, and more recently in Finland and Estonia. It is a more potent drug than fentanyl; its *cis-* isomer (4.7) is said to be 6,600 times more potent than morphine as an analgesic, with the *trans-* isomer (4.8) 1,000 times more potent.

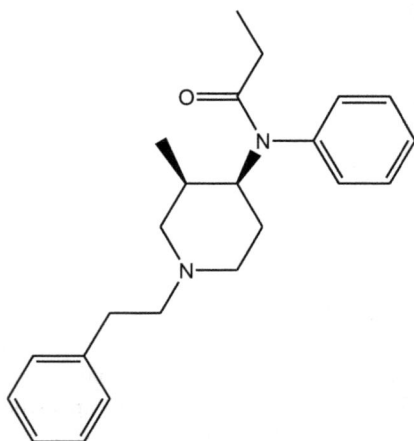

(4.7)

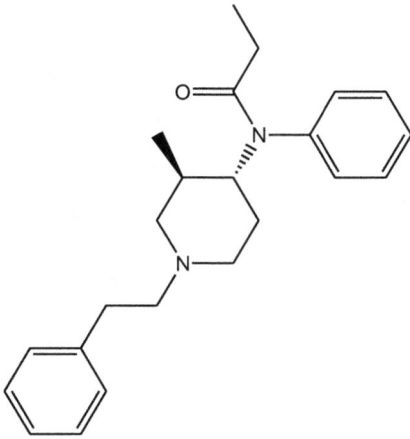

(4.8)

In Europe, eight fentanyl-related deaths were reported in Sweden in 1994; serious problems emerged from the late 2000s with reports of hundreds of overdose deaths in Estonia. More recently, it has been suggested that fentanyl and 3-methylfentanyl have been marketed as heroin substitutes, sometimes described as 'China White' again. Other European countries like Germany and the UK have reported outbreaks of fentanyl-related deaths for the first time.

This is part of a wider and much greater problem, as fentanyl has been a major contributor to a massive spike in overdose deaths caused by opioids. This has particularly affected the USA over the last 25 years (see Figure 4.1).

Around 2000, there was a significant rise in overdose deaths in the USA, driven by considerable increases in the prescribing of natural and semisynthetic opioids – neither heroin nor fentanyl use affected this. Then around 2010, fatalities caused by heroin started to rise very steeply, peaking around 2016. Young opioid users found that their increasing dependence meant that they could not obtain the quantities of pills that they needed, so they moved to the more available (and increasingly pure) heroin. Thirdly, deaths due to fentanyl (and other synthetic opioids) began a precipitous rise, exceeding the other two causes of death within 3 years. This attracted public attention, not least due to the death of Prince, the celebrated musician, from an accidental fentanyl overdose on 21 April 2016; shortly afterwards, the singer Tom Petty died from accidentally taking an overdose of a mixture of painkillers, including fentanyl, despropionylfentanyl and acetylfentanyl.

A factor in the 'third wave' is likely to be that the stronger fentanyls are being used to 'spike' heroin or even marketed as heroin. A 'fourth wave' has been linked particularly to 'polysubstance abuse', especially to a combination of fentanyls and stimulants, particularly methamphetamine and cocaine.

Annual drug overdose-caused deaths in the USA have risen from less than 20,000 in 1999 to over 106,000 in 2021, a figure including illicit drugs and prescription opioids.

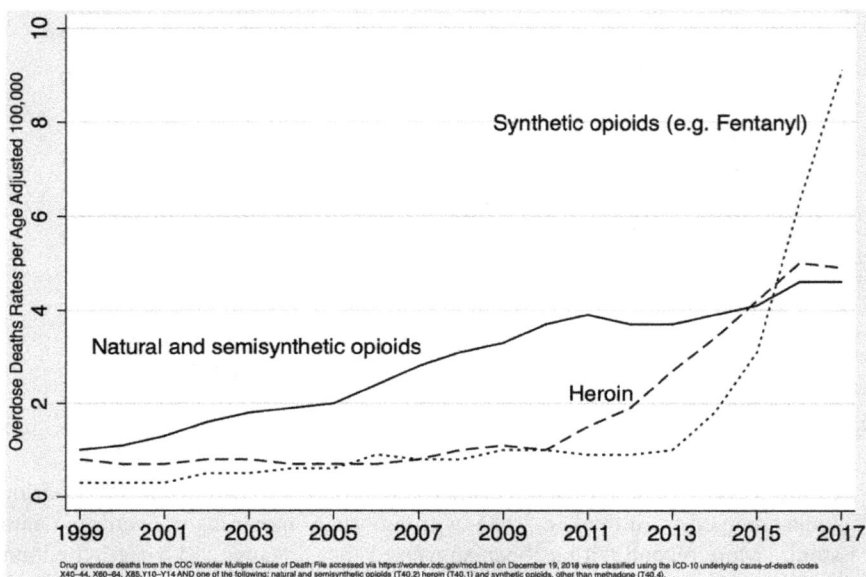

FIGURE 4.1 Reproduced with permission from Daniel Ciccarone, The triple wave epidemic: Supply and demand drivers of the US opioid overdose crisis, *International Journal of Drug Policy*, 2019, 71, 183–188.

Carfentanil and Other Powerful Fentanyls

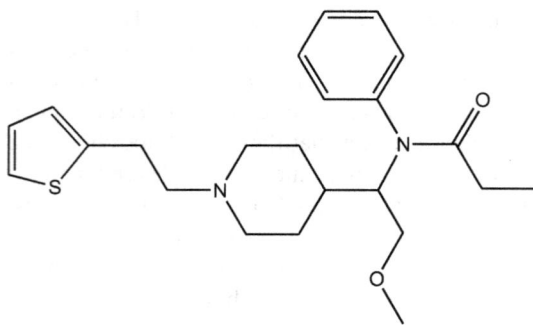

(4.9)

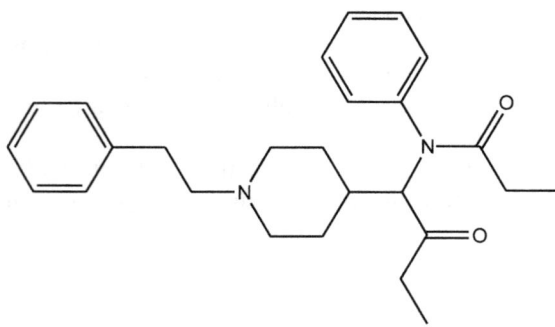

(4.10)

Paul Janssen didn't stop with the synthesis of fentanyl, of course, as he kept searching for better analgesics with fewer side effects and higher safety margins. In particular, he looked for more lipid-soluble molecules that would get into the central nervous system more quickly and thus afford faster analgesic action. They came up with derivatives including sufentanil (4.9), which is 5–10 times stronger than fentanyl. Then in 1974, they synthesised carfentanil (4.10), a good 100 times stronger than fentanyl – and since fentanyl is some 100 times stronger than morphine, this makes carfentanil the most potent commercially available opioid. It is so toxic that it has no approved use in humans.

The best-known use of carfentanil is to anaesthetise large animals like elephants, rhinos or 'big cats'; it is the active ingredient in the former anaesthetic *Wildnil*. The lethal dose of carfentanil in humans is unknown, but it must be similar to that of the incredibly toxic morphine derivative, etorphine, in the region of 20 µg or less. That is micrograms, not milligrams – too small to be seen. The vets anaesthetising the 'big game' shoot darts at the animals; the 'darts' are effectively syringes fitted with a hypodermic needle and are shot out of a gun by compressed gas. The darts are stabilised by a 'flight', a tuft of fibre, making them fly straight at the animal.

At the same time, vets have to take precautions to protect themselves, starting from when they put the 'darts' together until the animals are anaesthetised. That means they have to work in groups of at least two; wear full face coverings, so that they neither inhale the carfentanil nor let it come into contact with mucous membranes; wear latex gloves so that nothing gets into cuts or broken skin (some people believe that carfentanil is so toxic that it could be absorbed through the skin); wear clothing that covers the whole body (T-shirts and shorts are OUT); and finally, carry several vials of the antidote naloxone (Narcan), not just one. This sounds like overkill, but back in 2010, a vet was trying to sedate a wild elk. He had to remove a dart from a tree trunk, getting splashed in the face, eyes and mouth. He immediately washed his face but started to feel drowsy within a couple of minutes. Naloxone was not available, so his colleagues administered naltrexone (the 'reversal agent' used on animals) so that he was largely recovered by the time he got to A+E an hour later.

Carfentanil has become a significant part of the 'painkiller crisis' in the USA since around 2016, when there were about 400 seizures of the drug, largely centred around Cincinnati in Ohio. Since then, its use has spread. It is particularly associated with the 'Rust Belt' of the American Midwest. Internationally, it has reached countries as diverse as Canada, Australia, the United Kingdom and Latvia.

One source of carfentanil is China. It was estimated *circa.* 2016 that it could be obtained there for $2750 per kilogram (with 'no questions asked'). In June 2016, a squad of Royal Canadian Mounted Police officers (aka Mounties) descended upon Vancouver airport. They were dressed for business, wearing respirators, facemasks and hazmat suits, with the seams covered in tape, as they intercepted a consignment that had arrived from China. It was labelled 'printer accessories', but the blue cartridge labelled 'HP laserjet printer toner cartridges' contained a kilogram of carfentanil, reportedly sufficient for 50 million lethal doses (at $ 0.000055 per dose). The order was labelled for a man in Calgary – he was arrested. Shortly afterwards (March 2017), China introduced regulations restricting fentanyl and its derivatives, though 'underground' laboratories continue to operate over the 'Dark Web'.

One problem that has become more significant since the fentanyls, particularly carfentanil, came on the scene is the marketing of heroin cut with fentanyl or, even worse, carfentanil. The higher toxicity of carfentanil in particular means that there is a real chance of drug users unsuspectingly taking a lethal dose.

But the most public display of carfentanil's toxicity took place on another continent. On 23 October 2002, 40 Chechen terrorists occupied the Dubrovka Theatre in Moscow, taking around 900 people hostage. Three days later, the siege was brought to an end when Russian authorities pumped a 'chemical agent' into the building's ventilation system. Russian special forces killed the terrorists, but alongside that, around 125 of the hostages were killed by the 'chemical agent', whose identity was not stated. The Russian emergency services had been told to bring opioid antagonists to the scene but did not bring sufficient naloxone or naltrexone.

Two days after the siege ended, clothing and blood samples from British survivors held in the theatre arrived at the Defence Science and Technology Laboratory (Dstl) at Porton Down, near Salisbury, in the United Kingdom. They used chemical analysis, particularly mass spectrometry, to show that a mixture of carfentanil (4.10) and remifentanil (4.11) was used.

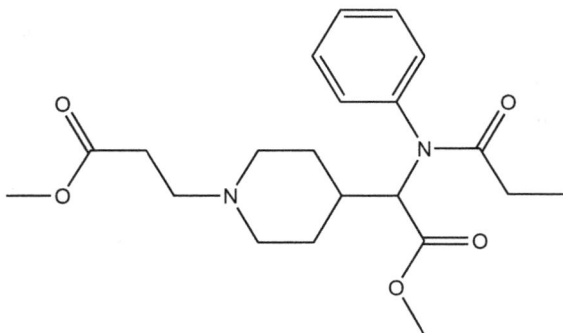

(4.11)

OXYCODONE AND OXYCONTIN

OXYCODONE

Trying to reduce pain is a natural human impulse. Chemists have been trying to develop alternatives to plant-produced remedies for over a century, but the results have not always been felicitous. Of course, the opium poppy, *Papaver somniferum*, produces two natural painkillers, codeine (4.12) and morphine, the active agents in opium consumed by humans for many years.

Oxycodone (4.13) doesn't occur in nature but has the same rather complex molecular backbone as codeine and morphine. Oxycodone is 'semisynthetic'. This means that, instead of being made 'from scratch', you start from a molecule that has the 'core structure' ready-made, then carry out just a few more steps. This is quicker and also gives a much better yield of the desired product.

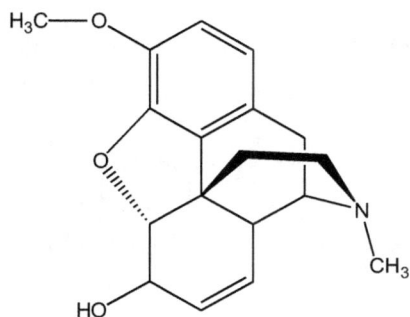

(4.12)

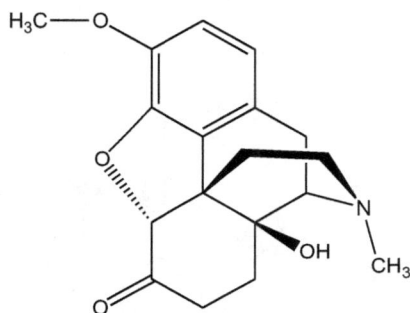

(4.13)

Oxycodone is usually made in two steps (4.14) starting from thebaine or codeine; thebaine is another alkaloid present in opium, which itself does not have any opioid properties. The first stage involves oxidation to 14-hydroxycodeinone, usually done with m-chlorobenzoic acid (m-CPBA) in acetic acid/trifluoroacetic acid, though hydrogen peroxide also works (but with lower yield). Stage two is the catalytic hydrogenation of 14-hydroxycodeinone; the benzene ring is unaffected, but the C=C bond in the other ring is reduced.

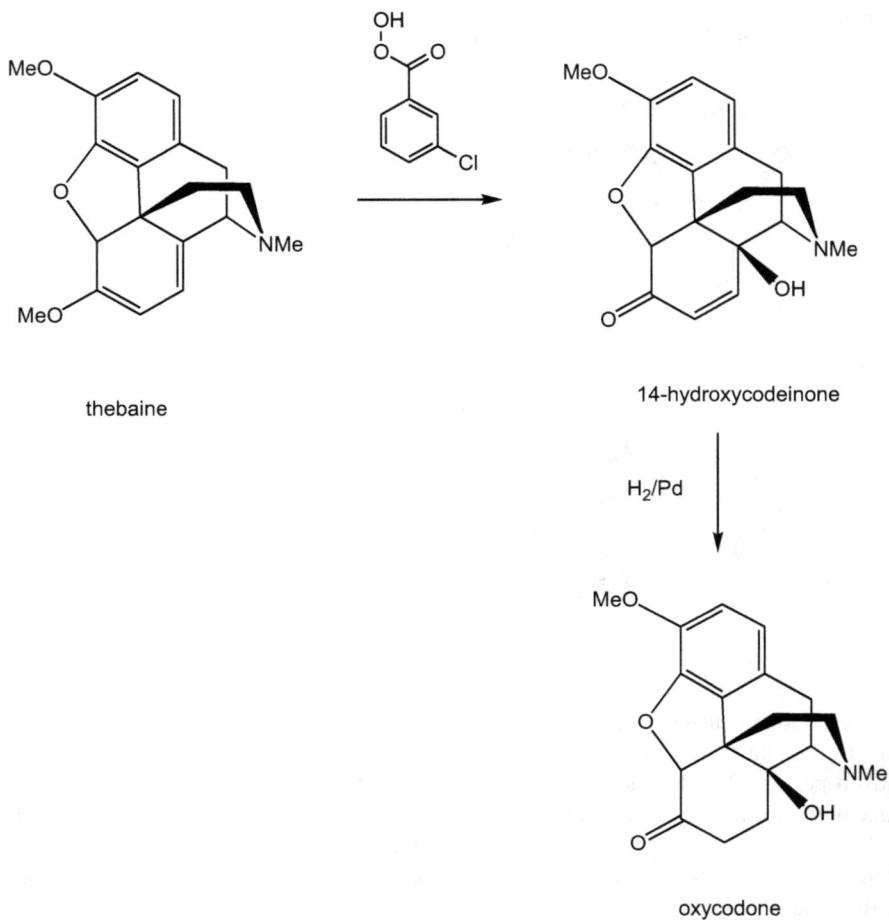

thebaine

14-hydroxycodeinone

H_2/Pd

oxycodone

(4.14)

Oxycodone was first synthesised in 1917 (German (DE) Patent 296916), two decades after chemists thought they had made a non-addictive improvement upon morphine as a painkiller; their diacetylmorphine molecule is commonly called heroin. We know that morphine achieves its painkilling effect by binding to the μ-(mu) opioid receptors in the body's central nervous system; how oxycodone achieves this is less certain.

In 1995, OxyContin® was introduced as a controlled-release formulation for oxycodone, in which the oxycodone content in the medication was released gradually into the patient, saying that they would need to take it roughly every 12 hours. Doctors in the United States began prescribing it widely as a painkiller, both for severe pain and for milder situations. It has been seen as a valid alternative to morphine and a first-line treatment for cancer pain.

People who took OxyContin for pain relief liked the feelings that they came to associate with this substance and started on the addictive pathway. Very quickly, people discovered that they could simply crush the OxyContin to a powder, releasing the oxycodone at once; they could then abuse it, either by 'snorting' it or injecting it as a solution. OxyContin became known as "Hillbilly heroin" around Virginia, West Virginia and Kentucky. OxyContin addiction became a serious problem, linked with thousands of deaths in the United States.

In late 2010, the formulation of OxyContin was changed to make the tablets much more difficult to crush and thus deter abuse. Ironically, it has been suggested that this reformulation has resulted in addicts changing their drug of abuse to heroin. As a cause of US drug overdose deaths between 2011 and 2016, the top-ranked drug in 2011 was oxycodone; from 2012 to 2015, it was heroin; and in 2016, it was fentanyl (from 2011 to the end of 2016, the rate of drug overdose deaths involving heroin more than tripled).

Painkiller addiction has become a real problem in the USA; among the celebrities linked with the issue are Steven Tyler, Courtney Love, Charlie Sheen, Nicole Richie, Winona Ryder, Matthew Perry, Chevy Chase, Kelly Osbourne and Heath Ledger. Besides oxycodone, other painkillers are involved, such as hydrocodone (4.15). This has a similar structure – and painkilling properties – to oxycodone.

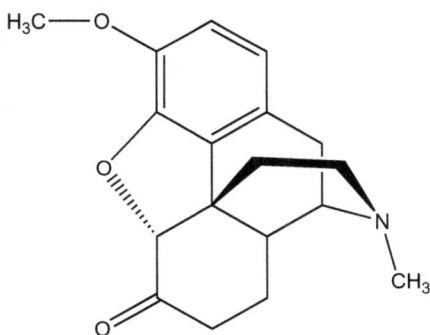

(4.15)

Hydrocodone is found, along with paracetamol (aka acetaminophen or Tylenol), in the popular US medication Vicodin®; another widely used painkiller, Percocet® similarly contains oxycodone and paracetamol.

KROKODIL

The use of some drugs like heroin and cocaine extends across the world. The drug known as 'Krokodil' is largely restricted to Russia, although its origins trace back to the USA. Its traditional chemical name is desomorphine (4.16), and you would not be wrong in suspecting that it bears a strong resemblance to morphine (4.17). It has come to be known as the 'flesh-eating drug'.

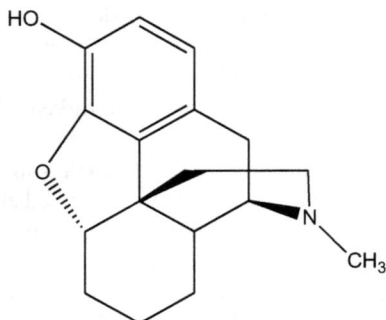

(4.16)

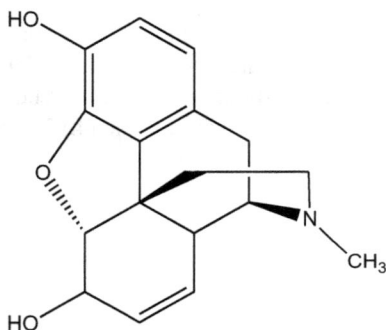

(4.17)

Back in the early 1930s, people were trying to improve upon morphine as a painkiller, something with fewer side-effects. People had tried this for many years; towards the end of the 19th century, diacetylmorphine was an attempt to create a non-addictive painkiller. This was unsuccessful, with that molecule today usually being referred to as heroin. In 1933, desomorphine was found to be a stronger pain-killer than morphine, with fewer side effects (e.g. depressing respiration), but it had a downside – it had a greater potential for dependence; therefore, it wasn't adopted as a medicine.

Its enhanced pain-killing effects can be explained by the fact that desomorphine has half as many –OH groups as morphine. This makes it less water-soluble than morphine, and conversely, more lipid-soluble, allowing it to cross the blood–brain barrier effectively. Thus, desomorphine should reach the brain better than morphine. In fact, desomorphine is roughly 10 times stronger as a painkiller than morphine, whilst its effects are felt sooner – and it's eliminated from the body faster, giving leading to a shorter 'high'.

Desomorphine can be made from codeine in a fairly straightforward synthesis (4.18). The first step is to replace the hydroxyl group with a halogen using thionyl chloride, then catalytic reduction removes the chlorine and also reduces the C=C bond – the one bit of unsaturation outside the benzene ring, which is more difficult to reduce. Finally, the CH_3 is removed from the methoxy group. This 'semisynthesis' means that the hard work of making the skeleton of the molecule, the core of the structure, has been done by Mother Nature; you just add the finishing touches.

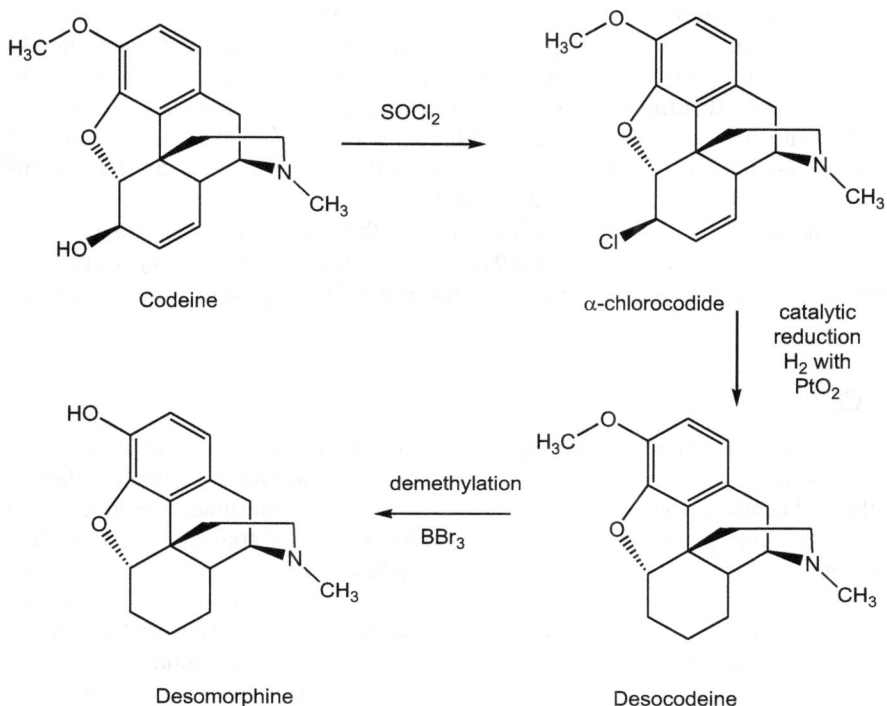

$$(4.18)$$

Its use was banned in the USA in 1936, where it is a Schedule I drug. It was used in Switzerland as 'Permonid' (Hoffman-LaRoche) from 1940 to 1952; otherwise, it hardly saw any use until it resurfaced in Russia, around the year 2000. The reason for this was that less heroin was being produced in Afghanistan, hence less was being supplied to Russia; therefore, Russian addicts were seeking an alternative drug. A new homemade (literally) drug that became known as 'Krokodil' came to fill the gap in supply. An alternative synthesis to the traditional way of making desomorphine made it cheaper than heroin, which it has tended to replace, as "the heroin of the poor". This synthesis involves codeine, at the time an easily obtained over-the-counter drug; iodine, a common antiseptic; and red phosphorus, obtainable from matchboxes.

People with no chemical training could make 'Krokodil' in their kitchens in about an hour – this was important, as the drug's short-lived 'high' meant that addicts needed to inject it every few hours. They first needed to extract the codeine from tablets, then heat powdered codeine with phosphorus and iodine. There is a very serious downside to all this. Addicts are injecting a crude and impure mixture straight from the reaction vessel, not a medicine purified by experienced chemists in a pharmaceutical laboratory. The result is that these abusers develop scaly skin with a green-black colour, resembling a crocodile, hence the name 'Krokodil'. Health problems develop, such as thrombosis, gangrenous damage, subcutaneous abscesses and ulceration, leading to rotting flesh and infection of the bone and muscle, which can result in amputations. Thus, Krokodil has become known as a "flesh-eating" drug. By around 2010, it was estimated that some 100,000 Russians had a serious addiction to Krokodil, and at that time, it was estimated that perhaps 5% of Russians had used it. The government tried to limit its availability by making codeine a prescription-only medicine, so that it was no longer available 'over the counter'. 'Krokodil' has continued to be a black-market drug in Russia, though.

And addicts face additional problems, such as the risks of sharing syringes whilst being intravenous drug users – AIDS/HIV and hepatitis C. If addicts try to come off Krokodil, detox takes around a month, compared with the week or so for recovering heroin addicts.

SPICE

Opium has been around for thousands of years; the opposite is true of 'Spice', one of the 'Novel Psychoactive Substances' that have been around since the first decade of the 21st century. The origins of 'Spice' can be traced to marijuana, obtained from the Indian hemp plant, usually *Cannabis sativa*, which, of course, has also been a drug of abuse for thousands of years. The synthesis of the psychoactive molecule in dried marijuana, Δ-9-tetrahydrocannabinol ((4.19), delta-9-THC or Δ^9-THC) was first reported in 1965; this is the molecule that is taken in the blood from the lungs to the brain. Δ^9-THC interacts with what have come to be known as 'cannabinoid receptors.' Two of these, CB1 (which stands for Cannabinoid receptor type 1) and CB2, were discovered in the early 1990s.

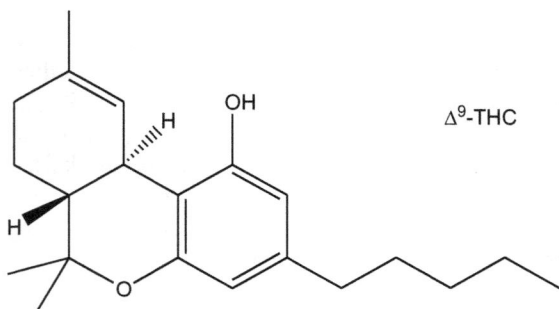

Δ^9-THC

(4.19)

Scientists became interested in synthetic cannabinoids, molecules that bind to these receptors but do not occur in nature, as they might shed light on how the receptors work, have possible medical uses against various diseases and disorders or provide insights into medical applications for cannabis. In the 1990s, John W. Huffman, a professor of chemistry at Clemson University, South Carolina, USA, was one of these scientists; his research team synthesised hundreds of new molecules, many of them belonging to the naphthaloylindole family. It was already known that these substances did not have to have structures like Δ^9-THC to bind to the receptor. The results were written up in scientific papers, and the compounds were referred to by tags like JWH-018 (4.20) for ease of reference, rather than their formal chemical names. The synthesis of JWH-018 first appeared in print in 1994.

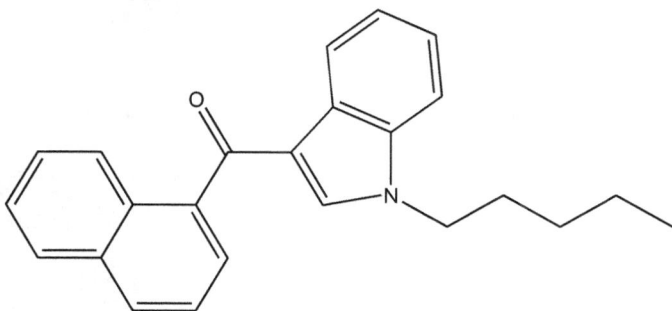

(4.20)

Over a decade later, in 2008, some German forensic chemists came across JWH-018 molecules in an unexpected context. They had been sprayed onto plant leaves – to give them a 'natural' context – and sold under the name of 'Spice'. This 'imitation cannabis' was, of course, designed to activate the 'cannabinoid receptor' in the brain. Some entrepreneur had been reading Huffman's 1994 paper, as its progenitor realised. This is not how he intended his research to be accessible to people on the streets (nor 'reaching an international audience', for that matter).

JWH-018 has cropped up in various products, also going under names like "Herbal spice" and "Herbal Gold". Reportedly, it is a very potent agonist for the CB1 receptor, with four times the potency of THC. Users are said to get a marijuana-like "buzz" similar to cannabis when they smoke it.

JWH-018 is not the only chemical of this type to be used in 'herbal drug' mixtures. Others include JWH-073 ((4.21), naphthalen-1-yl-(1-butylindol-3-yl)methanone), marketed as a "fertiliser" product called "Forest Humus", as well as JWH-250 ((4.22), 2-(2-methoxyphenyl)-1-(1-pentylindol-3-yl)ethanone) and JWH-200 ((4.23), (1-(2-morpholin-4-ylethyl)indol-3-yl)-naphthalen-1-ylmethanone) in "herbal" smoking blends.

Such synthetic psychoactive molecules are often known across the European Union as 'new psychoactive substances' (NPS), though across the world, 'designer drugs' or 'legal highs' are more often used. They are marketed via the Internet to provide alternatives to traditional illicit drugs, as they can circumvent the legislation in place for 'controlled substances'. These compounds are relatively straightforward for trained chemists to make and have been provided by labs all over the world, some in the Far East, some in Europe and some in America. Many have been banned in much of Europe, like JWH-018. A problem here is that it is easy for clandestine chemists to make similar molecules, perhaps differing by a methyl group (for example), which is not covered by the legislation, so are not 'illegal'.

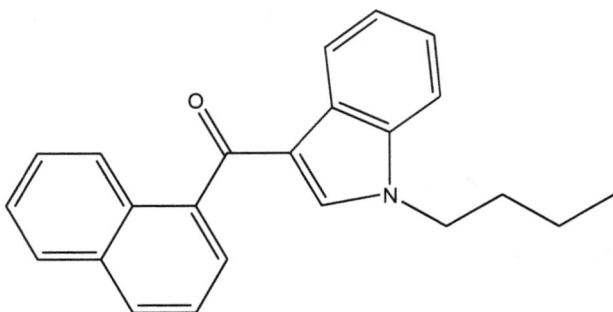

(4.21)

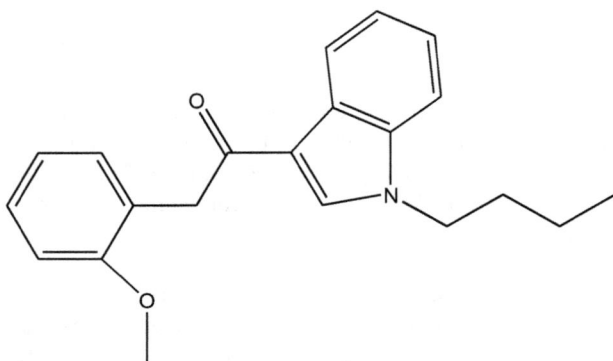

(4.22)

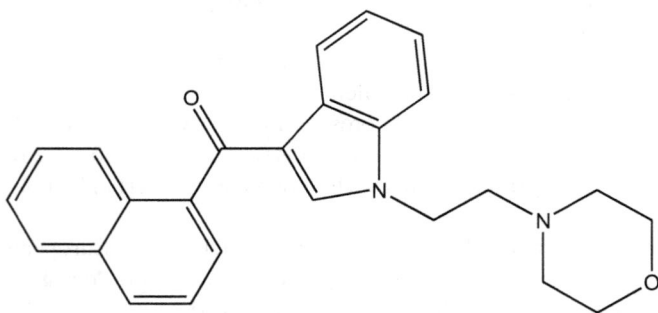

(4.23)

Around 2008–2010, the same time that these 'spice' drugs started to emerge, a different kind of drug came onto the streets, going by names like 'bath salts', 'pond cleaner' and 'plant food'. Of course, they were nothing of the kind. These had significantly different structures from the 'spice' family, resembling amphetamines. The first of these drugs to attract attention became known as Mephedrone (4.24); its chemical name is methylcathinone. It is closely related to cathinone (4.25), the main psychoactive substance in khat, the leaves of an East African evergreen shrub (*Catha edulis*), which is habitually chewed as a stimulant by many Yemenis as well as other East Africans (e.g. Somalis, Ethiopians and Kenyans). These are stimulants that do not bind to the cannabinoid receptor and thus are not 'spice' drugs.

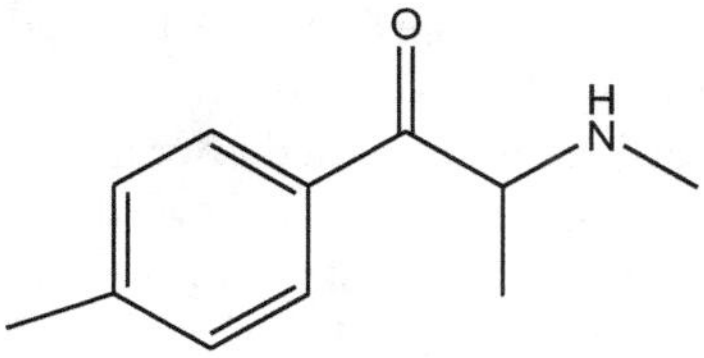

(4.24)

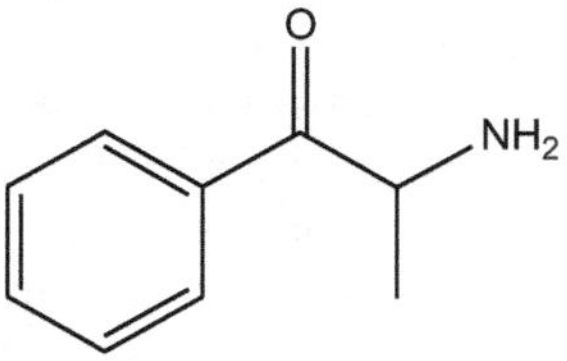

(4.25)

From around 2013, a different type of active ingredient has been found in 'Spice' products. The ADB-FUBINACA molecule was developed by Pfizer as a potential therapeutic medication and patented in 2009, but never marketed. Along with AMB-FUBINACA (aka MMB-FUBINACA) and related molecules, these indole carboxamides were found to be active agonists for the CB1 receptor.

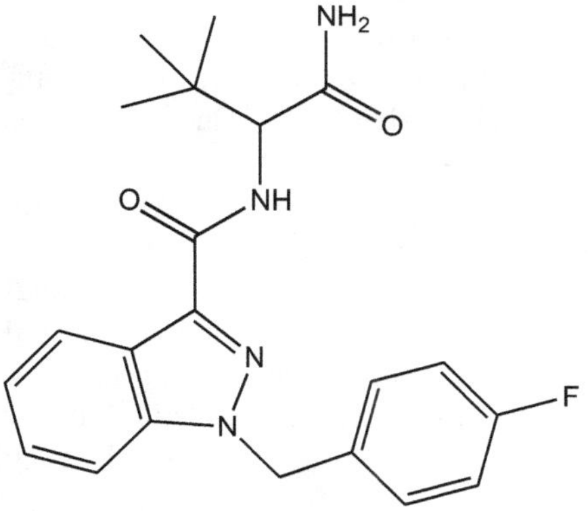

(4.26)

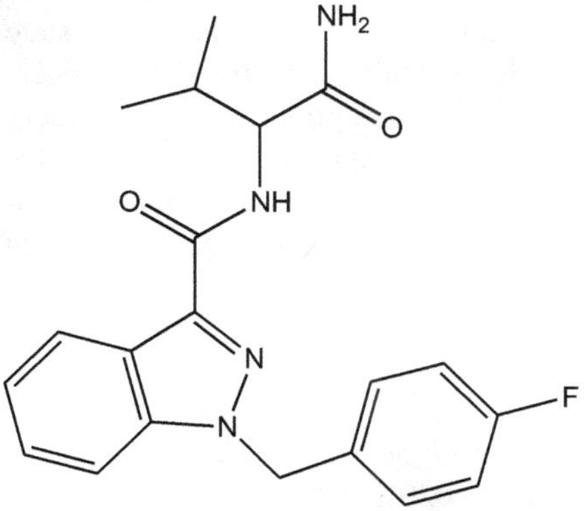

(4.27)

They have been detected as 'herbal drugs' all over the world. The first to be detected was ADB-FUBINACA (4.26) in Japan in 2013, followed by AMB-FUBINACA (4.27) in Sweden in 2015. The same year, a 41-year-old Louisiana woman died after smoking a synthetic cannabinoid product known as 'Mojo'; post-mortem analysis revealed the presence of ADB-FUBINACA. A 'mass casualty event' in Brooklyn on 12 July 2016, resulted in New York's Emergency Medical Services being called out to attend to 33 'zombie-like' casualties, of whom 18 were hospitalised. An herbal 'incense' product known as 'AK-47 24 Karat Gold' was recovered from one of the victims, who had been smoking it, and analysis revealed the presence of AMB-FUBINACA, whilst a metabolite of this molecule was found in blood and urine samples from eight of the patients (no common drugs of abuse were present). Though no fatalities were associated with this New York event, around 20 deaths were attributed to AMB-FUBINACA during 2017 alone in New Zealand.

AMB-FUBINACA and similar molecules have made their way into the United Kingdom. Sometimes the drugs have been made abroad and then shipped into the UK; in other cases, the precursor chemicals were imported from elsewhere, and the actual synthesis was performed in Britain. A case that came to court in 2018 involved the synthesis of AMB-FUBINACA in the kitchen of a 16th-floor flat in a high-rise block in Oldham, Greater Manchester. A gang of four received prison sentences totalling 19 years, with the gang leader getting 9 years. Using chemicals bought from Hong Kong, Shanghai and Poland, such as corrosive sulfuric acid and highly flammable diethyl ether, they appear to have carried out their syntheses without prior training. This placed the residents of neighbouring flats, including families with small children, in real danger.

'Spice' reaches places beyond the street, including prisons. People find ways of smuggling drugs in. Prisoners have time on their hands, and 'doing drugs' is one way to cope with this. It is believed that there's a higher percentage of synthetic cannabinoid users in prisons than in the country's population as a whole. One unexpected method of smuggling the synthetic cannabinoids in is to dissolve the drug in a solvent, which can be absorbed by writing paper, then used for writing a letter – either from a friend or prison visitor or maybe a 'lawyer's letter'; once in the jail, the drug can be extracted from the paper and consumed.

This 'cannabinoid' is much more potent than 'ordinary' cannabis. Users can be seen (and photographed) in a zombie-like state, slumped in public places in a state of semi-consciousness. It is a real public health issue. Quite apart from the dangers to consumers of drugs sold by pushers for human consumption, these substances have never been tested on animals, let alone humans. There is also the further issue of other, more potent drugs – including fentanyls – being mixed with synthetic cannabinoids or cutting ordinary cannabis with Spice. It is Russian roulette, but without the gun.

NITAZENES

Nitazenes are a new class of drug that has come to prominence in the last 5 years, particularly in Europe and North America. As of 1 May 2019, China put fentanyl-type drugs onto their list of controlled narcotic drugs and psychotropic substances with nonmedical use. This has affected their availability on the black market, and the nitazenes have come to fill the vacuum. They are μ-opioid receptor agonists with effects broadly comparable to others in the class, such as morphine, but stronger, with similar potencies to fentanyl.

Known as nitazenes, a series of 2-benzylbenzimidazoles was first synthesised in the late 1950s by CIBA Pharmaceuticals as potential analgesic medicines, rather more potent than morphine. They showed that it was possible to produce an opioid painkiller without the complicated core structure of morphine, which was not easy to synthesise. However, CIBA did not commercialise their discovery. In the subsequent decades, there were occasional reports of individual nitazenes being abused, but it was not until 2019 that these benzimidazole opioids came to light as new psychoactive substances, first with the detection of isotonitazene (4.28) on the 'drug market' in Belgium (sold as the homologue etonitazene), rapidly followed by others (4.29–4.31). That year, it was implicated in fatalities in the U.S. states of Illinois, Indiana, Minnesota, Pennsylvania and Wisconsin, as well as in cases from California and Tennessee. In some of the fatalities, it had been implicated along with other opioids (notably fentanyl) and benzodiazepines, such as etizolam and flualprazolam. Outside the USA and Belgium, isotonitazene also appeared at that time in Canada, Estonia, Germany, Latvia, Sweden, Switzerland and the United Kingdom. By the spring of 2020, it was reported that an average of 40 Americans were dying every month from consuming isotonitazene. Such fatalities are not just an American problem. It has been reported that in June and July 2023, there were over 30 deaths associated with nitazenes in the Birmingham area (UK) alone, whilst the UK National Crime Agency reported 54 deaths in which nitazenes were detected in post-mortems over a 6-month period in 2023. It should be remembered that nitazenes may well be mixed with other illegal agents like fentanyl.

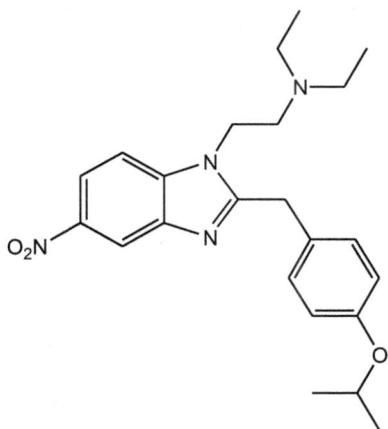

(4.28)

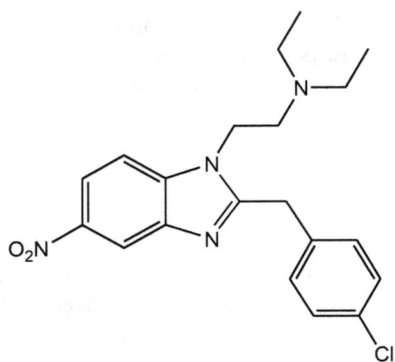

(4.29)

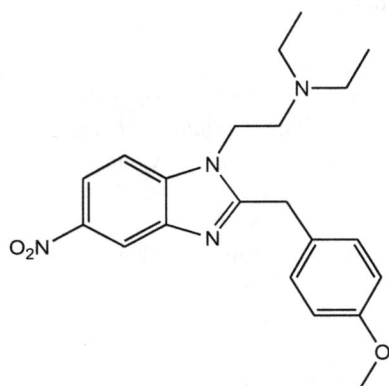

(4.30)

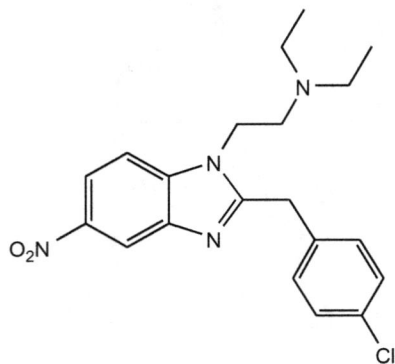

(4.31)

5 Nasty-Smelling Molecules

INTRODUCTION

Mention chemistry to non-chemists, and there is a good chance that the conversation will soon turn to 'smelly chemicals'. Quite a few of these smells involve sulfur compounds.

HYDROGEN SULFIDE

When it comes to 'bad smells', hydrogen sulfide (H_2S) is likely the first that comes to mind, with its smell reminiscent of rotten eggs. Hydrogen sulfide is formed when bacteria decompose proteins, as two of the natural amino acids, cysteine and methionine, contain sulfur. Although hydrogen sulfide is quite toxic (comparable to hydrogen cyanide), its smell usually provides ample warning at sub-lethal concentrations, around 0.005 ppm. Despite this, exposure to H_2S, notably in the sewage and petroleum manufacturing industries, led to deaths of 52 workers due to hydrogen sulfide poisoning in the United States over 7 years from 1993 to 1999; since H_2S is denser than air, it accumulates at the bottom of unventilated spaces. Hydrogen sulfide is also generated in the human body in very low concentrations, serving as a signalling molecule. In a different form of signalling, H_2S is the main odorant in human flatus; although CH_3SH and $(CH_3)_2S$ are also present in significant (but lesser) amounts, the smell of flatus correlates with the amount of H_2S. H_2S is also detectable as a malodorant in faeces, along with CH_3SH, ammonia, propanal, pyridine and carboxylic acids; levels of carboxylic acids can rise by a factor of up to 100,000 in cases of diarrhoea.

The H_2S molecule contains just two hydrogen atoms bound to the central sulfur; it is not linear but V-shaped, like H_2O (page 210). There are eight electrons in the 'outer shell' of the sulfur atom in the H_2S molecule, including two non-bonding electron pairs.

$$H \diagdown \underset{\cdot\cdot \overset{\displaystyle S}{} \cdot\cdot}{} \diagup H \tag{5.1}$$

These electron pairs around the central atom try to get as far away from each other as they can. So, just like water, H_2S has four central electron pairs, arranged approximately tetrahedrally. The four pairs are not in identical environments; the two non-bonded pairs (lone pairs) produce stronger repulsions than the electron pairs

DOI: 10.1201/9781003658115-5

in the H–S bonds, so the two bond pairs are squeezed together, and the H–S–H angle is well below 109½°, measuring 92.1° in fact.

A parallel example of the difference in smell between oxygen and sulfur-containing analogues is provided by acetone and thioacetone. Acetone (5.2) is a stable, if volatile, molecule with a 'nail varnish remover' smell. Thioacetone (5.3) is unstable, normally existing as a trimer. This can be cracked at around 500°C to yield the monomer, a substance with a quite nauseating smell, detectable even in traces from hundreds of yards away.

$$H_3C\diagdown\underset{\underset{O}{\|}}{C}\diagup CH_3$$

$$(5.2)$$

$$H_3C\diagdown\underset{\underset{S}{\|}}{C}\diagup CH_3$$

$$(5.3)$$

DIMETHYL SULFIDE

One smelly sulfur-containing molecule that crops up in all sorts of places is dimethyl sulphide (5.4).

$$H_3C\diagdown \underset{\ddot{S}}{}\diagup CH_3$$

$$(5.4)$$

Like the water molecule – and H_2S – dimethyl sulfide is V-shaped. Its bond angle of 99.1° is greater than the value of 92.1° in H_2S because the greater size of the methyl groups prevents them from approaching each other as closely as the two hydrogens in H_2S.

$(CH_3)_2S$ is a heavier molecule than H_2S, with stronger Van der Waals forces, accounting for its higher boiling point (37°C, compared with the value of – 60.3°C for H_2S).

The unpleasant smell of $(CH_3)_2S$, described as "rotting cabbage" or "cooked cauliflower", can be detected at concentrations as low as 1 ppb. Its formation in cooking can be attributed to the decomposition of either methionine or S-methylmethionine, which are present in large amounts in brassicas (5.5).

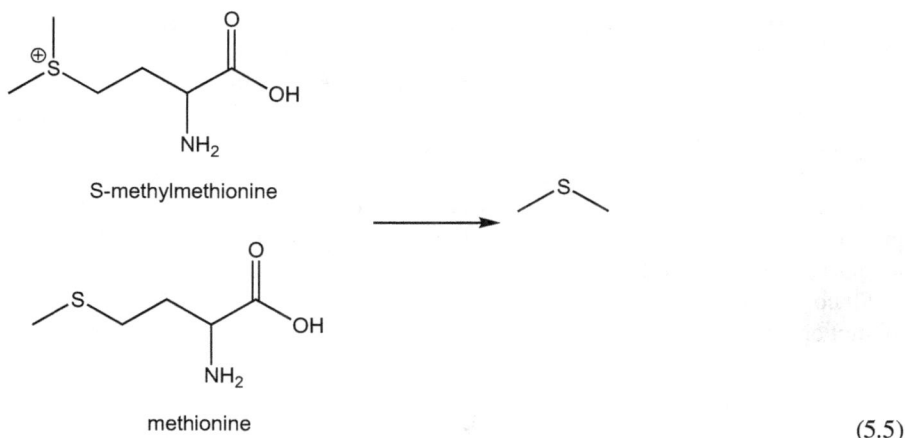

S-methylmethionine

methionine

(5.5)

So that is why Brussels sprouts in your Christmas dinner can have certain, er, consequences. At lower concentrations, dimethyl sulfide can smell more pleasant – it is found in some wines and can enhance fruity flavours. It is found in cheese like Cheddar and Camembert, again formed from methionine; enzymatic breakdown of proteins generates the free amino acids, and methionine decomposes into several small sulfur-containing molecules, notably $(CH_3)_2S$. It is found in some beers, believed to be formed from molecules like S-methylmethionine in germinating barley and drying malts. Certain English breweries, like Burton in Derbyshire, are credited with producing beers with an "eggy" note, which may derive from high sulfate concentrations in the breweries' water supply.

Dimethyl sulfide is associated with truffle-hunting, though it took a long time for this to be proved. Shortly before his death on 2 February 1826, the great French gastronome Jean Anthelme Brillat-Savarin published the book forever linked with his name, *La Physiologie du Goût* ('The Physiology Of Taste'). One memorable passage is: "Whoever says 'truffle' utters a great word which arouses erotic and gastronomic memories among the skirted sex and memories gastronomic and erotic among the bearded sex." He was talking principally about the black truffle of Périgord, *Tuber melanosporum,* produced by fungi associated with oak trees, a byword in the cuisine of the epicure. Italy is likewise associated with white truffles. Black truffles are traditionally hunted using pigs or dogs, the latter being more usual nowadays. The black truffle produces some 80 volatile molecules, including many aldehydes, ketones and esters, but for many years, it was believed that the snuffling pigs were interested in a steroid, which could be a pheromone, 5α-androst-16-en-3α-ol. In the 1980s, this idea was tested by a French chemist named Thierry Talou, who opined that dimethyl sulfide was responsible. What he did was devise a synthetic truffle aroma involving $(CH_3)_2S$ (but which did not contain the steroid); he then buried – at different locations – separate samples of the synthetic aroma, real truffles and 5α-androst-16-en-3α-ol. He then freed the pigs; they made for either the real truffles or the synthetic truffle aroma (i.e. $(CH_3)_2S$). The steroid was ignored.

Two decades after Talou published his findings, a group of over 50 European researchers spent 5 years decoding the genome of *Tuber melanosporum*. A key *discovery* was that the genes for making dimethyl sulfide were involved, showing that $(CH_3)_2S$ was actually produced by the truffle, not by soil microbes, as some had suggested. The truffle's synthesis of $(CH_3)_2S$ begins with sulfate ions in the soil; a series of enzymatic reactions generates a whole family of organic sulfur compounds, including the amino acid methionine, which then forms 3-(methylthio)propanal and methanethiol, precursors of dimethyl sulfide (5.6). The Italian white truffle (*Tuber magnatum*) also produces $(CH_3)_2S$, as well as $CH_3SCH_2SCH_3$ (2,4-dithiapentane).

Of course, these sulfur-containing odorants are present at such low levels that they are not obnoxious.

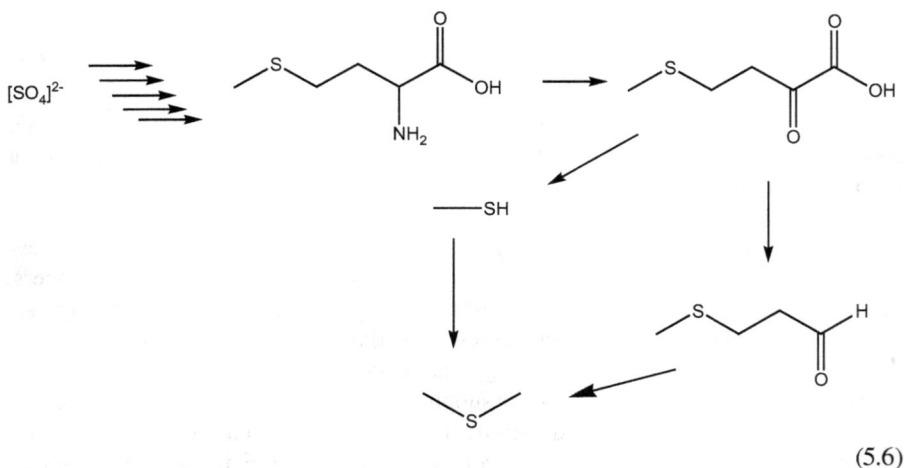

(5.6)

Last of all, the major role of dimethyl sulfide in our well-being must be considered: its role in the climate. It was the great eco-guru James Lovelock (1919–2022) who suggested that $(CH_3)_2S$ was key to regulating the global climate cycle, and in fact, dimethyl sulfide is responsible for three quarters of the global sulfur cycle (Figure 5.1). And it starts with plankton in the oceans, where phytoplankton, at the very bottom of the oceanic food web, are the ultimate food supply for everything living therein, from tiny zooplankton up to the big whales. Phytoplankton make dimethylsulfoniopropionate (DMSP); when zooplankton or small fish eat the phytoplankton, DMSP is converted to $(CH_3)_2S$, which attracts more zooplankton to feed. Some 90% of the $(CH_3)_2S$ gets decomposed before reaching the atmosphere; nevertheless, annually some 50 million tonnes reach the air. This $(CH_3)_2S$ is successively oxidised to SO_2, $(CH_3)_2SO$, $(CH_3)_2SO_2$, CH_3SO_3H (sulfonic acid) and sulfuric acid (Figure 5.1), with sulfuric acid being largely responsible for the cloud-forming aerosols above the oceans. Clouds reflect energy coming from the sun back into space. Because this energy does not reach the Earth, dimethyl sulfide is responsible for cooling the planet.

A wide range of predators is attracted by $(CH_3)_2S$ to the feeding grounds rich in zooplankton; apart from fish, the range encompasses foraging seabirds, like Leach's

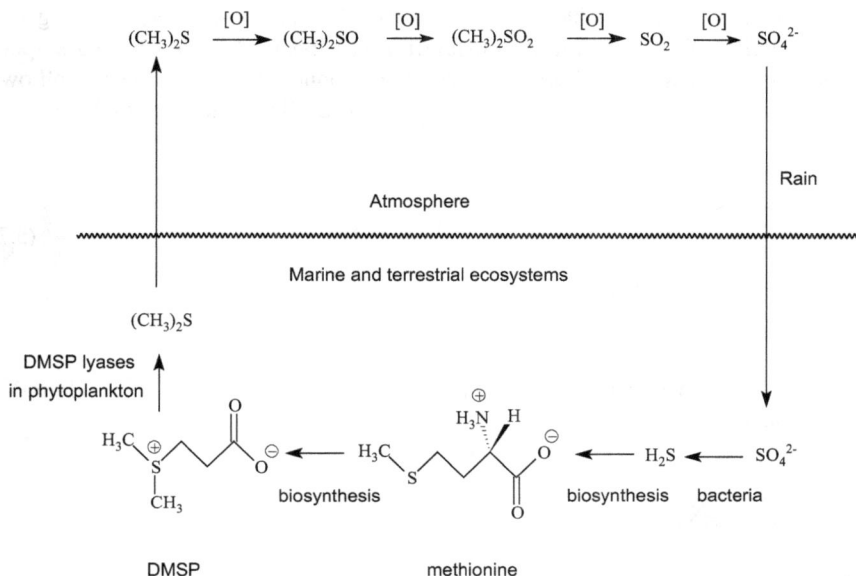

FIGURE 5.1   Global sulfur cycle [Based on Fig. 1 of T. G. Chasteen and R. Bentley, *Journal of Chemical Education*, 2004, **81**, 1524].

storm-petrels (*Oceanodroma leucorhoa*) and the wandering albatross (*Diomedea exulans*). There are also penguins and seals, which can be seen poking their noses out of the water to sniff the dimethyl sulfide. Professor Andy Johnston's research group at the University of East Anglia at Norwich discovered the gene responsible for turning DMSP into dimethyl sulfide. After inserting this gene into *E. coli* bacteria, they grew these bacteria on DMSP, whereupon they smelled the "rotten cabbage" smell of $(CH_3)_2S$. They took a bottle full of these engineered bacteria out to the salt marshes on the North Norfolk coast, whereupon the research team was set upon by hungry seabirds, who thought that the smell of dimethyl sulfide spelled 'food'.

And one more thing: low concentrations of $(CH_3)_2S$ in the air cause the "smell of the seaside" – not ozone.

DIMETHYLDISULFIDE AND THE TITAN ARUM

In contrast to the not unpleasant smell of the low concentrations of dimethyl sulfide in truffles, we move to the smelliest plant in the world, the titan arum *Amorphophallus titanum*. Discovered in 1878 in the rainforests of central Sumatra, some examples have since been cultivated elsewhere. It flowers irregularly, every few years and then just for 2 or 3 days. But the flowering can't be missed if you are in the neighbourhood, as its rotten flesh smell can be detected within a half-mile radius. The smell helps attract the kind of insects that like to feed on decaying flesh – flies and carrion beetles – whilst its deep red inflorescence looks like meat. These insects pollinate it.

The molecules responsible for the smells of various species of the genus *Amorphophallus* (Araceae, the Arum family) have been identified by mass spectrometry. The titan arum and several others that produce the rotting flesh smell owe their odour to mixtures of dimethyl disulfide (5.7, DMDS) and dimethyl trisulfide (5.8, DMTS).

(5.7)

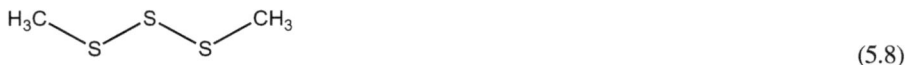

(5.8)

But there are other nasty smells – *Amorphophallus eichleri* has a faecal odour due to the presence of indole (5.9).

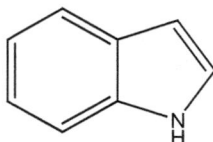

(5.9)

There are other nasty-smelling plants. Another arum found on the islands of Sardinia and Corsica, *Helicodiceros muscivorus*, is known as 'dead-horse arum' because of its rotting flesh smell. It uses dimethyl sulfide (DMS) as well as DMDS and DMTS to attract blowflies.

THIOLS

Replace one of the hydrogen atoms in hydrogen sulfide with an alkyl group, and you form a thiol, the sulfur analogue of an alcohol. Alcohols generally have very weak smells, but thiols are another matter, with quite disagreeable smells. The simplest thiol, methanethiol, CH_3SH, is associated with the smell of bad breath, body odour and farts.

A major use of thiols, unsuspected by most people, is in the home. We take the smell of domestic gas for granted today, but that only came into being after a disastrous explosion on 18 March 1937; a leak of odourless gas ignited and destroyed a school building in New London, Texas. There were around 300 fatalities; as a result, the state decided to make natural gas odourisation obligatory, and the practice spread. Organic sulfur compounds are the most widely used odorants for this purpose, with a number of blends in use. The cyclic compound tetrahydrothiophen (5.10, THT) is employed in many European countries; in contrast, tert-butylthiol (5.11), also known as tert-butyl mercaptan (TBM), is more generally employed in the USA. The United Kingdom uses a mixture of dimethyl sulfide ((5.4), DMS) and tert-butylthiol (5.11) (TBM), whilst ethylmercaptan or ethanethiol (5.12), (EM) is used in Romania.

(5.10)

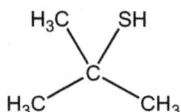

(5.11)

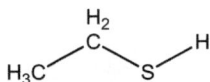

(5.12)

(5.13)

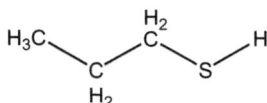

(5.14)

Thiols appear in foods. Perhaps the most notorious example is the durian fruit from Asia. It combines the reputation of having a wonderful taste with a quite nauseating smell, as a result of which it is widely banned in Thailand, Japan, Hong Kong and Singapore. You won't be able to take it on airlines or the mass transit systems in Singapore or Bangkok, and very likely it is banned from your hotel. 'No Durian' signs are in widespread evidence in that part of the world. There is no single molecule responsible for this – analysis has identified around 50 molecules that contribute, including the three simplest thiols: methanethiol (5.13) described as having a rotten cabbage smell; ethanethiol (5.12), which smells like rotten onion and propane-1-thiol (5.14), which has a rotten durian smell.

On the other hand, it is worth recalling that the smell of many mercaptans depends on their concentration. Thus, benzylmercaptan ((5.15), phenylmethanethiol) has a strong, unpleasant smell when undiluted, but it also occurs in certain wines made from *Vitis vinifera* grapes (e.g. Sauvignon), where it contributes a 'smoky' note at low concentrations to the wine.

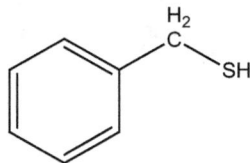

(5.15)

The grapefruit owes its smell mainly to a mercaptan (5.16). Its full name is (*R*)-2-(4-methylcyclohex-3-enyl)propane-2-thiol, so it is commonly referred to as 'grapefruit mercaptan' ('for short') or 1-p-menthene-8-thiol. It has a chiral carbon and exists as two enantiomers. The (S)-enantiomer (5.17) has a much weaker smell that does not relate to grapefruit, a reminder that enantiomers can smell very different, as in the case of the isomers of carvone (see *Every Molecule Tells a Story*, pp. 60–61). So not all sulfur compounds smell bad.

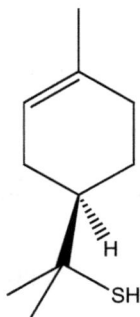

(5.16)

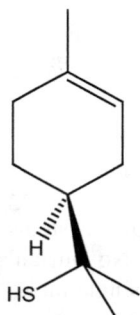

(5.17)

Another thiol that is rather important to peoples' everyday lives is (furan-2-yl) methanethiol (5.18), a key molecule in the smell of roasted coffee.

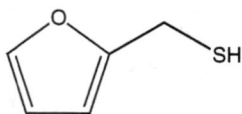

(5.18)

The sulfur-containing aldehyde methional (5.19) contributes to the smell of boiled potatoes and also appears in other foods, like cheddar cheese and potato crisps. It is formed through the breakdown of the amino acid methionine (5.20).

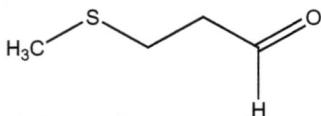

(5.19)

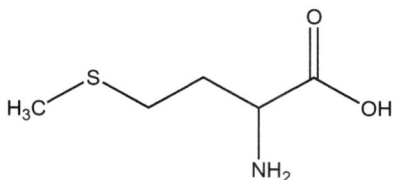

(5.20)

SKUNKS

Of course, skunks make it onto any 'awful smell' list, being described as worse than sewers or like a combination of rotten eggs, garlic and burnt rubber. A threatened skunk first adopts a defensive posture, giving a warning by hissing or stamping its feet (sounds like it has a career in politics). If the threat remains, it raises its tail and directs a foul-smelling yellow liquid spray from its hindquarters, over a distance of three metres or more from its anal glands. Back in 1896–7, the American chemist Thomas Aldrich split the liquid into two fractions, the lower boiling range being 100°C–130°C. Noting that butane-1-thiol, butyl mercaptan (5.21), had a boiling point of 97°C, and with sulfur analyses to hand, he suggested that this was an important ingredient in skunk spray.

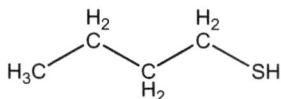

(5.21)

More recent investigations, aided by modern techniques like mass spectrometry and NMR, have looked at the spray of the striped skunk (*Mephitis mephitis*), finding that it was a mixture, whose main constituent was (*E*)-2-butene-1-thiol (5.22). Its M_r is 88, compared to 90 for butane-1-thiol; thus, they have very similar sulfur analyses. They also found that the spray contained significant amounts of 3-methyl-1-butanethiol (5.23). Butane-1-thiol has not been found.

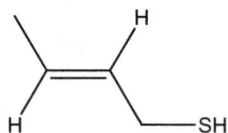

(5.22)

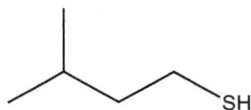

(5.23)

2-methylquinoline (5.24) is also present in significant amounts in the spray of the striped skunk, whilst small amounts of 2-quinolinemethanethiol (5.25) are also found.

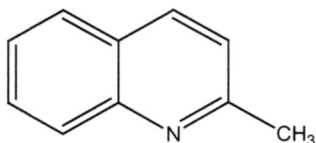

(5.24)

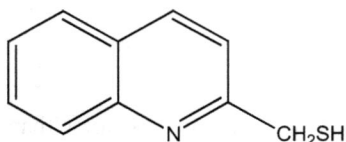

(5.25)

Different species of skunks have different compositions of their spray. In addition to the above thiols, the spray of the striped skunk (*Mephitis mephitis*) also contains the thioacetates of the already-discovered thiols, such as *S*-2-butenyl thioacetate (5.26) and *S*-3-methylbutyl-1-thioacetate (5.27).

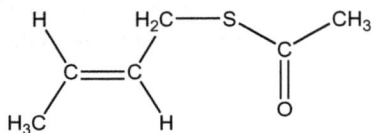

(5.26)

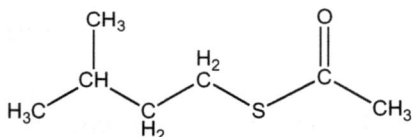

(5.27)

Though the thioacetates are less smelly than the parent thiols, upon hydrolysis by water, they release the nauseating thiols. That is why domestic pets that get sprayed by a skunk and are then cleaned can give off a skunk smell again a few days later if they get wet again. The spray of the spotted skunk (*Spilogale gracilis*) also contains both (*E*)-2-butene-1-thiol and 3-methyl-1-butanethiol; the thioacetates are not present here, but there is a significant amount of 2-phenylethanethiol (5.28).

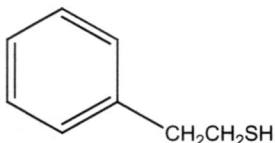

(5.28)

The spray from a fourth species, the hog-nosed skunk (*Conepatus mesoleucus*), is again different, containing (*E*)-2-butene-1-thiol and (*E*)-2-butenyl thioacetate, but neither 3-methyl-1-butanethiol nor its thioacetate. Nor is there any 2-phenylethanethiol.

Unsaturated thiols crop up in other places as well. Beer that has been exposed to light ('lightstruck') acquires an off-flavour due to 3-methylbut-2-ene-1-thiol, which has also been reported in a Spanish wine.

Several unsaturated thiols, including 2-methyl-1-propene-1-thiol, (Z)-3-methyl-1-butene-1-thiol (5.29), (E)-3-methyl-1-butene-1-thiol (5.30), (Z)-2-methyl-1-butene-1-thiol and (E)-2-methyl-1-butene-1-thiol, have been found in roasted sesame seeds.

(5.29)

(5.30)

And most remarkably, in 2021, concentrated extracts of cannabis flowers were found to contain a number of sulfur-containing odorants, especially 3-methyl-2-butene-1-thiol. Skunk, anyone?

PERSONAL HYGIENE – AND WINES

Perhaps it is not too surprising that several sulfur compounds are associated with body odour, particularly 3-mercapto-2-methylbutan-1-ol (5.31), 3-mercapto-3-methylhexan-1-ol (5.32) and 3-mercaptohexan-1-ol (5.33), which are generated when skin microorganisms decompose molecules formed by sweat glands (especially around the armpits).

(5.31)

(5.32)

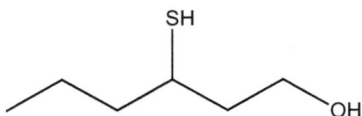

(5.33)

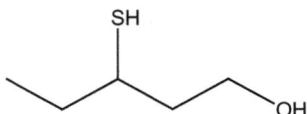

(5.34)

A similar molecule, 3-mercapto-2-methylpentan-1-ol (5.34), which is only CH_2 different from 3-mercapto-2-methylbutan-1-ol, has a very intense smell, described as sweaty and oniony. It is actually found in both raw and cooked onions, as well as in beef and pork vegetable gravies.

In a recent experiment, 3-mercapto-2-methylpentan-1-ol was tested as one of about 200 key food odorants; it was found to be the only one of these to trigger a response from one particular human odorant receptor, OR2M3. The 3-mercapto-2-methylpentan-1-ol/OR2M3 pairing enables 'our highly specific and sensitive perception of this onion-related key food odorant'.

Many oenophiles would be surprised, or just unhappy, to find out the role of compounds of this type in giving their favourite wines their *bouquet*. So ethyl 3-mercaptopropionate is found in the Concord grapes used to make Sauvignon blanc wines; it's also been identified in aged champagne wines – and in Munster and Camembert cheeses. Neat ethyl 3-mercaptopropionate has a 'strong skunk-like aroma'; however, when very dilute, at the parts per million level, it changes to a 'fresh and fruity grape character'.

It's natural to wonder where these molecules originate. It is believed that ethyl 3-mercaptopropionate is biosynthesised starting from the amino acid homocysteine (which is probably formed from the essential amino acid methionine). On transamination, the amine group is replaced by a carbonyl, forming 2-oxo-4-mercaptobutanoic acid. Its decarboxylation, losing a carbon atom, results in the aldehyde 3-mercaptopropanal. In turn, the aldehyde undergoes facile oxidation to 3-mercaptopropanoic acid, which can be esterified with ethanol to afford ethyl 3-mercaptopropanoate (5.35).

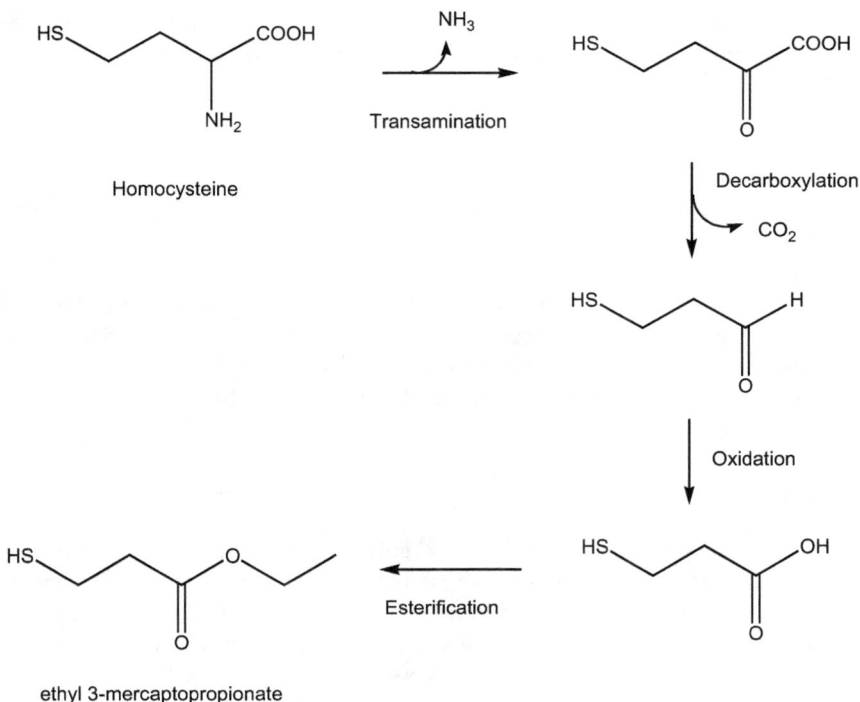

Homocysteine

ethyl 3-mercaptopropionate

(5.35)

3-mercapto-3-methylbutan-1-ol (5.36), as well as 3-mercaptohexan-1-ol (5.37) and its ester 3-mercaptohexyl acetate (5.38), all contribute to the 'tropical fruit' aroma of Sauvignon blanc wines too.

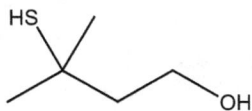

(5.36)

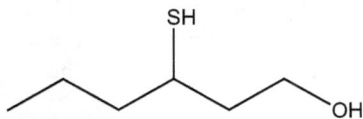

(5.37)

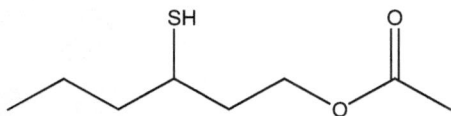

(5.38)

For a different sort of drink, 3-mercapto-3-methylbutan-1-ol has been found in roasted coffee: its formate (5.39) and acetate esters are important contributors to the odour of coffee.

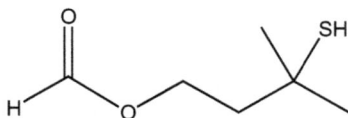

$$(5.39)$$

A very different source of some of these thiols is certain felines. 3-mercapto-3-methylbutan-1-ol, described as 'intensely odorous', has been found to occur in the urine of various carnivorous cats like leopards and bobcats (as well as domestic cats). It is thought to cause the smell of cat urine, whilst it may also be a 'warning' to potential prey.

TRIMETHYLAMINE

Many of the compounds described here are not associated directly with humans. Here is one that, sadly, sometimes is. Trimethylamine (5.40) is a molecule whose structure can be described as based on ammonia, NH_3, with each of the three hydrogen atoms replaced by methyl, CH_3, groups. Like ammonia, it is a base.

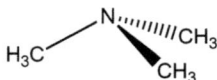

$$(5.40)$$

Both ammonia and trimethylamine have trigonal pyramidal structures. The C–N–C bond angle in $(CH_3)_3N$ is 110.9°, compared with 107.2° in NH_3, for the corresponding H-N-H angle presumably due to greater repulsions between the methyl groups.

The human body contains enzymes called flavin monoxygenases (FMOs), which have the job of breaking down toxic molecules. FMO_3 is one such enzyme; it decomposes Me_3N by oxidising it to odourless Me_3NO ((5.41), trimethylamine N-oxide, TMAO), which is then excreted. One source of trimethylamine is choline ((5.42), $Me_3N^+CH_2CH_2OH$), found in eggs, liver, legumes and some grains; bacteria break down choline into Me_3N. Mutations of FMO_3 cannot decompose Me_3N, so it builds up in the body.

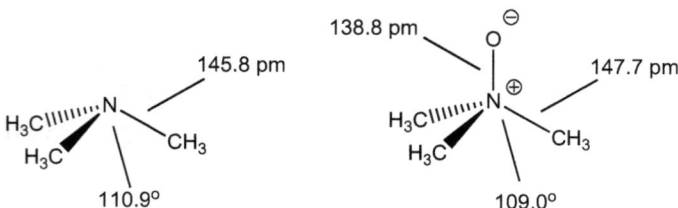

$$(5.41)$$

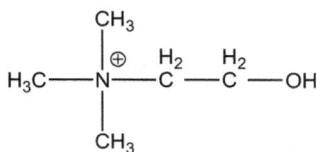

$$
\begin{array}{c}
CH_3 \\
| \oplus \quad H_2 \quad H_2 \\
H_3C \text{——} N \text{——} C \text{——} C \text{——} OH \\
| \\
CH_3
\end{array}
$$

(5.42)

There are also methanogenic (methane-making) bacteria in the gut, such as *Methanobrevibacter smithii*, which turn $(CH_3)_3N$ into CH_4, which may be responsible for flatulence.

The consequence of this build-up of trimethylamine is its smell, a very unpleasant, fishy one, known as 'fish odour syndrome' or 'fish-breath syndrome'. Pharmacogenetic screening indicates that the condition is inherited as an autosomal recessive trait. Everyone has two copies of the FMO_3 gene, one inherited from each parent, and perhaps 1% of people have a defective copy of this gene. If both copies are defective, then the result is 'fish-breath syndrome', affecting perhaps one person in 10,000.

It is obviously a very unpleasant condition to live with. Sufferers get ostracised, children are called names at school; dealing with it is a daily problem. Some people even start smoking to disguise the smell.

THE SMELL OF THE LIVING AND THE DEAD

Every person has a different smell. You could say that we have our unique molecular cocktail. Traditionally, dogs have been employed to distinguish between people and track them, as they have many more receptors in their noses for odorants. This was famously employed by Arthur Conan Doyle as a plot device in his novel *The Hound of the Baskervilles*, where the character Stapleton gives the hound Sir Henry Baskerville's boot for the dog to recognise Sir Henry's smell.

To begin with, our fresh sweat has no smell. The lipids in sweat are released especially by the hands and feet as well as from the armpits. Bacteria found on the skin decompose the esters of long-chain carboxylic acids found in the lipids, generating the free acids. In turn, these long-chain carboxylic acids can be broken up into smaller molecules, short-chain carboxylic acids. These smaller molecules are more volatile and thus have stronger smells.

Carboxylic acids do not have pleasant smells. The C_1 compound, methanoic acid, is associated with ants, many of which use it in their venoms; it formerly took its name, formic acid, from *formica*, the Latin word for ants. Similar historical associations account for the old, non-systematic names of several other carboxylic acids. Thus, the C_2 acid, ethanoic acid, which can be obtained by oxidation of ethanol, was known as acetic acid, since it is found in vinegar, for which the Latin noun is *acetum*. The C_4 acid, butanoic acid (5.43), was known as butyric acid. Butter (Latin *butyrum*) contains the ester of glycerol and butanoic acid, which on breakdown forms butanoic acid, with a rancid butter smell. The clue to the smell of the C_6 acid, hexanoic acid (5.44), is found in its traditional name, caproic acid, *caper* being the Latin noun for goat. Several carboxylic acids are associated with 'body odour' smells.

(5.43)

(5.44)

Branched-chain acids are also involved in body odour. 3-methylbutanoic acid ((5.45), traditionally called isovaleric acid) is often described as having a 'cheesy' or 'sweaty feet' odour. Just as sweaty feet attract African mosquitoes, so do some cheeses containing it, like Limburger. The isomer, 2-methylbutanoic acid (5.46), differs in having a chiral carbon atom (*) and thus has two optical isomers.

(5.45)

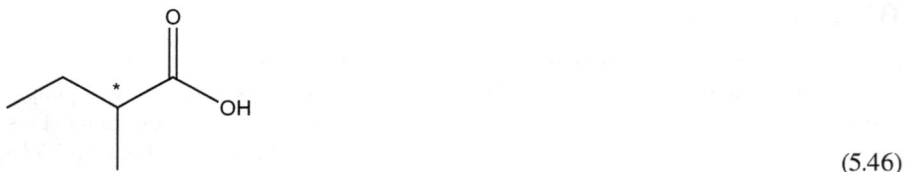

(5.46)

As with many pairs of optical isomers, these isomers have significantly different smells (5.47). Thus, the *R*-isomer of 2-methylbutanoic acid smells 'cheesy' or 'sweaty'; on the other hand, the *S*-isomer smells fruity and sweet.

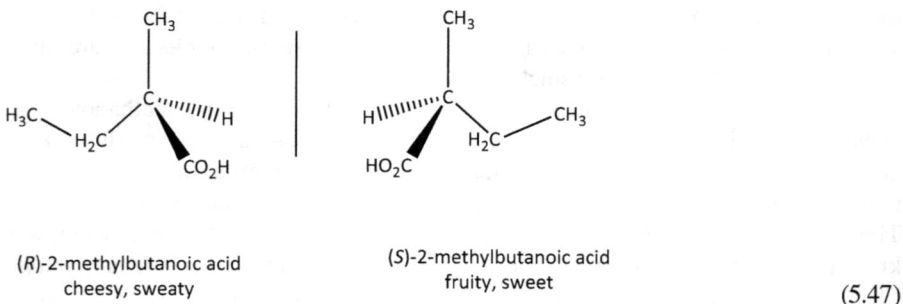

(R)-2-methylbutanoic acid
cheesy, sweaty

(S)-2-methylbutanoic acid
fruity, sweet

(5.47)

Several other molecules contribute to body odour. Research has led to (E)-3-methyl-2-hexenoic acid (5.48) being called "the predominant olfactory contributor in axillary sweat".

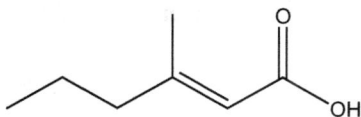

(5.48)

It cannot be smelled in fresh sweat, as it (and other smelly molecules) are present, not as free molecules, but linked to glutamine residues (as 'conjugates'). The skin bears *Corynebacteria*, which supply enzymes that catalyse the hydrolysis of an enzyme link, splitting the conjugate into glutamine and the free acid, the latter supplying the smell (5.49).

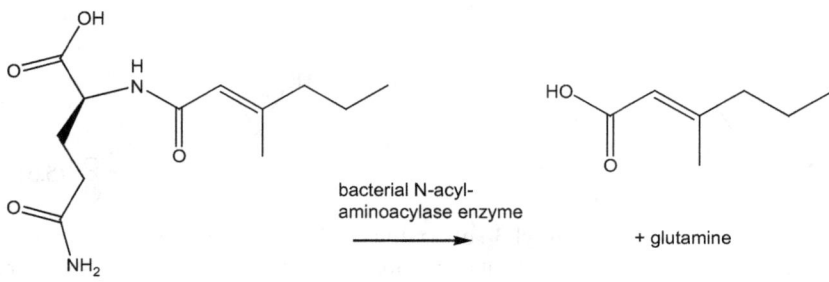

bacterial N-acyl-
aminoacylase enzyme

+ glutamine

(5.49)

Mosquitoes are the most important global vectors of disease, with *Anopheles gambiae* being the principal African vector of malaria, whilst *Aedes aegypti* is highly efficient at spreading viruses, including yellow fever, dengue, chikungunya and Zika among humans. Less than half of the species of mosquitoes mediate the malaria parasite.

Female mosquitoes feed on humans and other vertebrates; their attraction to humans may well be due to humans having more abundant free fatty acids on their skin surface than non-human animals. There has thus been sustained interest in discovering chemicals that attract mosquitoes to humans, such as L-lactic acid (1968) and butanoic acid (1997). Some of these chemicals are produced by the breakdown of lipids by skin bacteria. A number of carboxylic acids, particularly those with short chains, including propanoic acid, butanoic acid, 3-methylbutanoic acid, pentanoic acid, heptanoic acid, octanoic acid and tetradecanoic acid, have been identified as promoting attraction to the malaria mosquito *Anopheles gambiae* (though some of them are not attractive to humans!), with butanoic acid and 3-methylbutanoic acid identified in more than one study.

Some people are 'mosquito magnets'. A recent study involving the yellow fever mosquito *Aedes aegypti* compared people who are exceptionally attractive or unattractive to mosquitoes, finding that highly attractive people produce significantly greater amounts of the carboxylic acids from their skin. In the case of *Aedes aegypti*, 'highly attractive' subjects generate notably higher levels of pentadecanoic, heptadecanoic and nonadecanoic acids (as well as some unidentified ones).

In contrast, scientists have examined compounds that seem to reduce attractiveness to mosquitoes (in contrast to repellents). Three of these are the C_8–C_{10} aldehydes octanal, nonanal and decanal; another is the ketone 6-methyl-5-hepten-2-one ((5.50), sulcatone)

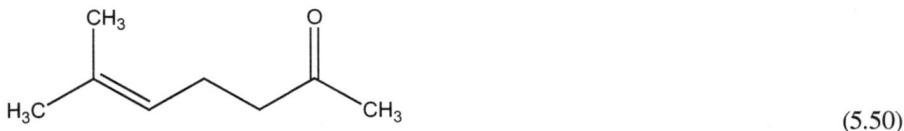

(5.50)

As mentioned earlier (page 125), certain sulfur compounds have been isolated from human sweat, such as 3-hydroxy-3-methylhexanoic acid and 3-methyl-3-sulfa-nylhexan-1-ol. These molecules both have a chiral carbon atom (*), so each exists as two optical isomers (5.51).

(5.51)

Of the two isomers of 3-methyl-3-sulfanylhexan-1-ol isolated from sweat, the major isomer, the (S)-form, has a "sweat and onion-like" smell; in contrast, the odour of the (R)-isomer is described as fruity and grapefruit-like (5.52).

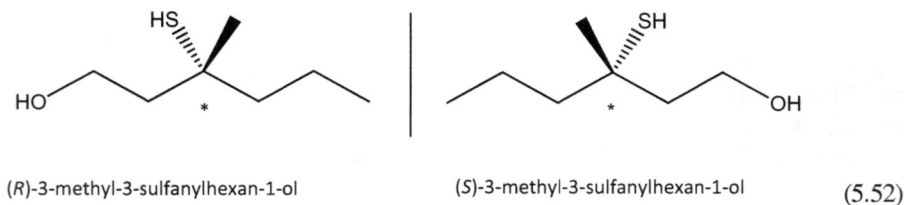

(R)-3-methyl-3-sulfanylhexan-1-ol   (S)-3-methyl-3-sulfanylhexan-1-ol   (5.52)

THE SCENT OF DEATH

Tissue decay starts in bodies soon after death. Carbohydrates, proteins, lipids and nucleic acids all decompose into smaller molecules, covering most types of simple organic compounds – aliphatic and aromatic hydrocarbons; aldehydes, ketones, alcohols and carboxylic acids, as well as organic nitrogen and sulfur compounds.

Two of the molecules with unpleasant-sounding names are putrescine (rotten fish smell) and cadaverine (decomposing body smell). Both of these result from enzyme-catalysed decarboxylation of amino acids, specifically orthinine (5.53) and lysine (5.54).

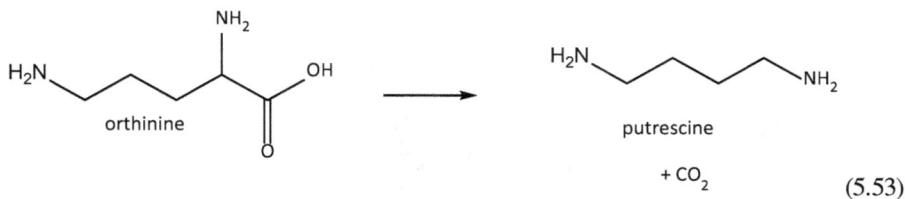

orthinine → putrescine

+ CO$_2$

(5.53)

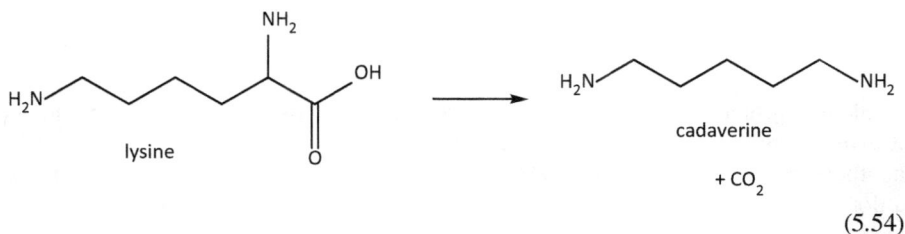

lysine → cadaverine

+ CO$_2$

(5.54)

These are not, of course, the only nitrogen-containing compounds that contribute to the smells of decomposition. Others include indole (5.55) and 3-methylindole ((5.56), skatole), which also contribute to the smell of faeces; these are breakdown products of the amino acid tryptophan (5.57). The first stage of the breakdown is catalysed by the enzyme tryptophan deaminase, forming indoleacetate, which in turn is converted into indole by another enzyme, indoleacetic acid decarboxylase (5.58).

(5.55)

(5.56)

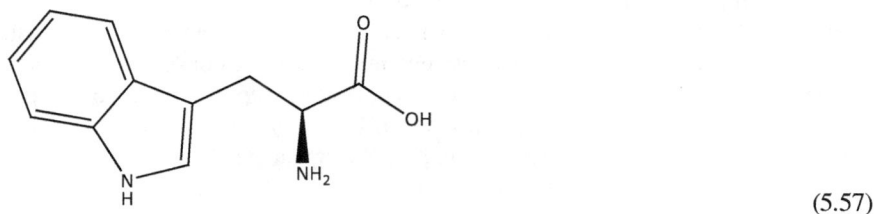

(5.57)

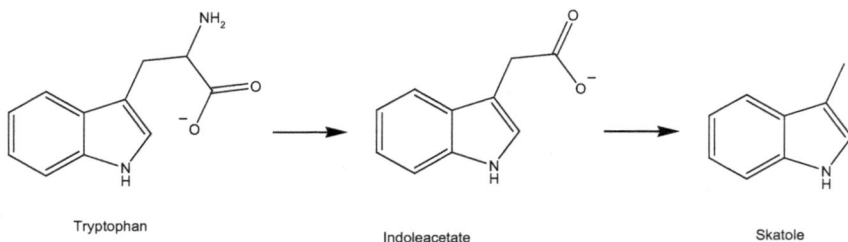

Tryptophan   Indoleacetate   Skatole

(5.58)

Skatole gets its name from the Greek word σκωρ or σκατ (*skat*), meaning dung. It gives meaning to the adjective scatological, which means filthy or obscene. It is another compound whose smell depends upon concentration; in tiny amounts, it has a floral smell (jasmine and orange blossom), being used as a perfume additive – as well as in strawberry ice cream.

The smelly sulfur compounds most associated with human (and pig) remains are the sulfides: dimethyl sulfide (DMS), dimethyl disulfide (DMDS) and dimethyl tri-sulfide (DMTS).

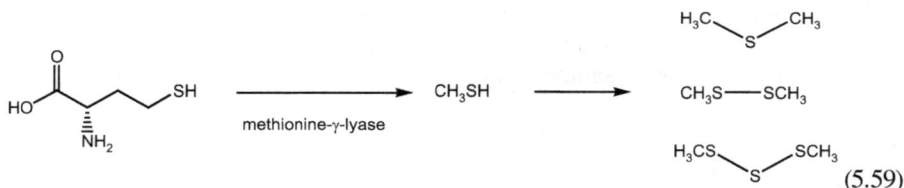

methionine-γ-lyase

(5.59)

These are generated through the decomposition of compounds like the amino acid methionine (5.59). They are also associated with many types of truffles (page 117), along with some of the world's smelliest flowers, like the Titan arum (page 119).

Forensic scientists want to be able to detect human remains. Traditionally, because dogs have many more odorant receptors than humans, they have been used to track bodies. Tests have shown that trained dogs can detect, with 98% accuracy, pieces of carpet that have been in contact with newly dead bodies for as little as 10 minutes. Training tracker dogs can be fraught with problems. There are ethical reasons for not using human remains, even if they could be obtained, so pigs are sometimes used as substitutes, though some question their suitability.

Finding 'human marker' molecules, particularly ones that can be detected instru-mentally, has proved difficult. An instrument can give a consistent and reliable response and not have 'off days', unlike a dog. A study reported in 2015 compared decomposing human remains with those of different types of animals. It was found that some esters, ethyl propanoate (5.60), propyl propanoate (5.61), propyl butanoate (5.62) and ethyl pentanoate (5.63), were commonly found in human and pig remains rather than in other animal remains.

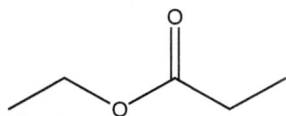

(5.60)

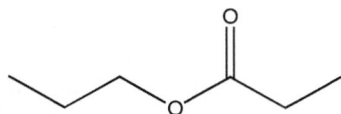

(5.61)

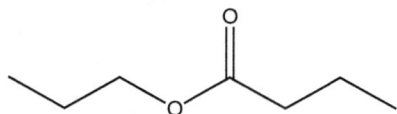

(5.62)

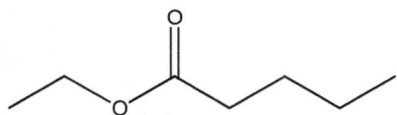

(5.63)

The presence of five more esters, 3-methylbutyl pentanoate (5.64), 3-methylbutyl 3-methylbutanoate (5.65), 3-methylbutyl 2-methylbutanoate (5.66), butyl pentanoate (5.67) and propyl hexanoate (5.68)), could distinguish between pig remains from human remains.

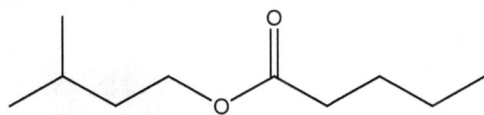

(5.64)

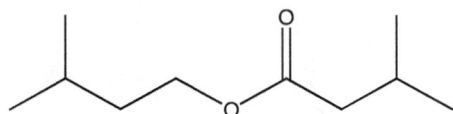

(5.65)

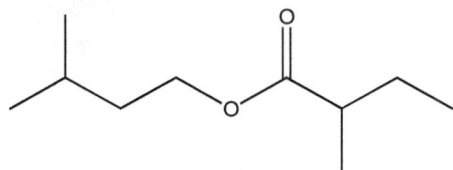

(5.66)

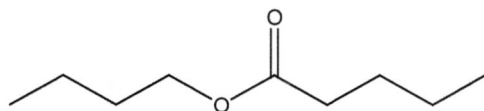

(5.67)

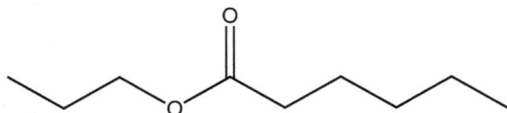

(5.68)

EUGLOSSINE BEES

Not everyone dislikes 'strong' smells. Euglossine bees (Euglossini) are a species of long-tongued tropical bee with unusual pockets on their hind legs, used to collect volatiles from a range of sources (including flowers, decaying plant and vegetable matter, and faeces) that they use in courtship. Many are attracted to a range of pleasant esters (e.g. methyl benzoate, benzyl acetate, methyl salicylate and methyl cinnamate), not to mention skatole! (see 'War and Peace in Nature' chapter, p. 154).

6 War and Peace in Nature

In the eyes of some, chemistry has become associated with warfare, whether in the form of 'Greek fire' used by the Byzantine Empire a thousand years ago, its successor, napalm, or the poison gases of World War I like phosgene and mustard gas, as well as nerve agents, most recently in the March and June 2020 Novichok incidents in Salisbury (UK). However, plants and insects have been using chemical warfare for millions of years.

Despite its title, this chapter is not written by Leo Tolstoy. Unlike Alexander Porfiryevich Borodin, whose day job was 'chemist,' Tolstoy focused on literature. Aldehydes were one of Borodin's specialties, appropriate for the study of green grass, as will emerge.

CUT GRASS

The smell of freshly cut grass is very evocative. Chemists know that for a molecule to be smelled, it must have a molecular mass of less than about 300. But what small, smelly molecules are involved, and why are they generated? Some interesting chemistry is involved, and it is not confined to grass or to the action of the mower. These same molecules are formed when an insect chews on plant material. This smell is known as 'green leaf volatiles' (GLV) or 'green odour' and is caused by a blend of six-carbon (C_6) aldehydes, alcohols and esters formed from the breakdown of lipids (chemically, esters) present in the blades of grass, in a series of enzyme-controlled reactions.

Cutting the grass breaks cell membranes, allowing the lipids to come into contact with an enzyme that hydrolyses the ester linkages, a hydrolase enzyme. This generates free unsaturated long-chain carboxylic acids like linoleic acid (9Z, 12Z-octadeca-9,12-dienoic acid) and α-linolenic acid (9Z, 12Z, 15Z-octadecatrienoic acid). Both are C_{18} molecules, with linoleic acid having one less C=C bond. 'Green grass' volatiles are formed in the next step, where a different type of enzyme, known as a lipoxygenase, uses atmospheric oxygen in an oxygenation process. The enzyme grips the linoleic acid molecule so that an oxygen molecule oxidises the C=C bond, between atoms 12 and 13 in the chain (Figure 6.1), with the oxidation process being assisted by the presence of a Fe^{3+} ion in the enzyme, which is important for electron transfer in the redox process.

The initial product is a hydroperoxide (6.1), where a –OOH group is attached to carbon 13 of the chain. Next, a hydroperoxide lyase enzyme cleaves the carbon chain, generating a C_6 molecule, hexanal (as well as another aldehyde group at the end of the rest of the carbon chain). The hexanal can be released directly into the atmosphere, but some will also be reduced by an alcohol dehydrogenase enzyme to the primary alcohol hexan-1-ol; this alcohol can subsequently be acetylated by an acetyltransferase enzyme, forming the ester hexyl acetate (6.2). Both hexan-1-ol and hexyl acetate can also escape into the surrounding air.

DOI: 10.1201/9781003658115-6

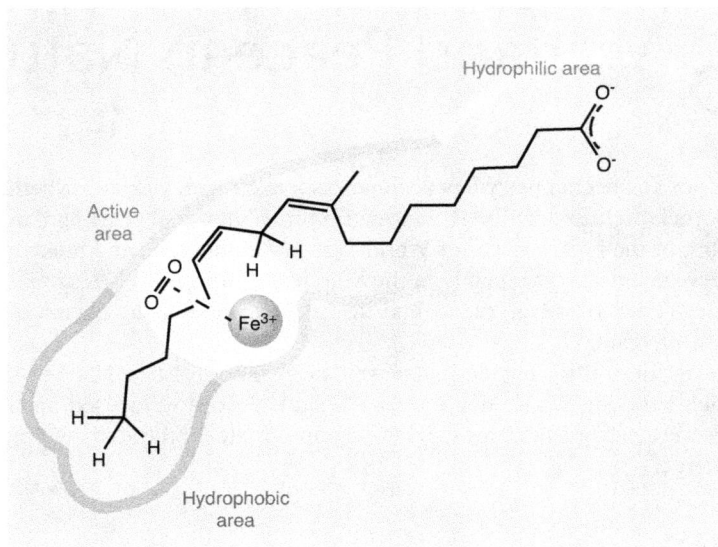

FIGURE 6.1 Linoleic acid binding to the active site of a lipoxygenase enzyme. From S. A. Cotton, *Chemistry Review*, November 2024, **34**(2), 20–25; based on A. Hatanaka, *Phytochemistry*, 1993, **34**(5), 1201–1218.

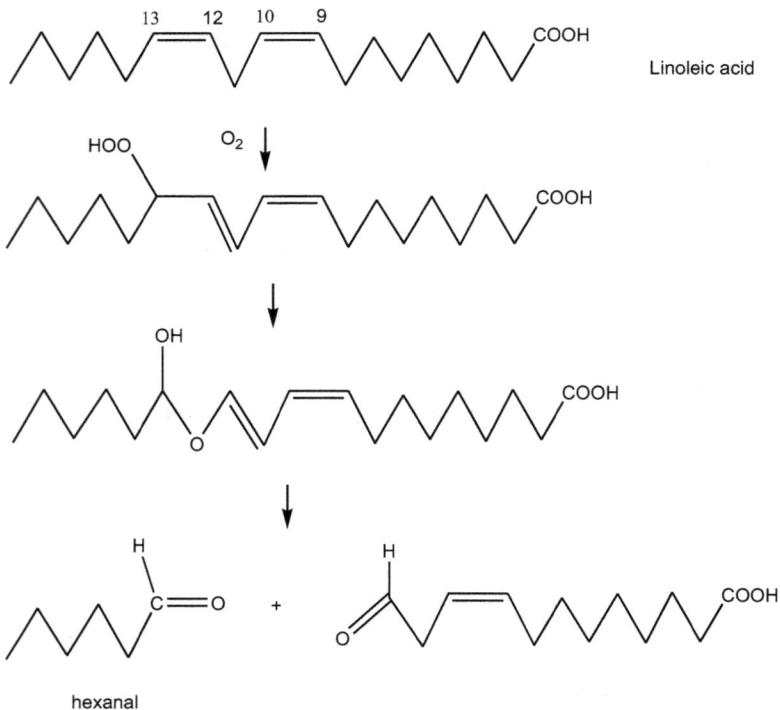

$$(6.1)$$

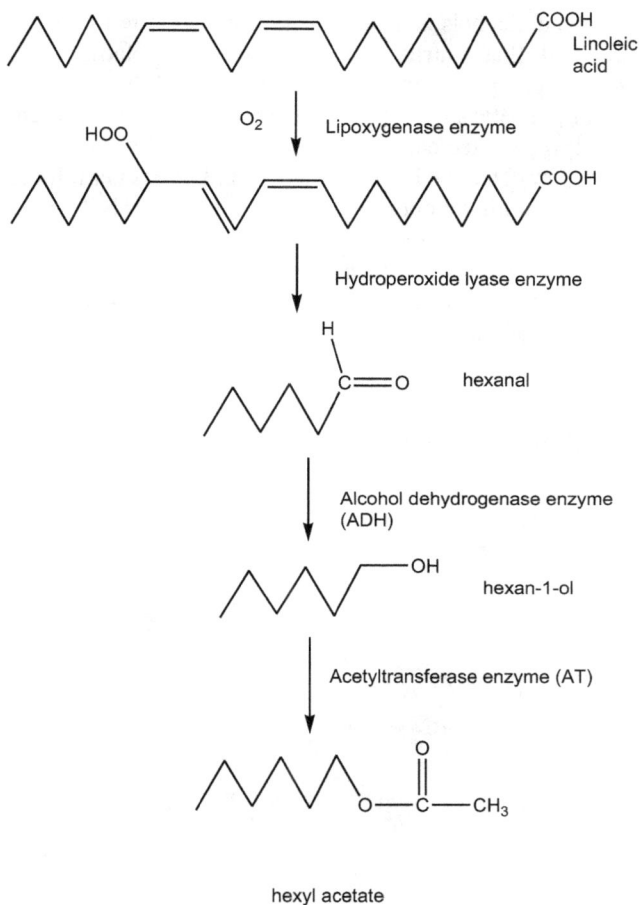

hexyl acetate

(6.2)

These are not the only 'green grass' volatiles. The breakdown of α-linolenic acid broadly follows a similar route (6.3). The lipoxygenase enzyme grips α-linolenic acid similarly to linoleic acid, so that the carbon chain here also gets cleaved between carbons 12 and 13. Compared to linoleic acid, the carbon backbone in linolenic acid contains another double bond, between carbons 15 and 16, so that this C=C double bond is incorporated into the structure of the C_6 product, the unsaturated aldehyde (Z)-3-hexenal (6.3). This can undergo similar reactions to those described for hexanal, including reduction to the unsaturated alcohol (Z)-3-hexen-1-ol, and then acetylation by acetyltransferase to form the ester (Z)-3-hexenyl acetate.

However, there is an important difference in the reactions involving (Z)-3-hexenal. This compound is unstable, undergoing two transformations catalysed by 'isomerisation factor' enzymes. First, (Z)-3-hexenal rapidly changes to its geometric isomer (E)-3-hexenal; this can in turn isomerise to (E)-2-hexenal, when the double bond moves one position nearer to the carbonyl group. These two aldehydes can both undergo analogous reactions to those described for (Z)-3-hexenal, including reduction to their corresponding unsaturated alcohols and subsequent acetylation to esters. So there are quite a few C_6 compounds that contribute to the 'Green Leaf Volatiles'!

What does this mean for the 'cut grass' smell? The short-lived (Z)-3-hexenal is the main contributor to the characteristic smell of fresh-cut grass. Even when present at very low concentrations, it can be detected, as it has an exceptionally low odour threshold – the amount required for detection by our noses – of 0.25 parts per billion. The more stable (E)-2-hexenal has a higher odour threshold, 17 parts per billion. Its smell is less 'fresh' than that of its unstable precursor, and this may contribute to the change in the smell of cut grass that you notice after a period of time. (E)-2-hexenal is known as 'leaf aldehyde'. (Z)-3-hexen-1-ol ("leaf alcohol") has a weaker green smell, with an odour threshold of 70 ppb, but is still used in perfumery on account of this.

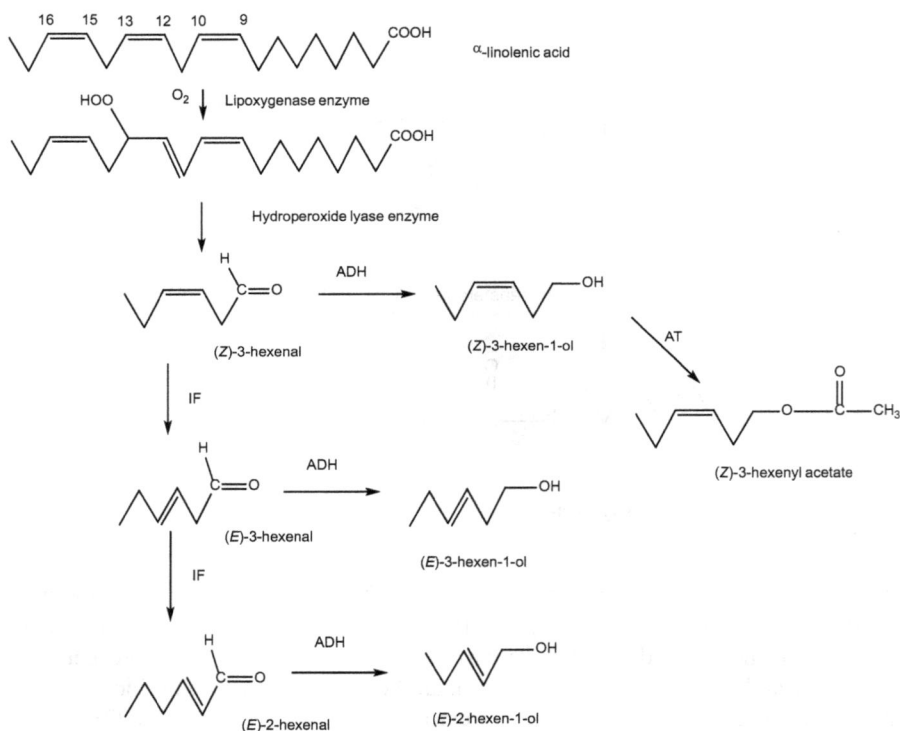

(6.3)

Scientists have analysed the vapour emitted from grass, both before and after cutting. A majority of the emissions from ordinary (uncut) grass is made up of five simple, rather small molecules – methanol (11%–15%), ethanol (16%–21%), ethanal (13%–16%), propanone (11%–16%) and hexanal (3%–4%). Together, they make up 59%–68% of the total emissions, with no other one compound making up more than 1% of the rest, including isoprene (ca. 0.1%) and monoterpenes making up 1%–1.5%. Upon cutting the grass, emissions of volatile organics increased immensely, with a very different profile, being greatly enhanced in C_6-related compounds in particular. (Z)-3-hexenyl acetate made up nearly 40% of the emissions, whilst two other

C_6 species, (Z)-3-hexen-1-ol (leaf alcohol) and (Z)-3-hexenal, accounted for another 20%, along with (E)-2-hexenal (7%). The levels of ethanal and propanone dropped to only about 4% and 2%, respectively (in contrast to uncut grass), though there was still a significant amount of methanol (10%). The C_6-related emissions in cut grass total some 70%, rather than the 1% from uncut grass.

Apart from their smell, what role do these molecules play in the grass? One important function is believed to be promoting the healing of 'cut ends'. These C_6 molecules are generally produced by green plants almost instantly when the plants are stressed, including during herbivore or pathogen attacks. They spread through the grass and diffuse to nearby plants, signalling to them to 'switch on' their defence mechanisms. Many unsaturated aldehydes have bactericidal and antifungal properties; both (E)-2-hexenal and (E)-3-hexenal act against a range of bacteria, including *Salmonella choleraesuis*, a very common cause of septicaemia.

These C_6 GLVs have other roles. There is an interesting report of what happens when tobacco hornworm larvae (*Manduca sexta*) feed on tobacco plants (*Nicotiana attenuata*). The plants naturally produce a mixture of the two isomeric hexenals, (Z)-3-hexenal and (E)-2-hexenal; when the larvae feed on the plants, their saliva causes a considerable increase in the amount of (E)-2-hexenal in the mixture. This change in the ratio of the isomers recruits predatory bugs, *Geocoris* spp., which attack the larvae, a 'plant defence' that operates faster than the plants can manufacture their own defence compounds. So, these GLVs act as messengers.

Apart from cut grass, we recognise (Z)-3-hexenal in the flavours of fruits such as watermelon, strawberries and tomatoes, whilst (E)-2-hexenal contributes to the freshness of apple juice. They may also benefit us in other ways, as "green odour" mixtures involving (E)-2-hexenal and (Z)-3-hexen-1-ol have been shown to relieve the psychological damage caused by stress in monkeys, as well as stress in rats and pain sensation in humans.

PLANT DEFENCE

Plants are at the bottom of the food chain. Though it sounds like confused biology to use the expression, they seem like sitting ducks. Vulnerable, yes, but they have a range of tricks up their sleeves to defend themselves against predators.

To the eye, the most obvious are physical defences. These range from barriers – the thicker the leaf, the better – to surfaces, where a waxy coating can stop insects from getting a grip, or latex in rubber plants that becomes tacky and traps insects, or the resins in conifers that perform a similar function, the latter containing chemical repellents like β-thujaplicin (6.4) in cedars. Thorns and spines deter larger predators. Stinging nettles use another defensive chemical.

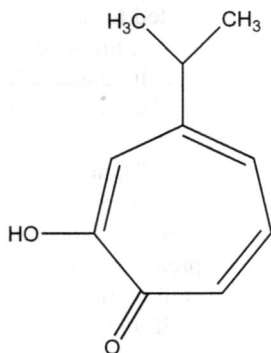

(6.4)

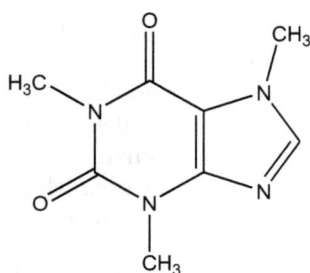

(6.5)

Though the public does not usually associate these chemicals with plant defence, almost everyone consumes one of these every day – the caffeine (6.5) in their tea (*Camellia sinensis*) or coffee (e.g. *Coffea Arabica* and *Coffea robusta*). Some members of the public are smokers, which adds nicotine (6.6) from tobacco (*Nicotiana tabacum*) to the list. These molecules are known as secondary metabolites – compounds made by plants that are not essential to their basic functioning. Other molecules made by plants for a similar reason are morphine (6.7) in poppies (*Papaver somniferum*) and cocaine (6.8) in coca plants (*Erythroxylum coca*). These molecules are toxic to insects and thus have a protective role in plants – for example, it has been shown that nicotine protects wild tobacco plants from herbivores. These molecules are all basic (due to the nitrogen atoms they contain) and are collectively known as alkaloids. Of course, molecules like morphine, cocaine and nicotine can cause fatalities in humans too, though morphine in particular has legitimate medical uses.

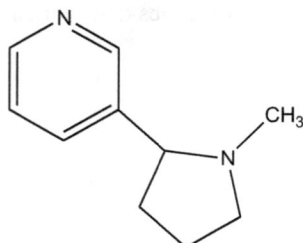

(6.6)

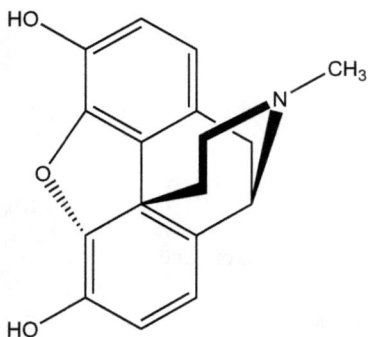

(6.7)

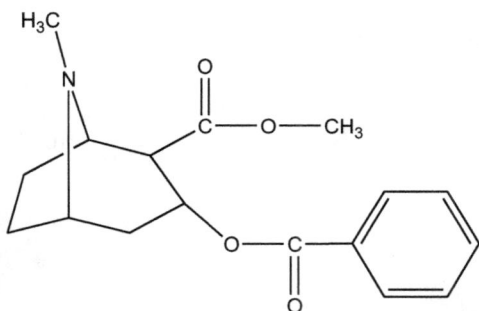

(6.8)

There are other molecules that protect plants against insects. One family of them is the pyrethrins (6.9), made by plants of the chrysanthemum family. They are fatal to insects in concentrations that are not toxic to humans. Pyrethrins are environmentally friendly insecticides because they are easily degraded in the air or sunlight. Urushiols (6.10) are produced by poison ivy (*Toxicodendron radicans*) as a defence against herbivores. They are not well tolerated by *Homo sapiens*, as contact with the skin leads to a rather unpleasant form of dermatitis.

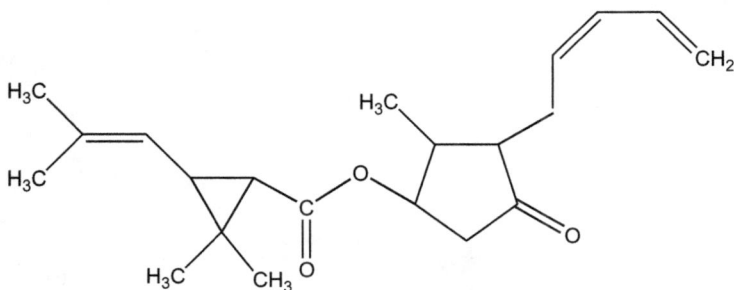

(6.9)

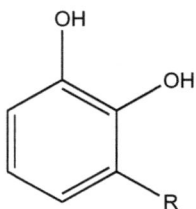

R e.g. (CH$_2$)$_{14}$CH$_3$

(6.10)

Capsaicin (6. 11) is familiar to us as an ingredient in chillies, enhancing our enjoyment of food (pp. 74). Its role in chillis is protective, against microbial fungi. Although
capsaicin irritates us, birds are not affected, which enables them to spread chilli seeds.

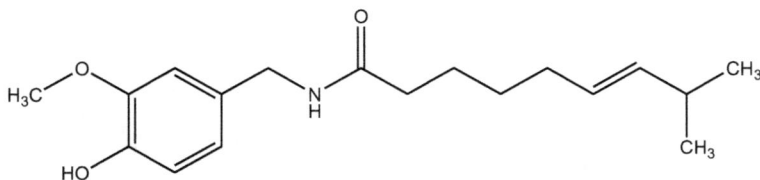

(6.11)

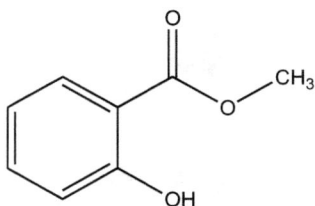

(6.12)

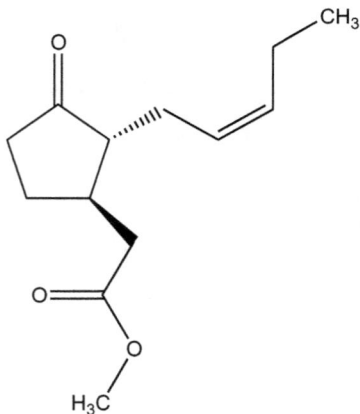

(6.13)

Other chemicals play a more indirect role. Take the volatile ester methyl salicylate
(6.12), familiar as the smell of 'wintergreen' embrocation, which appears in several
situations, notably as a 'messenger molecule'. It sends a 'not welcome' message when
it repels aphids from beans and cereals. In a different situation, spider mites have a

taste for lima beans. When the mites attack the beans, the damaged plants release methyl salicylate, which attracts predatory mites that feed on the spider mites.

Sometimes the message is directed to other plants, warning them of the attack, to 'tell the neighbours'. So when a tobacco plant is attacked by the tobacco mosaic virus, the affected plant emits methyl salicylate vapour, which is detected by other plants, who 'get the message' to activate their disease resistance genes. This phenomenon is not confined to tobacco plants. American scientists examined a forest of walnuts (*Juglans californica*×*Juglans regia*) experiencing significant day–night temperature fluctuations, finding that the stressed plants were emitting considerable quantities of methyl salicylate, which could be absorbed by other plants, often several feet away. The receiving plant uses an enzyme to hydrolyse the ester, converting it into salicylic acid, which is much less volatile than methyl salicylate. Salicylic acid is a plant hormone. It signals the walnut plant to respond – to stress, recover from disease or resist microbial pathogens. The plants use methyl salicylate as the messenger because salicylic acid is too involatile for that purpose – it has a melting point over 150°C higher than methyl salicylate and a correspondingly higher boiling point.

So, the sequence is: plants turn salicylic acid into methyl salicylate, which makes an aerial journey to another plant, where it is converted back to salicylic acid, signalling the plant to produce antimicrobial proteins.

Methyl salicylate isn't the only ester to do this. Methyl jasmonate ((6.13), MeJA) is another one. If MeJA is applied to the leaves of tomato plants, defensive proteinase inhibitor proteins are formed, both in the treated plants and in their neighbours. Furthermore, sagebrush, *Artemisia tridentata,* another plant containing MeJA in its leaves, communicates the MeJA to neighbouring tomato plants, again generating proteinase inhibitors in the tomato plants.

These proteinase inhibitors are a key part of the plant's defence, interfering with the digestive system of insect pests and deterring them from feeding on the plant.

WEAPONISED INSECTS

Most insects are quite small, which makes them vulnerable to predators, and they have developed several ways to overcome this. One obvious strategy is using defensive chemicals, some of which are poisonous whilst others have unpleasant smells. Many insects have striking colours that act as a warning ('aposematic') not to mess with them or eat them, as seen in ladybirds, whose colouring hints at their bitter taste.

Perhaps the most obvious examples of insects with a chemical defence are ants (Latin *Formica*). Around one-third of them are formicine ants, which have venom containing up to 60% methanoic acid ((6.14), traditionally formic acid), which they discharge as a spray from their abdomens.

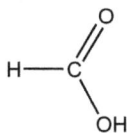

$$H\!-\!\underset{\displaystyle OH}{\overset{\displaystyle O}{\|\,C}}$$

(6.14)

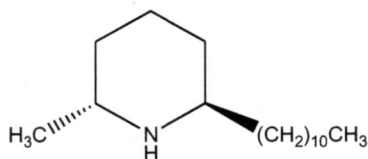

(6.15)

This acidity has more than one use. *Solenopsis invicta* fire ants have a multi-component venom that contains toxic proteins as well as basic piperidines like (*E*)-2-methyl-6-undecanylpiperidine (6.15). The latter causes an intense burning pain when humans are stung (which is the origin of the name fire ants), triggering an allergic reaction in many of the people they sting, which may be fatal. *Nylanderia fulva* worker ants use their venom against *S. invicta* fire ants, where the acid may neutralise the basic piperidines as well as denature enzymes.

It's not just about acids. The bombardier beetle (e.g. *Stenaptinus insignis*) employs hydrogen peroxide (H_2O_2), which it produces itself. It also synthesises hydroquinone ((6.16): benzene-1,4-diol). When the insect feels threatened, it mixes these substances in a reaction chamber, whose walls produce two types of enzymes – catalases and peroxidases – which catalyse two different reactions. The role of the catalase is to decompose hydrogen peroxide into water and oxygen:

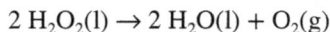

$$2\ H_2O_2(l) \rightarrow 2\ H_2O(l) + O_2(g)$$

The peroxidase oxidises hydroquinone to benzoquinone (6.17) in a very exothermic reaction (around 100°C).

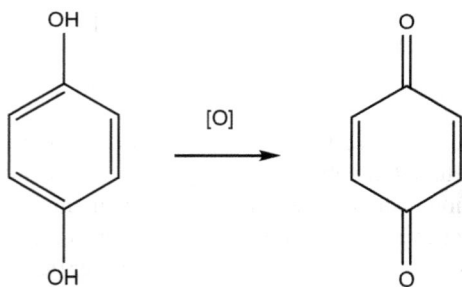

(6.16, 6.17)

The pressure from the generated oxygen forces the hot mixture out of a tip at the rear of the abdomen, which the beetle can aim at its target, with the mixture being very damaging to the target's eyes and respiratory system.

The bombardier beetle isn't the only spraying beetle. The red-lined carrion beetle, *Necrodes surinamensis*, like the bombardier beetle, has a moveable jet at the tip of its abdomen. Instead of hydrogen peroxide, it uses two saturated acids, hexanoic acid (6.18) and octanoic acid (6.19); two unsaturated acids, (*E*)-3-hexanoic acid (6.20) and (*E*)-4-hexanoic acid (6.21); and several terpenoids, including the unsaturated alcohol lavandulol (6.22), also found in lavender oil.

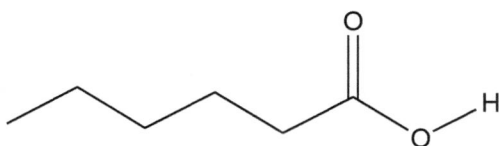

(6.18)

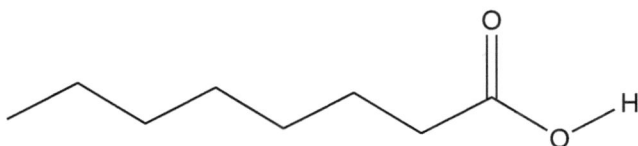

(6.19)

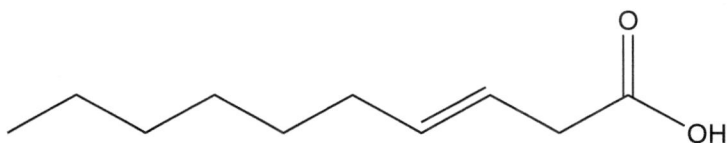

(6.20)

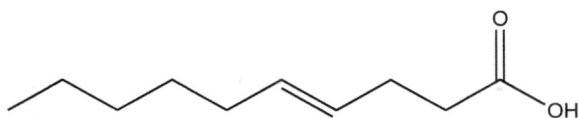

(6.21)

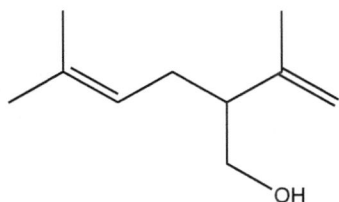

(6.22)

Less acidic than carboxylic acids, phenols are highly corrosive. *Romalea guttata* (the Lubber grasshopper) emits frothy bubbles from its thorax and mouth when threatened. The bubbles burst, forming a very unpleasant mist containing phenol itself (6.23), para-cresol ((6.24): 4-methylphenol), verbenone (6.25), isophorone (6.25) and 1,4-benzoquinone (6.17), the last of which is also emitted by bombardier beetles.

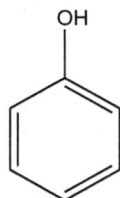

(6.23)

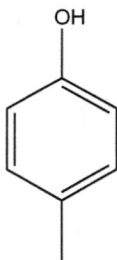

(6.24)

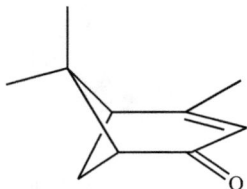

(6.25)

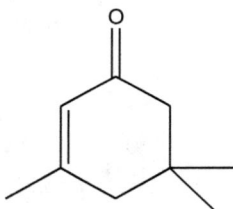

(6.26)

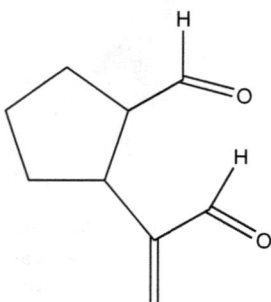

(6.27)

The American walkingstick insect *Anisomorpha buprestoides* is another insect that produces a defensive spray from its abdomen. In this case, the defensive compound is anisomorphal ((6.27): also known as dolichodial). This is particularly painful to the eyes of humans and is also irritating when inhaled.

Soldier termites of the Australian species *Nasutitermes exitiosus* do something even more complicated; they have been described by the authors of the book '*Secret Weapons*' as 'mobile artillery units'. They fire their weapon from a gland on their heads, not their abdomens, but it can be directed ahead, to the sides and even behind them. It is a mixture composed of odorous terpene hydrocarbons, particularly limonene (6.28), α-pinene (6.29) and β-pinene (6.30), as well as larger molecules such as kempene (6.31) and trinervitine (6.32). The mixture emerges from the gland as a fine filament, usually directed at ants that have invaded the mound inhabited by the termites. The terpenes irritate the ant targets, but the other components are larger and more viscous molecules, which serve to stick the invaders to the walls of the mound. The smell of the volatile limonene and pinenes also acts as an alarm pheromone, summoning other soldiers who surround their prey, adding their secretions to the target, which then expires.

(6.28)

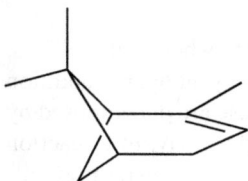

(6.29)

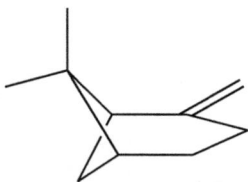

(6.30)

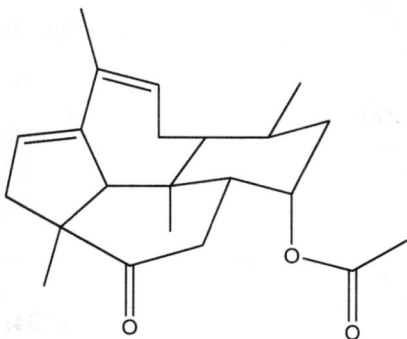

(6.31)

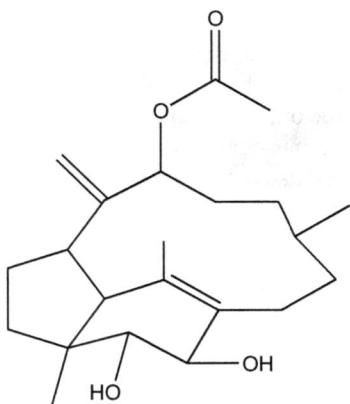

(6.32)

OUT OF DETECTIVE NOVELS

You do not have to be a fan of Agatha Christie or another traditional crime writer to know that cyanide is toxic to humans. It is generally toxic to animal life, as it interferes with cytochrome oxidase enzymes, stopping respiration. A number of insects make use of it.

They don't store hydrogen cyanide (HCN) itself; it is generated when needed. Thus, soil centipedes of the order *Geophilidia* carry internally the cyanohydrin addition products mandelonitrile (6.33) and benzoyl cyanide (6.34). These are decomposed by enzymes into HCN and benzaldehyde and benzoic acid, respectively, in a reaction chamber 'on demand', then discharged from the flanks of the insect. Polydesmid millipedes like *Aphelporia corrugata* defend themselves in a similar way.

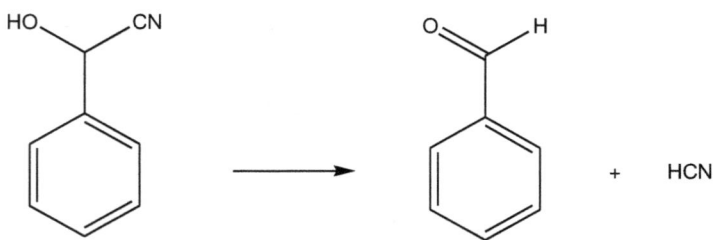

(6.33)

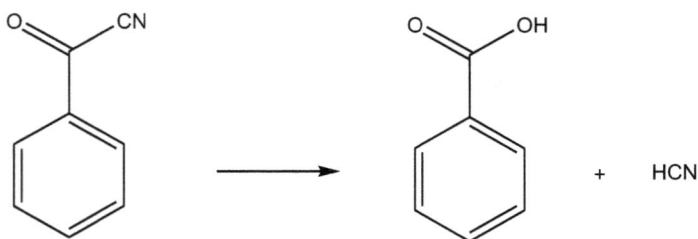

(6.34)

Some arthropods, including butterflies and moths, use HCN to defend themselves. A well-known example is the six-spot burnet moth (*Zygaema filipoendulae*), which has a black and red wing pattern to advertise its unpalatability, again aposematism. They utilise the compounds linamarin (6.35) and lotaustralin (6.36), which contain the cyanohydrins of propanone and butanone respectively, joined to glucose (Glc). The moths can obtain them from a host plant, *Lotus corniculatus*, as well as produce them themselves.

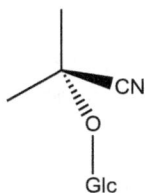

(6.35)

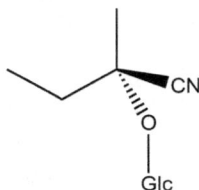

(6.36)

BLISTER BEETLES

Another example of toxic beetles, infamous in human history, is provided by the so-called blister beetles of the family Meloidae, which use cantharidin (6.37).

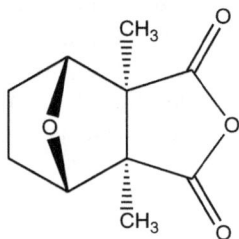

(6.37)

This is produced by male beetles as a defensive agent and passed on to females. When the beetles are disturbed or attacked, they bleed. If they are held by the leg, they bleed from the knee of that leg, whilst if they are held by the body, they bleed from all their knees and sometimes from other joints too; the blood contains cantharidin. Cantharidin is toxic to predators – spiders and ants are known to reject food containing cantharidin.

Dried and powdered remains of these beetles are known as 'Spanish fly' and were used hundreds of years ago as an aphrodisiac by human males, including the Italian libertine Giacomo Casanova. They do induce male erections, but cantharidin also causes severe damage to the renal and reproductive systems. The lethal dose for humans is around 100 mg. Infestation of hay by these beetles is not unknown; this can cause poisoning in horses that eat the hay. On the skin, cantharidin causes blisters, which is where the beetles get their common name.

There are beetles (*Neopyrochroa flabellata*) where cantharidin is used by males to attract females for mating. The females use it to protect their eggs from ants (for example).

PLANTS ATTRACTING INSECTS

Nature is a brilliant synthetic chemist, and that includes plants. Obviously, they make chemicals to build their structures and to maintain their living systems, but there's more to it than that. They also have to reproduce, which includes attracting pollinators. This is achieved with coloured substances and patterns in their leaves, which appeal to pollinators. Scented molecules are also used as attractants. Additionally, there's also the matter of self-protection from their enemies; they use molecules to repel insects that would otherwise feed upon them, either through their bitter taste or by being toxic (see pp 142–144).

The transport of pollen to eggs, the pollination of flowering plants, which is carried out by insects in around two-thirds of cases, enables the plants to reproduce and is absolutely vital to the world's agriculture. Scented molecules are an alternative to visual cues and have the advantage of working after dark. Bees are the most important pollinators, providing spin-off benefits like honey as food (and a source of sweetness, especially in the days before sugars were widely available) and wax.

Many of the molecules used by plants to attract bees are terpenoids, which are built up of multiples of the unsaturated hydrocarbon isoprene (C_5H_8). These attractants include the hydrocarbons trans-β-ocimene (6.38) and trans-β:(3*E*)-3,7-dimethylocta-1 ,3,6-triene); alcohols such as linalool (6.39), (*E, E*)-farnesol (6.40), geraniol (6.41) and nerol (6.42); the aldehyde (*Z*)-citral ((6.43), geranial) and the acids geranic acid (6.44) and nerolic acid (6.45), which are synthesised in the plants from precursors like carbohydrates and amino acids. Aromatic compounds are another class of widely used attractants, with methyl salicylate (6.12), methyl benzoate (6.46), phenylethanol (6.47) and phenylacetaldehyde (6.48) are among the most common honeybee attractants.

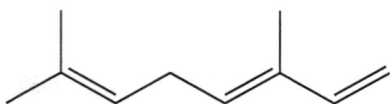

(6.38)

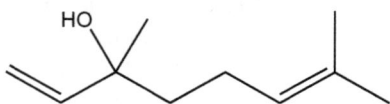

(6.39)

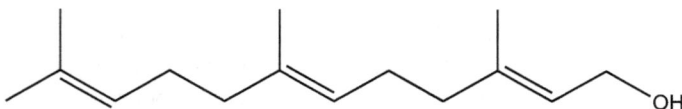

(6.40)

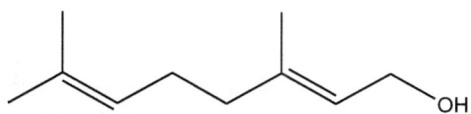

(6.41)

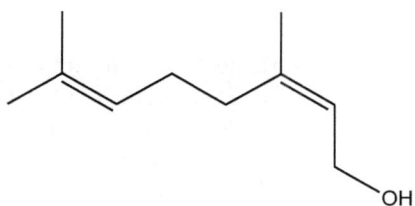

(6.42)

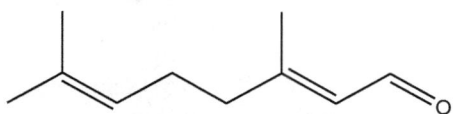

(6.43)

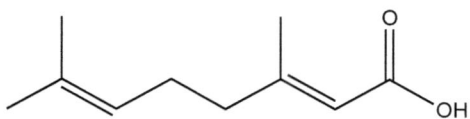

(6.44)

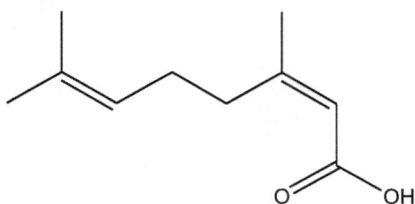

(6.45)

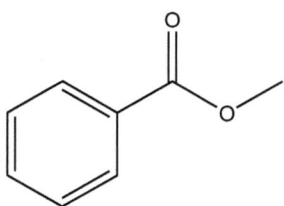

(6.46)

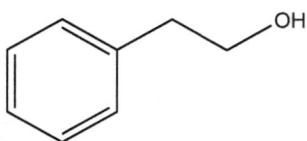

(6.47)

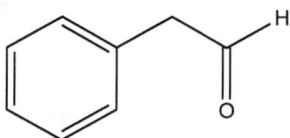

(6.48)

One type of pollinating bee stands out: Euglossine bees (Apidae: Euglossini) are found in South and Central American countries, including Brazil, Costa Rica, Guatemala, Ecuador, Mexico and Panama, but some *Euglossa viridissima* have become recently naturalised in Florida (USA). These solitary bees have iridescent green, blue or bronze colours. Male euglossine bees recognise the flowers whose nectar they feed on through chemicals, but they also *collect* chemicals from sources including decaying wood, faeces and even insecticides! They are important pollinators of some orchids. They rub brushes on their front tarsi against the flower's surface to suck up chemicals (using their own lipids) and then transfer them to internal storage pouches in their hind legs via combs on their middle legs. These chemicals are subsequently used by the bee to form a 'perfume' to attract females for breeding. Euglossine bees can detect a very wide range of organic molecules, though only some are attractants, and those vary between species. The five most popular odorants are cineole ((6.49), also known as eucalyptol), methyl salicylate (6.12), skatole ((6.50), the smell of faeces), vanillin (6.51) and methyl cinnamate (6.52). Most remarkably, one species of euglossine bee, *Eufriesea purpurata*, collects the insecticide DDT (6.53) with no ill effects, from DDT levels averaging 2,000 µg per bee!

(6.49)

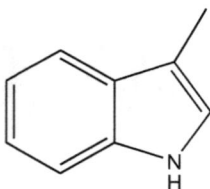

(6.50)

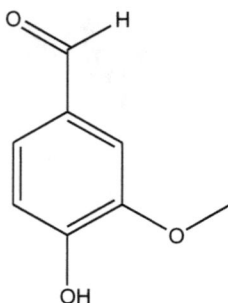

(6.51)

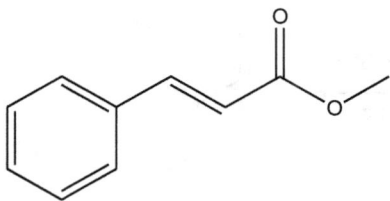

(6.52)

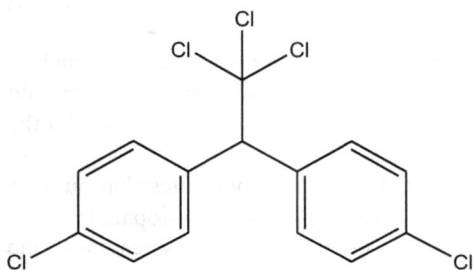

(6.53)

INSECT REPELLENTS

As we have seen, plants have been ahead of us in making chemicals like nicotine to combat insects. For thousands of years, bruised plant materials, which can emit repellent molecules, have been hung in homes. In some parts of the world, like the United Kingdom, insects are just a nuisance. Elsewhere, it is another matter, especially where mosquitoes are concerned. *Anopheles* mosquitoes are the only mosquito genus that carries malaria, with *Anopheles gambiae* transmitting the lethal Plasmodium falciparum parasite. Malaria kills well over half a million people a year. The *Aedes* mosquito transmits other diseases, including dengue fever, yellow fever and Rift Valley fever, the latter being among those carried by *Culex* mosquitoes.

Humans have been well aware of insect-borne diseases for thousands of years and have long sought ways of to combat their transmission. The ancient Egyptians are credited with using repellents, whilst the celebrated Cleopatra is said to have slept under mosquito nets, which nowadays are often loaded with insecticides.

World War II can be seen as giving an impetus to the development of synthetic insecticides. British and American forces were operating in many parts of the world where diseases like malaria were rife, at a time when access to established antimalarial drugs like quinine was severely reduced. The best-established synthetic repellent, DEET, was developed then and came into use in the US Army in 1946, subsequently becoming available to civilians. DEET is shorthand for its chemical name, N, N-diethyl-meta-toluamide; its systematic name is N, N-diethyl-3-methyl-benzamide. DEET is straightforward to make (6.54). The organic starting material is 3-methyl-benzoic acid, which is first turned into the corresponding acyl chloride; this is a more reactive compound, which, when it reacts with diethylamine, forms DEET.

(6.54)

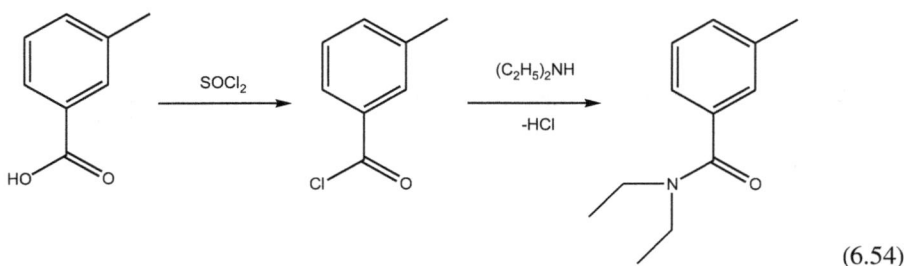

DEET is widely used as a mosquito repellent, with some minor side effects, including irritation of the skin. However, reports of DEET poisoning are very rare, and
these have generally resulted from someone deliberately taking an overdose. All the
evidence suggests that it is very safe. So how does it work?

It is understood that mosquitoes are attracted to several molecules that humans
emit, notably carbon dioxide (6.55), lactic acid ((6.56), 2-hydroxypropanoic acid),
1-octen-3-ol (6.57) and 6-methyl-5-hepten-2-one ((6.58), known as sulcatone). One
theory suggests that repellents may form an invisible 'blanket' of vapour that effectively acts as a barrier between the insect and the human emitting these molecules,
which means that the insects cannot sense them.

$$O{=}C{=}O$$

(6.55)

(6.56)

(6.57)

(6.58)

Experiments have shown that female *Aedes aegypti* mosquitoes are attracted to
humans even when DEET is present, but after they contact skin bearing DEET, they
are repelled. With *Drosophila melanogaster* flies, DEET interferes with food ingestion, linked with bitter taste neurons in their proboscis; although mosquitoes have
similar receptors in their proboscis, tests on Aedes aegypti mosquitoes show that
they sense DEET through their feet, not their mouths or proboscis. A study of *Culex
quinquefasciatus* mosquitoes shows that they dislike the smell of DEET.

And what about natural repellents? One that has found favour is citronella oil, which suffers from the disadvantage of being quite volatile, meaning that it has to be reapplied every hour or so. Among its ingredients are citronellal, found in a number of plant oils, and lemon eucalyptus. The most promising 'natural' repellent is p-menthane-3,8-diol (PMD, or citriodiol), which is found in lemon eucalyptus oil. This is less volatile than citronella oil, so it stays on the skin for several hours, meaning that it is longer before another application is needed. p-menthane-3,8-diol has been approved for use in malaria-afflicted areas. Apart from being a naturally occurring chemical, it can be made by more than one route. One of these is a hydration reaction with dil. H_2SO_4 catalyst. This reaction not only turns the carbonyl group into an OH group but also generates another –OH group, as well as a C–C bond, forming the necessary ring system (6.59). Temperature affects the product obtained, as PMD exists as four diastereoisomers. 1R-(+)-cis-PMD is obtained at −78°C, but 1R-(−)-trans-PMD is the product at +70°C.

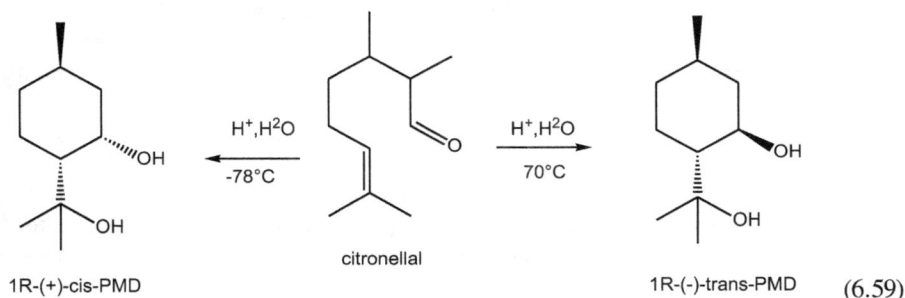

1R-(+)-cis-PMD   citronellal   1R-(-)-trans-PMD   (6.59)

This reaction suffers from the disadvantage of low yields, but there is an alternative. A similar hydration reaction (6.60), this time using isopulegol as the organic starting material, has yields of up to 94%.

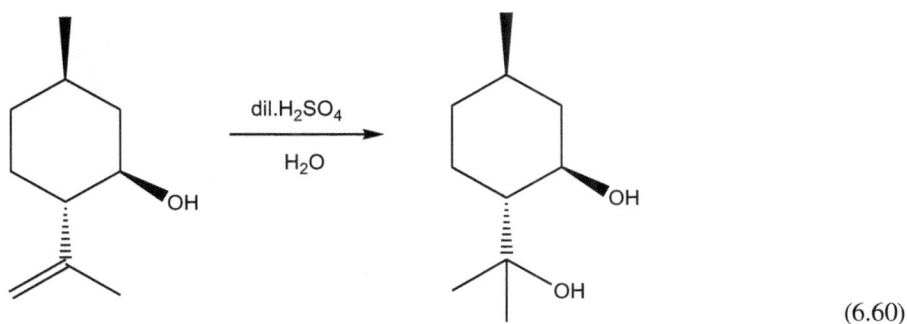

(6.60)

The four diastereoisomers of PMD (6.61) have been tested separately for 'repellency' against *Aedes albopictus* mosquitoes and have shown significant differences. (1R)-(+)-cis-PMD is the strongest repellent, more repellent than (1S)-(−)-cis-PMD and (1S)-(+)-trans-PMD; (1R)-(−)-trans-PMD is the least repellent of the four.

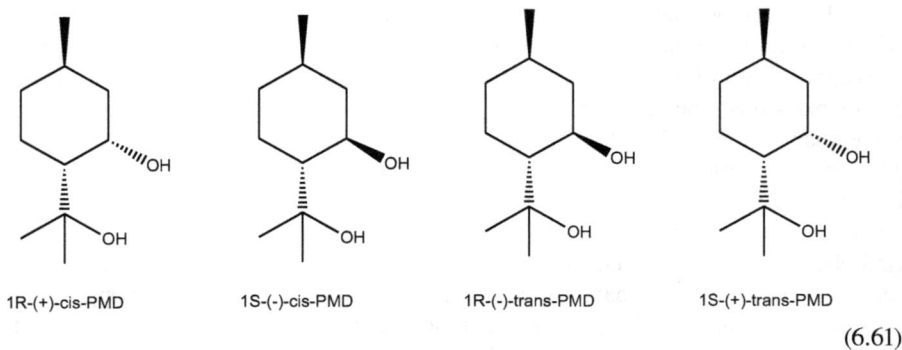

1R-(+)-cis-PMD   1S-(-)-cis-PMD   1R-(-)-trans-PMD   1S-(+)-trans-PMD

$$(6.61)$$

As diastereoisomers can behave differently in living systems, most evidently in their smells (e.g. carvone, see *Every Molecule Tells a Story*, pp. 60–61) as well as in drug action (e.g. thalidomide, see *Every Molecule Tells a Story*, pp. 115–117), these findings are not surprising. Different stereoisomers have different spatial orientations of their atoms, so they engage receptors differently, both in the human body and in mosquitoes.

In recent years, some people have favoured using devices such as bracelets, sonic repellers or citronella candles as their mosquito repellents. Scientific tests have indicated that many of these products have absolutely no effect as mosquito repellents. One comparative test that examined DEET and p-menthane-3,8-diol alongside alternatives found that the two first-named chemicals were by far the most effective mosquito repellents.

7 Organochlorine Compounds

There is a lot of chlorine present in the world, not least in seawater. Around 3.5% of seawater is dissolved solids (rising to 27% in the Dead Sea), and about 86% of that is NaCl, sodium chloride ('inorganic' chloride).

Elemental chlorine is, of course, too reactive to be found in nature. However, it has one very important everyday use: water supplies have been chlorinated for nearly a century (since 2001). This eliminated water-borne diseases like cholera, typhoid and meningitis. Chlorine is also cheap, and the process works at low levels of chlorine.

Chlorine reacts with dissolved organic matter, forming organochlorine compounds. These could be toxic, albeit only at very high levels. Some people moved to drinking bottled water, which, of course, brings its own problems. So in 1991, Peru decided to stop chlorinating their water supply. The result was outbreaks of cholera, with 10,000 fatalities. They resumed chlorinating water, but it took a long time to eliminate the cholera.

Greenpeace even took the view that chlorine should be banned. Really.

ORGANIC CHLORINE COMPOUNDS

Organic chlorine compounds have the public image of being environmental contaminants introduced into the environment by human activity ('pollutant' being the usual term used), with chlorofluorocarbons (CFCs) being the best example that comes to mind. In fact, large numbers of organic chlorine compounds are produced naturally. Back in 1954, fewer than a dozen natural organic compounds containing chlorine were known; this number is now well in excess of 2,500 as more research has been carried out and better instruments are available to detect them.

The most abundant of these organochlorine compounds is CH_3Cl (7.1), chloromethane (b.p. −24°C). Its atmospheric concentration is around 0.6 parts per billion. The great eco-guru James Lovelock detected CH_3Cl in the air over southern England in 1974–1975.

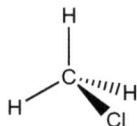

$$\begin{array}{c} H \\ | \\ H - C \cdots''''H \\ \quad\quad Cl \end{array}$$

(7.1)

About 5×10^6 tonnes of CH_3Cl are released globally each year, with well over 90% of this, possibly as much as 99%, coming from natural sources. The atmosphere contains CH_3Cl at a concentration of some 0.6 ppb, so it contributes much more chlorine

DOI: 10.1201/9781003658115-7

than CFCs like CF_2Cl_2, but the greater reactivity of CH_3Cl ensures a shorter atmospheric residence time (1.4 years compared with 100 years for CF_2Cl_2). Compared to CFCs, CH_3Cl has very low global warming potential, but it is decomposed by UV radiation, forming ozone-depleting chlorine radicals, in a similar way to CFCs (p. 174).

$$CH_3Cl \rightarrow \cdot CH_3 + \cdot Cl$$

Some 20 years ago, it was estimated that chloromethane contributed around 15% of ozone-depleting emissions; this proportion will increase as anthropogenic emissions of CFCs decrease.

It was once believed that the main source of chloromethane was the oceans, but research since the 1990s has shown that some 90% (or more) comes from a range of terrestrial sources. It is believed that there are 200,000 lightning-triggered fires across the world each year – biomass burning, forest and brush fires – which result in chloromethane. Fires created by human activity contribute much less. Wood-rotting fungi also produce chloromethane. Salt marshes, mangroves and peat bogs are major emitters of CH_3Cl; thus, a halophytic plant that occurs widely in salt marshes called *Batis maritima* produces chloromethane. Above all, it is calculated that tropical plants emit around half the chloromethane produced worldwide.

One of the reactions involved is believed to use pectin (7.2), a polysaccharide widely found in plant cell walls; this is believed to be the source of methyl groups, through an S_N2 reaction between chloride ions and a methoxy group in pectin.

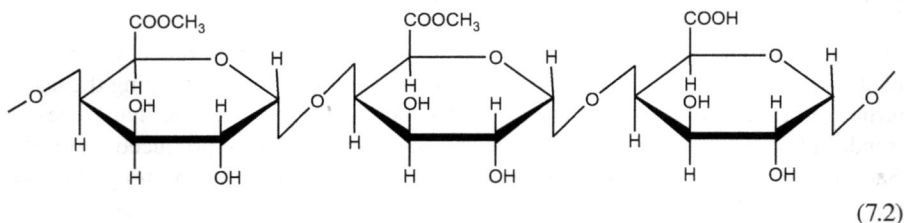

$$(7.2)$$

Two unexpected places where chloromethane has been found – not contributing significant amounts – are volcanoes and outer space. The fumes from several volcanic eruptions have released CH_3Cl, such as the 1969 eruption at Santiaguito in Guatemala, the 1980 eruption of Mount St. Helens in Washington, USA (which also released CH_3Br and CH_3I), and more recently at Etna in Italy. A report in 2017 revealed that chloromethane was the first organohalogen to be detected in outer space, in the coma of comet 67P/Churyumov–Gerasimenko (67P/C-G) and, some four hundred light-years away, within several young star clusters. In fact, both isotopologues ($CH_3{}^{35}Cl$ and $CH_3{}^{37}Cl$) were detected in the comet.

Although chloromethane has significant industrial uses, such as making methylchlorosilanes, like $(CH_3)_2SiCl_2$, which is important in making silicone polymers, and as a starting material in the synthesis of methanethiol (CH_3SH) and methylamines (CH_3NH_2, $(CH_3)_2NH$ and $(CH_3)_3N$), the industrial production of around a million tons a year is well below the amount made in nature.

Significant amounts of CH_3Cl are produced by nucleophilic substitution (using Cl^- present in seawater) of naturally produced CH_3Br and CH_3I. Among other halomethanes, bromomethane and iodomethane are also found in the atmosphere, but at much lower levels. Compared to CH_3Cl (ca. 600 ppt), the levels of CH_3Br and CH_3I are around 9 and 2 ppt, respectively. The polychlorinated methanes are also much less abundant than chloromethane. $CHCl_3$ can be formed by the degradation of humic acid.

PESTICIDES

We should start this section on organochlorine compounds with the 'elephant in the room'. Compared with great works of literature, scientific books are generally less read and less influential. Possibly the most famous book with scientific content is Charles Darwin's magnum opus, *'On the Origin of Species'* (1859). A close contender must be Rachel Carson's book *'Silent Spring'*, which appeared in 1962, a book where DDT is prominently featured.

Malaria, traditionally associated with marshes and swamps, has plagued much of the world throughout recorded history. It was not until 1897–8 that a British doctor, Ronald Ross, discovered that the *Anopheles* mosquito was involved in transferring the malaria parasite. Quinine had been used for several hundred years, originally in South America, as both a treatment and a prophylactic, but in the early 20th century, insecticides, such as pyrethrum sprays started to be employed, with limited success.

In 1873, a German doctoral student named Othmar Zeidler, a PhD student of the great Adolf von Baeyer, was the first person to synthesise dichlorodiphenyltrichloroethane (DDT) (7.3), though the systematic name is (2,2-bis(p-chlorophenyl) 1,1,1-trichloroethane). However, it was not until 1939 that the Swiss chemist Paul Müller discovered its insecticidal properties. The Swiss government carried out exhaustive tests and found that DDT was effective in destroying the Colorado potato beetle, which was severely damaging the Swiss potato crop. DDT was also effective against lice and other insects. It exerts its effects upon the nerve cells in insects (but not humans), causing Na^+ ions to flow in unchecked, leading to rapid firing of nerves, so that the insect dies of exhaustion ('excited to death').

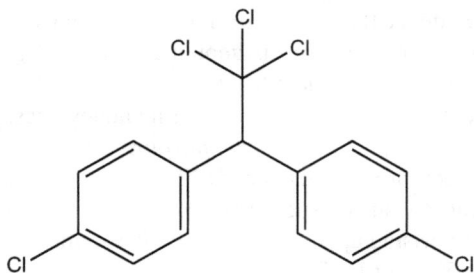

(7.3)

The Allies, especially the Americans, used DDT widely to control outbreaks of typhus and malaria in Europe from the latter part of World War II. In the best-known instance, in September 1943, German forces retreated from the Italian city of Naples, destroying some of the city's infrastructure as they went. There was no running water, gas or electricity, making the city an extremely unhygienic place. That month, 19 cases of typhus were recorded; by December there was a typhus epidemic, with 341 new cases. At the end of the year, the Allies began a mass campaign of dusting the population with DDT; by March, 1.3 million people had been dusted at a rate of 50,000 a day; the epidemic was effectively over. DDT was widely used by Allied forces in the Pacific campaign against malaria and other diseases like dengue fever. The British Prime Minister, Winston Churchill, described DDT as an 'excellent powder'.

Paul Herman Müller was awarded the 1948 Nobel Prize in Physiology or Medicine for his discovery of the effectiveness of DDT. After the end of the war, DDT was used widely across the world in campaigns run by the World Health Organization to eliminate malaria in areas like Taiwan and the Balkans, as well as across the Caribbean, North Africa, Australia, Sri Lanka and India. In this, DDT saved millions of lives. For example, in the island of Sri Lanka (then known as Ceylon), there were 2.5 million cases of malaria a year; a campaign began in 1948 that involved spraying every home regularly with DDT. By 1962, the annual number of cases had been reduced to 31 [sic]. Spraying ceased in 1964. By 1969, the annual number of cases had again reached 2.5 million. Why had spraying ceased? Well, *Silent Spring* had been published.

DDT had become a victim of its own success, with some people evidently thinking, 'It kills pests? Great, let's use more to kill more pests'. Concerns had been growing about the effects of DDT upon wildlife due to its indiscriminate spraying in many areas. Rachel Carson, a professional marine biologist, produced a passionately written book that encapsulated those fears. She did not call for a ban on all pesticides but warned against their indiscriminate use. She pointed out that DDT-resistant strains of insects were emerging.

As a result of the increasing outcry, DDT has been widely banned – for example, in 1973, the United States prohibited its agricultural use. However, the World Health Organization allowed its reintroduction only for the control of vector-borne diseases in some tropical countries in 2006.

There is certainly a place for the sensible use of DDT in clearly defined situations, though most definitely not for indiscriminate spraying. It is safer than some claims have suggested; for example, there is no evidence that it is a carcinogen in humans. It also has relatively low toxicity – the lethal dose in humans is quoted at around 30 g. Compared to some more recent insecticides, it appears less toxic.

Additionally, some insects can tolerate it better than others, including honey bees. More remarkably, male Euglossine bees of the species *Eufriesia purpurata* are well known to be attracted to DDT and to collect large amounts of DDT sprayed on the walls in their native Brazil, making repeated visits on successive days. DDT has no adverse effects on these bees, despite it comprising up to 4% of their body weight (the equivalent of a human containing some 1,500 g of DDT).

DIELDRIN AND ALDRIN

Dieldrin ((1aR, 2R, 2aS, 3S, 6R, 6aR, 7S, 7aS)-3,4,5,6,9,9-hexachloro-1a,2,2a,3,6,6a,
7,7a-octahydro-2,7:3,6-dimethanonaphtho[2,3-b]oxirene)

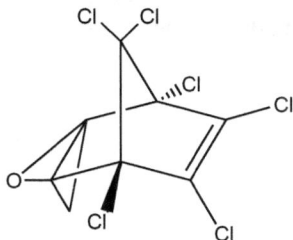

$$(7.4)$$

Dieldrin and Aldrin are synthetic insecticides developed in the wake of the success of DDT. They had early success but soon fell out of use because of their toxicity.
Aldrin is formed by the reaction of hexachloro-1,3-cyclopentadiene (a conjugated
diene) with norbornadiene (7.5), an example of the "Diels-Alder reaction" first
described by Otto Diels and Kurt Alder in 1928. In turn, aldrin undergoes oxidation
with a per-acid (e.g. peracetic acid), forming the epoxide dieldrin. Both aldrin and
dieldrin take their names from the reaction that makes them.

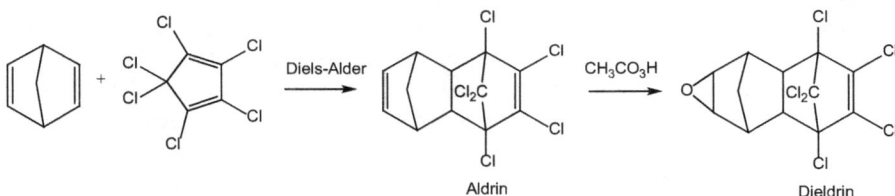

$$(7.5)$$

Aldrin was first prepared in 1948 in the wake of the initial success of DDT, by a
laboratory in Denver, Colorado. It was used primarily against insects in soil, notably
termites, where it was effective in amounts considerably lower than those of DDT.
It is oxidised in the insect to Dieldrin, which is the active insecticide (Aldrin is the
insecticidal version of a 'pro-drug'). Because of its lower volatility, Dieldrin was
used on crops (e.g. cotton, corn and citrus) and their foliage, in addition to soil and
seed dressing applications, both in considerably lower amounts than those of DDT.

The downside was their persistence, with a half-life of around 5 years for Dieldrin;
their bioaccumulation along the food chain; and their toxicity to a wide range of animals, including humans. Consequently, they were soon banned. For example, they
were first employed in Britain in the mid-1950s, but within a few years, they were
observed to kill birds that came into contact with treated seed, and they went out of
use in about a decade.

ANAESTHETICS – AND OTHERS

Perhaps the first successful synthetic organochlorine compounds were anaesthetics. Trichloromethane (7.6) – at the time called chloroform – was used as an inhalation anaesthetic from the 1840s. Prominently, it was employed by the British doctor John Snow when he delivered Queen Victoria's last two children during the 1850s. It was replaced by better anaesthetics, like halothane (7.7), which in turn has been superseded by others, mainly fluorine-based (see 183–184.).

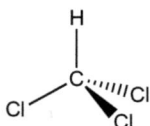

(7.6)

(7.7)

Various synthetic sweeteners have been deployed in the past century or so. The organochlorine compound sucralose (7.8), an artificial sweetener is both heat-stable (so it can be used in cooking) and zero-calorie. The latter follows from the fact that it is not metabolised, so most of it is excreted unchanged, and also because it is used in such small quantities that very small amounts are needed to produce a sweet response, as it binds so strongly to the sweet receptor compared to sugar molecules.

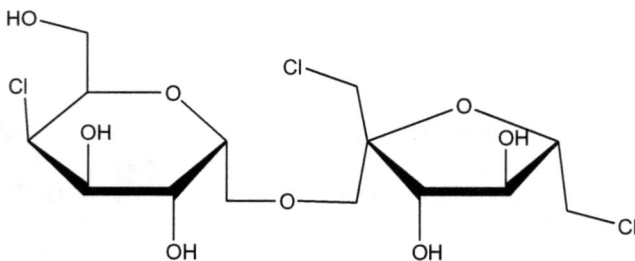

(7.8)

Sertraline (7.9, *Zoloft*) is another synthetic organochlorine compound in everyday use as a widely used antidepressant medication. Like Prozac and many others, it works as a Selective serotonin reuptake inhibitor.

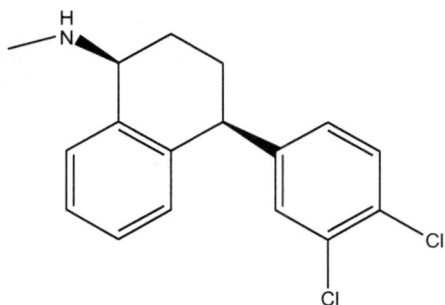

(7.9)

CHLORINATED PHENOLS

Some wit might suggest that a teacher's favourite insect is the tick, but this is unlikely, given their role in spreading numerous diseases, the best known of which is Lyme disease, along with others including Colorado tick fever, Rocky Mountain spotted fever, meningoencephalitis and typhus.

Females of many species of tick release the pheromone 2,6-dichlorophenol (7.10), which prompts excited males to search for and approach the female. The first report, in 1972, concerned the isolation and identification of 2,6-dichlorophenol from the female lone star tick, *Amblyomma americanum*. Within less than 5 years, others had identified the same molecule as the pheromone in the brown dog tick, *Rhipicephalus sanguineus*, the Rocky Mountain tick, *Dermacentor andersoni*, and the American dog tick, *Dermacentor variabilis*. The list has continued to mount, spreading to other continents. One interesting facet lies in the identification of 2,4-dichlorophenol (7.11) in females of the American dog tick, suggesting that this isomer may be a component of its sex pheromone.

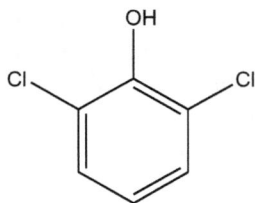

(7.10)

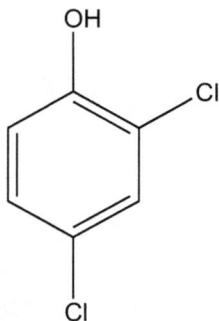

(7.11)

Chlorine compounds are widely used in antiseptics and disinfectants, such as chloroxylenol ((7.12), 4-chloro-3,5-dimethylphenol), better known as the active ingredient in Dettol (though a number of other products go under this name now).

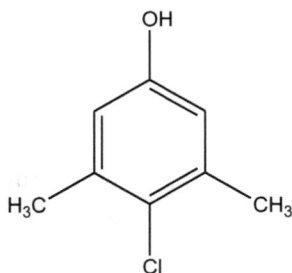

(7.12)

Introduced in 1933 (by the British firm Reckitt and Colman), it has a broad spectrum of activity; for example, it works against both gram-positive and gram-negative bacteria.

Dettol isn't the only well-known chlorine-containing antiseptic commonly used in the UK. TCP was marketed from 1918. The derivation of the brand name is uncertain. Many chemists would have assumed that TCP stood for trichlorophenol, but 'officially', the original ingredient was trichlorophenylmethyliodosalicyl, though its formulation was changed during the 1950s to a mixture of phenol and halogenated phenols. The chemist Jim Clark, the authority behind the *Chemguide* website, has tracked down several conflicting stories behind the name.

NATURAL GERM-KILLERS FROM THE EARTH

As so often, Mother Nature got there first, in this case by many millions of years. Following the discovery that penicillin was made by a fungus, scientists searched, especially in the soil, for other fungal-derived antibiotics. In 1945, the American Benjamin Duggar (a retired professor of plant physiology) isolated a molecule called chlorotetracycline (aureomycin) from some Missouri soil. This founded the family of tetracycline antibiotics.

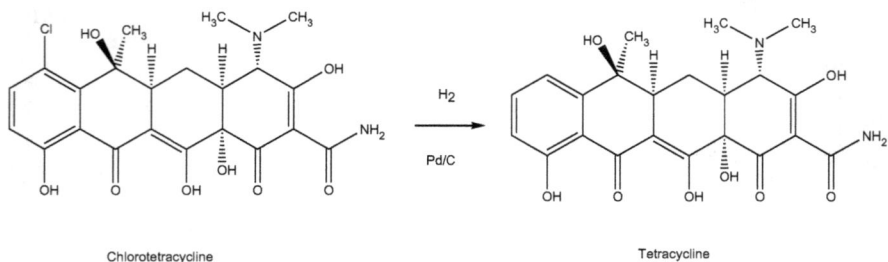

Chlorotetracycline   Tetracycline

(7.13)

Chlorotetracycline can be converted to the parent of the series, tetracycline (7.13), by catalytic hydrogenation, whilst another member of the family, tigecycline (7.14), is active against resistant microorganisms, notably methicillin-resistant *Staphylococcus aureus* (MRSA). Tetracycline is a broad-spectrum antibiotic commonly used to treat severe forms of acne.

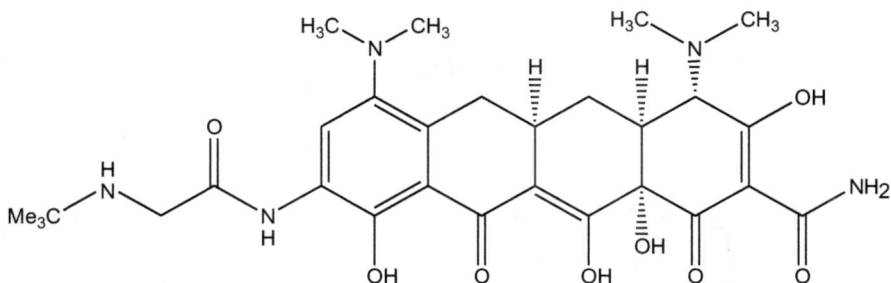

(7.14)

Currently the most widely used drug in the tetracycline family is doxycycline (7.15), first used in the 1960s, which combines high efficacy, rapid oral absorption, few side effects and low cost. It is used to treat a range of infections, including anthrax, Lyme disease, plague, scrub typhus, P. falciparum malaria and methicillin-resistant Staphylococcus aureus, as well as urinary tract infections, including chlamydia and syphilis.

It transpires that tetracycline itself was unknowingly used over 1,500 years ago as a health-giving ingredient in Nubian beer (*Every Molecule Tells a Story* pp. 171–172)

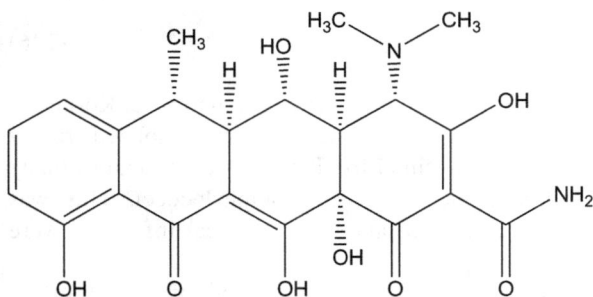

(7.15)

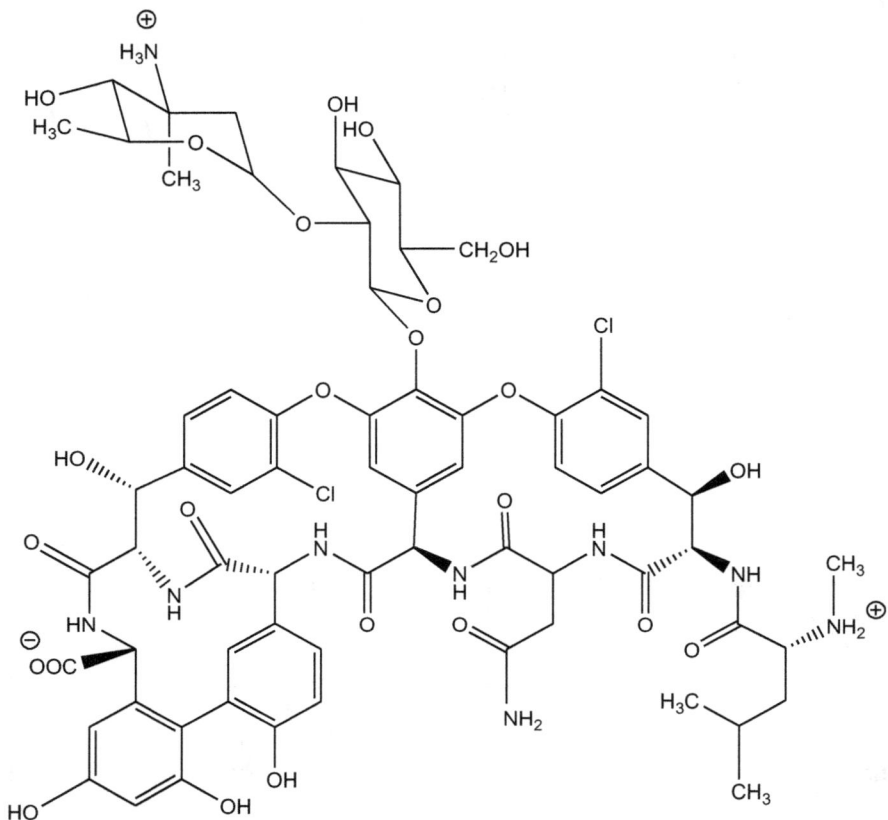

(7.16)

In 1952, a sample of dirt was sent (by a missionary in Borneo) to E. C. Kornfield, a chemist working for the pharmaceutical company Eli Lilly, who isolated from it an antibiotic that we now know as vancomycin (7.16). It proved active against many gram-positive organisms, including penicillin-resistant staphylococci. This was a most important discovery, as penicillin-resistant staphylococcal infections were beginning to be a problem.

Vancomycin is a tricyclic glycopeptide with a molecular mass of 1485. It exerts its activity through the formation of hydrogen bonds between its NH groups and the bacterial cell wall precursor D-Ala-D-Ala, thus inhibiting bacterial cell wall formation (Figure 7.1). Penicillin also inhibits bacterial cell wall synthesis, but in a different way, by binding to the enzyme that controls the formation of peptidoglycan crosslinks.

Over time, bacteria developed resistance to vancomycin, just like other antibiotics, due to increased (and partly unnecessary) use of antibiotics in both healthcare and agriculture, but it is still in demand as the front-line agent for treating complaints like methicillin resistant infections such as MRSA.

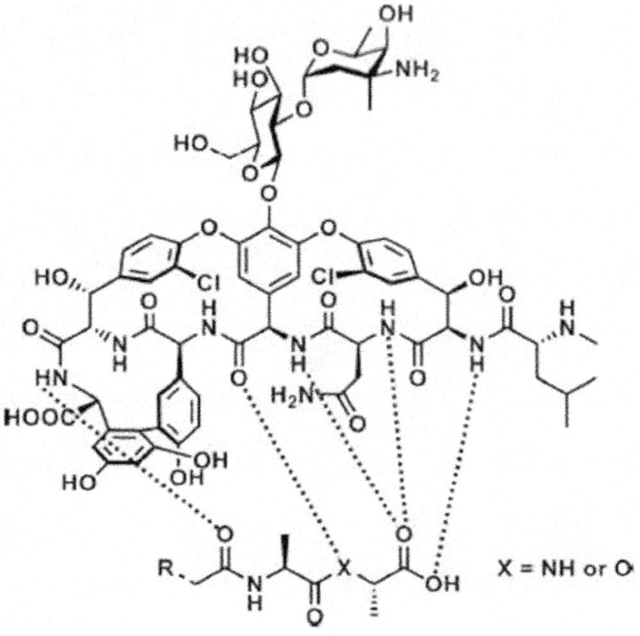

FIGURE 7.1 Interaction of vancomycin with the bacterial cell wall precursor D-Ala-D-Ala. Vancomycin forms hydrogen bonds (dotted lines), enabling inhibition of cell wall synthesis. E. Mühlberg et al., *Can. J. Microbiol.*, 2020, **66**, 11–16.

TEICOPLANIN

Teicoplanin (7.17) is a lipoglycopeptide antibiotic similar to vancomycin. It was first isolated from the bacterium *Actinoplanes teichomyceticus* in a soil sample from Nimodi village, Indore, India, and reported in 1978. Like vancomycin, it binds to bacterial cell wall precursors, inhibiting bacterial cell wall peptidoglycan synthesis, and is used particularly to treat serious infections caused by gram-positive bacteria, such as staphylococci (including MRSA), streptococci, enterococci and anaerobic gram-positive bacteria, including *Clostridium* spp.

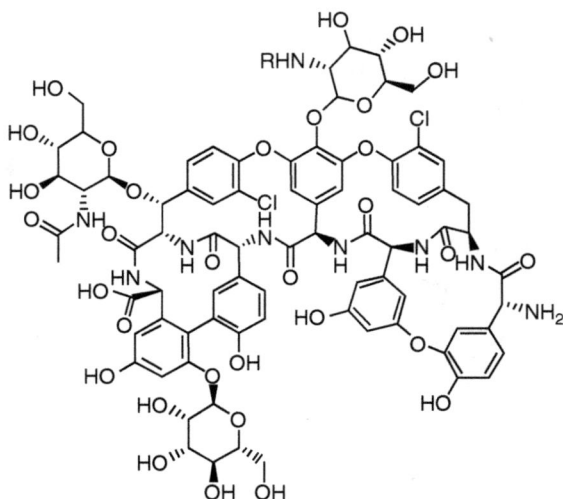

(7.17)

https://commons.wikimedia.org/wiki/File:Teicoplanin_core_and_major_compo-nents.svg#file

CHLORAMPHENICOL

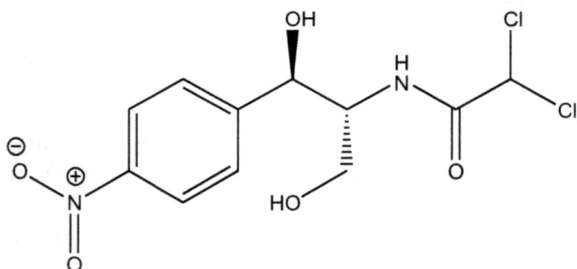

(7.18)

Chloramphenicol ((7.18), first known as chloromycetin), was first isolated from the culture fluid of the bacterium *Streptomyces venezuelae*, which was found in the soil near Caracas in Venezuela, and was first reported in 1947. Shortly after-wards, an independent report appeared of its isolation from compost on a farm at the University of Illinois at Urbana, Illinois. Within 2 years of the initial reports, its structure had been worked out, laboratory syntheses devised and successful clini-cal trials conducted. It was marketed by Parke-Davis from 1949. Chloramphenicol is a broad-spectrum antibiotic that works by interfering with substrate binding in the ribosome, preventing the synthesis of proteins. Soon, however, problems arose from its interaction with the bone marrow, decreasing the synthesis of blood cells and platelets, resulting in a form of aplastic anaemia. It has been widely used in eye ointments to treat conjunctivitis, but its use as an oral or injected medication for the treatment of diseases like bacterial meningitis, typhoid and Rocky Mountain spotted fever is limited to cases where 'safer' medications are not effective.

HALOGENATED COMPOUNDS FROM MARINE FUNGI

An area being investigated is that of chlorine-containing natural products from the sea and other marine sources. A paper on marine microbial natural products stated that natural products from marine fungi account for 63% of marine microorganisms. Given the large amounts of halide ions, especially chloride, present in the sea, it is expected that a significant proportion of these natural products will contain halogens.

ANOTHER KILLER

Back in the 1970s, some scientists were in Ecuador studying the small yellow frog *Epipedobates tricolor*. It is typical of the very small frogs of the region in bearing a colour warning that it contains defence chemicals. This frog was found to have a venom that, when tested upon laboratory mice, produced a painkilling response some 200 times stronger than morphine. Surprisingly, it did not operate through the opioid receptor to which morphine and other addictive opiate painkillers bind, but through the receptor used by nicotine; in retrospect, this is not a total surprise given the structural resemblance between nicotine (7.19) and the molecule from the frog, eventually named epibatidine (7.20) after the frog.

The frog, of course, would run the risk of self-poisoning from the venom. Research shows that in the frog, epibatidine shares a binding site with acetylcholine, and through a single amino acid mutation that has changed the configuration of the acetylcholine receptor, it has a sufficiently reduced sensitivity to epibatidine.

Epibatidine was the subject of considerable research interest for several years, as its mode of action, avoiding the opioid receptor, suggested that it – or a molecule derived from it – could be a non-addictive painkiller, but clinical trials have failed to produce a viable drug.

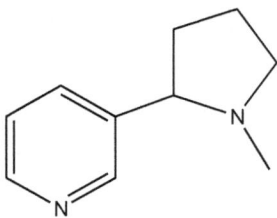

(7.19)

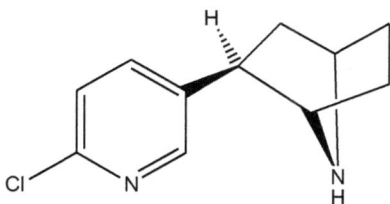

(7.20)

Abbot Laboratories developed a synthetic analogue of epibatidine, ABT-594 ((7.21), Tebanicline or Ebanicline), which also acted at the nicotine receptor. Animal and human tests showed that it was an effective analgesic, comparable with opioids, whilst being less toxic than epibatidine.

Unfortunately, results from Phase II clinical trials showed gastrointestinal side effects, such as nausea, dizziness and vomiting, which were sufficient to prevent its further development, but research continues in this area of analgesics.

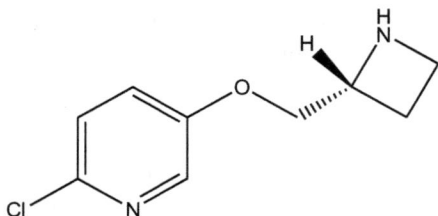

(7.21)

The world contains an amazing variety of organochlorine compounds – some are 'natural', some are wholly synthetic. Some of these compounds are toxic or harmful in other ways, but others are not only useful substances but quite safe.

Molecules are 'morally neutral'; they do not display their good or bad sides until they come into contact with people.

8 Organofluorine Compounds

Fluorine is a remarkable chemical element. It is the most electronegative element known and also the most reactive, reacting directly with virtually all the known elements (except helium, neon and argon). This is due to a combination of two factors: the weak F–F bond in the F_2 molecule and the strong bonds that fluorine forms with other elements. Compared to the Cl–Cl bond energy of 242.6 kJ/mol in the Cl_2 molecule, the F–F bond energy in the F_2 molecule is only 158.8 kJ/mol, due to repulsions between non-bonding electron pairs. The small size of a fluorine atom contributes to good overlap of its orbitals with those of other atoms, leading to strong bonds; its small size also means that many fluorine atoms can surround another atom, leading to the formation of multiple bonds.

What does this mean for its organic compounds? Well, the strength of C–F bonds means that organofluorine compounds are very unreactive. The C–F bond is the strongest single bond known, with an average bond energy (enthalpy) of 484 kJ/mol, compared with values for other carbon–halogen bonds: C–Cl 338, C–Br 276, and C–I 238 kJ/mol, as well as C–C 348, C–H 413 and C–O 351 kJ/mol.

Henri Moissan was the first person to isolate fluorine itself in 1886, but aspects of fluorine chemistry, like the etching power of HF, go back to the 17th and 18th centuries. The French chemists Jean-Baptiste Dumas and Eugène Péligot synthesised the first aliphatic organofluorine compound, fluoromethane (CH_3F), in 1835, through the reaction between heated dimethyl sulfate and potassium fluoride

$$(CH_3O)_2SO_2 + 2\ KF \rightarrow 2\ CH_3F + K_2SO_4$$

A much more famous organic chemist synthesised a pioneering aromatic organofluorine compound in 1862; Alexander Porfiryevich Borodin used KHF_2 to carry out a nucleophilic substitution reaction upon benzoyl chloride to make benzoyl fluoride (8.1). And yes, I had heard that Borodin wrote some music in his spare time; however, his day job was as a professor of chemistry at the Imperial Medical-Surgical Academy in Saint Petersburg.

(8.1)

An important reaction described in 1892 by the Belgian chemist Frédéric Swarts was the 'Swarts reaction', using SbF_3Br_2 as a fluorinating agent, for example, with chloroform to give $CHFCl_2$. The reaction was most famously used in the first synthesis of chlorofluorocarbons (CFCs) and hydrochlorofluorocarbons (HCFCs) (pp. 174–178).

Some significant developments occurred in the inter-war period in the first half of the 20th century. In 1926, two French chemists, Lebeau and Damiens, made CF_4 by the reaction of fluorine with wood charcoal; the properties of this compound were clarified by Otto Ruff and Rudolf Keim in 1930. An important step was taken in 1927 with the synthesis of aromatic fluorine compounds by Günther Balz and Günther Schieman, in what has logically become known as the Balz-Schiemann reaction (1926–7). The diazotisation reaction, in which aromatic primary amines react with nitrous acid at temperatures below 5°C to form diazonium ions, key to producing azo-dyes, was established in the late 19th century. It was already known that warming diazonium ions in the presence of iodide ions introduced iodine into the aromatic ring; now it was discovered that diazonium salts of the tetrafluoroborate ion (BF_4^-) were stable enough to be isolated, and that when dried and heated, in many cases, aryl fluorides were generated in good yield (8.2).

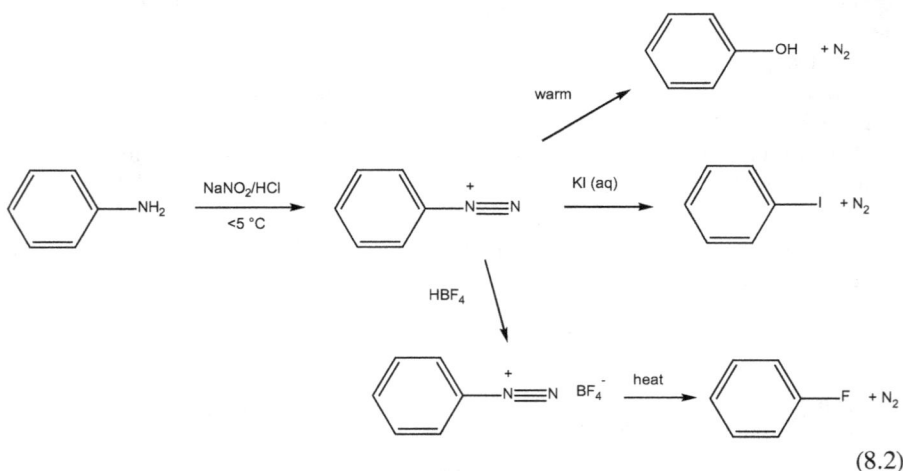

$$(8.2)$$

CFCs, SERENDIPITY AND A SERIOUS PROBLEM

We take domestic refrigerators for granted now, but the type that we have in every home goes back in essence to the early 1920s, when they were coming into widespread use in the USA. As already noted, back in the 1890s, Frédéric Swarts found a convenient way to make the unreactive CFCs from CCl_4, and 30 years later, the American Thomas Midgley Jr. (a mechanical engineer by training) popularised CFCs, compounds of C, F and Cl), like CF_2Cl_2 (CFC-12) as refrigerants. In an article published in 1937, Midgley told the story of his 'discovery'. In 1928, the most widely used materials as coolants in refrigerators were ammonia (NH_3), sulfur dioxide (SO_2) and chloromethane (CH_3Cl), which all posed problems of toxicity. Working for the

General Motors Corporation, Midgley and colleagues were looking for compounds that were substantially non-flammable and considerably less toxic than the refrigerants then commonly employed. CFCs (8.3) filled the bill, as they were hard to ignite and non-toxic and had the right volatility. HCFCs, also known as freons, containing C, F, Cl and H, can also be made.

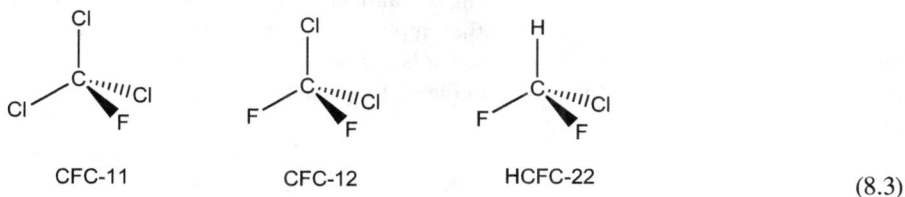

CFC-11   CFC-12   HCFC-22   (8.3)

BOX

The numbering system for CFCs, CFC-xyz (and similarly HCFCs), works like this:

x is the number of carbon atoms in the molecule, minus 1. If x is zero, it is omitted; y is the number of hydrogen atoms plus one; z is the number of fluorine atoms.

You can work back from CFC-xyz, by adding 90 to xyz, where the first digit is the number of carbons, the second is the number of hydrogens and the third is the number of fluorines (the rest are chlorines). So if you do that for CFC-12, you get 102 – that is one carbon, no hydrogens and two fluorines – and, by difference, two chlorines.

They wanted a compound combining the properties of a boiling point between 0°C and −40°C, stability, non-toxicity and non-flammability.

They looked at databases for suitable organic compounds and saw that CF_4 had a boiling point of −15°C. This started them thinking about fluorine compounds. They looked at the data again and doubted it – correctly, it was subsequently found that carbon tetrafluoride boils at −129°C. They thought that it would be toxic, so Midgley's colleague Albert L. Henne suggested that replacing some of the fluorine with chlorine would improve it, so they chose CF_2Cl_2, dichlorodifluoromethane. Synthesising these compounds had been worked out using SbF_3 some 40 years earlier by the Belgian chemist Swarts, so Midgley got on the 'phone' and ordered five 1-ounce bottles of antimony trifluoride (apparently 'all there was in the country at the time').

When the SbF_3 arrived, a bottle was chosen at random, and a few grams of dichlorodifluoromethane was prepared. A guinea pig was placed under a bell jar with it and, 'much to the surprise of the physician in charge, didn't suddenly gasp and die.' They repeated the synthesis and the guinea pig tests. The animal died. For the first trial, they chose the only bottle containing pure SbF_3. If they had chosen any other, the guinea pig would have died, and they wouldn't have bothered to do more tests.

So they proceeded with the use of CF_2Cl_2 (CFC-12) as a refrigerant material, with a slightly modified synthesis.

$$3 \, CCl_4 + 2 \, SbF_3 \rightarrow 3 \, CF_2Cl_2 + 2SbCl_3$$

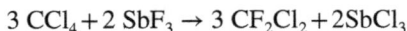

Famously, Midgley gave a dramatic demonstration of the new material by filling his lungs with it, then extinguishing a lighted candle at a 1930 meeting of the American Chemical Society.

More and more fluorocarbons were made, and their applications spread. Their properties also made CFCs useful in other applications, like solvents for grease in the electronics industry; propellants in aerosols; foam-blowing agents in the polymer industry for making polyurethanes, 'inert blankets' to exclude air for fire extinguishers and air conditioning for cars.

Everything in the garden seemed to be lovely, but reality caught up in the 1970s. In 1973, James Lovelock, best known as a British eco-guru but who was also a chemist and inventor, used his recently invented electron capture detector to show that $CFCl_3$ and other CFCs were present in the atmosphere, notably while travelling all the way from England to Antarctica on a voyage with the R.V. Shackleton. And in 1974 Sherwood Rowland and Mario Molina – and separately Paul Crutzen – linked the presence of CFCs to damage to the ozone layer. These three scientists shared the 1995 Nobel Prize in Chemistry for their work. Many think that Lovelock should have shared the award.

The exceedingly low reactivity of CFCs – they have atmospheric lifetimes of up to hundreds of years – meant that they were not decomposed in the lower part of the atmosphere, instead spreading to the stratosphere, where they underwent photochemical decomposition generating chlorine atoms. The • Cl radical decomposes ozone, O_3, into normal 'dioxygen', O_2, in a chain reaction. The chlorine atoms are regenerated in the process, which means both that they act as catalysts and are regenerated, so that one chlorine atom can decompose thousands of ozone molecules. Because C–Cl bonds (bond energy ~ 346 kJ/mol) are weaker than C–H bonds (~ 413 kJ/mol) or C–F bonds (~ 485 kJ/mol), when UV light of sufficient energy falls on CH_3Cl or a CFC molecule, the C–Cl bond undergoes homolytic fission, forming two free radicals.

$$CH_3Cl \rightarrow \bullet CH_3 + \bullet Cl$$

$$CCl_2F_2 \rightarrow \bullet CClF_2 + \bullet Cl$$

$$\bullet Cl + O_3 \rightarrow \bullet ClO + O_2$$

$$\bullet ClO + O \rightarrow \bullet Cl + O_2$$

In addition, it was realised that CFCs contributed some 20% of the man-made 'greenhouse effect' – in comparison, CO_2, CH_4, N_2O and tropospheric ozone contribute 50%, 15%, 5% and 7%, respectively.

By the late 1970s, the United States was one of several countries banning the use of CFCs as aerosol propellants. The best-publicised example of damage to the ozone layer was the development of a 'hole' in the ozone layer over Antarctica. This led to a global agreement known as the Montreal Protocol (1987, strengthened in 1990) to eliminate the use of CFCs, which has led to a general decrease in atmospheric CFC levels from a peak in the mid-1990s, though it will take many decades for a return to pre-CFC ozone concentrations. It was decided to completely ban CFC production in industrial countries by the year 2000 and in developing countries by 2010.

Obviously, other compounds had to be found to replace CFCs, compounds that had similar physical properties but did not damage the ozone layer. The first substitutes used were hydrochlorofluorocarbons (HCFCs); they have C–H bonds, making them more reactive, so they have shorter atmospheric lifetimes and are thus weaker ozone depleters. The best-known HCFC (8.4) is chlorodifluoromethane (HCFC-22), used as a substitute for CFC-12.

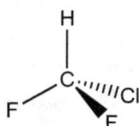

$$(8.4)$$

However, they are still ozone depleters, so they are only a short-term CFC replacement. Developed countries are set to stop using HCFCs by 2030, with a deadline extended to 2040 for developing countries.

The next candidates in line were hydrofluorocarbons (HFCs), such as CF_2H_2 (HFC-32) and CHF_3 (HFC-23) (8.5). As they contain no chlorine, they aren't ozone depleters.

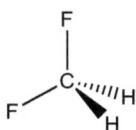 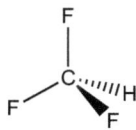

HFC-32   HFC-23

$$(8.5)$$

CF_3CH_2F, 1,1,1,2-tetrafluoroethane ((8.6), HFC-134a), in particular, has come to be employed in car air conditioners in place of CF_2Cl_2. However, CF_3CH_2F still has significant **greenhouse** potential (GWP100), so it is again only a short-term solution. Though CH_3CHF_2 ((8.7), HFC-152a) has quite a low GWP100 value, it is flammable because it contains more hydrogen, so it cannot replace CFC-12 in applications such as air conditioning or as a refrigerant.

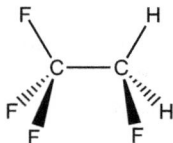

HFC-134a

$$(8.6)$$

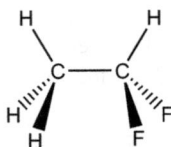

HFC-152a   (8.7)

HFCs are also scheduled to be phased out. In 2016, a new agreement was signed, the Kigali Amendment to the Montreal Protocol; this commits signatories to reduce the production and consumption of HFCs. Thus, the successors to HFCs are hydrofluoroolefins (HFOs), which are fluorine-substituted alkenes. Not only do they not contain chlorine, but they also contain a double bond, making them more reactive and thus shorter-lived in the atmosphere, which gives them a very low global warming potential (GWP100).

Examples of HFOs are the isomers E-1,3,3,3-tetrafluoropropene ((8.8), E-HFO-1234ze) and 2,3,3,3-tetrafluoropropene ((8.9), HFO-1234yf).

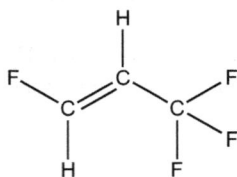

HFO-1234ze   (8.8)

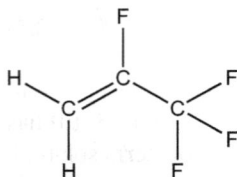

HFO-1234yf   (8.9)

Regulating the reduction in CFC levels has not always been straightforward. There are sites in different parts of the world that monitor atmospheric levels of these compounds, including the South Pole, Colorado (USA), Hawaii, Samoa, Alaska, Japan and South Korea. A report in 2018 indicated that emissions of CFC-11 (8.10) had suddenly increased since 2012, even though none was supposed to have been produced since 2006. One theory is that some 'unreported' CFC-11 manufacture has been taking place after 2010, possibly in Eastern China.

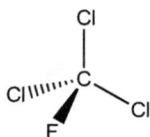

CFC-11

(8.10)

However, more recent studies have found that CFC-11 emissions from this source declined in 2017–2018. Another study, reported in 2023, looked at a wider timescale. It showed that atmospheric abundances and emissions of five CFCs increased overall between 2010 and 2020, the compounds being studied were CFC-13 (8.11), CFC-112a (8.12), CFC-113a (8.13), CFC-114a (8.14) and CFC-115 (8.15).

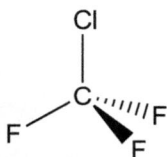

CFC-13

(8.11)

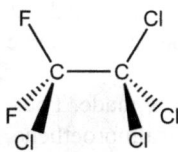

CFC-112a

(8.12)

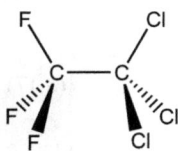

CFC-113a

(8.13)

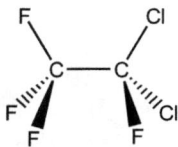

CFC-114a

(8.14)

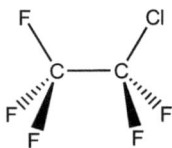

CFC-115   (8.15)

Some of this may be due to their being 'undesirable by-products' in the manufacture of HFCs, such as HFC-125 (8.16), which is widely used in air conditioning and refrigeration.

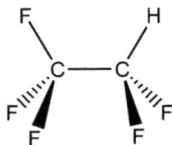

HFC-125   (8.16)

Another study that emerged in 2023 indicated that HFC-23 emissions from eastern China had almost doubled, from $5:0\pm0:4$ Gg/yr in 2008 to $9:5\pm1:0$ Gg/yr in 2019. Analysis suggested that the rise in HFC-23 emissions could be due to ineffective controls at known HCFC-22 production sites.

Although HCFCs are being eliminated, some are still allowed to be made. Thus, HCFC-22 is allowed, as it is used (8.17) to manufacture 1,1,2,2-tetrafluoroethene, the monomer for producing poly(tetrafluoroethene), also known as PTFE or Teflon.

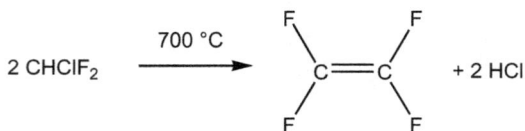

$$2\ CHClF_2 \xrightarrow{\ 700\ ^\circ C\ } \quad + 2\ HCl$$

(8.17)

Rising atmospheric concentrations of HFC-23 (8.18) could thus be a consequence of increased manufacture of HCFC-22 (8.19).

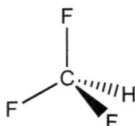

HFC-23   (8.18)

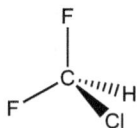

HCFC-23   (8.19)

Therefore, this continuous monitoring must persist if the environmental improvements resulting from the Montreal Protocol are to be maintained.

OTHER ORGANOFLUORINE COMPOUNDS

A more fortuitous serendipitous discovery was made at the American firm DuPont in 1938 by a chemist named Roy J. Plunkett. He found that Freon 114 (CF_2ClCF_2Cl) reacted with zinc, forming the gaseous substituted alkene, tetrafluoroethene, $F_2C=CF_2$, which he stored in cylinders.

$$CF_2ClCF_2Cl + Zn \rightarrow F_2C=CF_2 + ZnCl_2$$

Sometime later, he emptied one cylinder and was surprised to find that it contained some 11 g of a white solid. This was poly(tetrafluoroethene), PTFE; in the interim, the tetrafluoroethene had fortuitously polymerised (8.20).

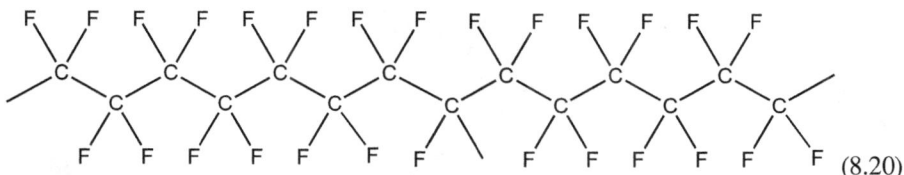

(8.20)

PTFE proved to be a wonder material. Chemically, it was inert; it did not dissolve in any solvents and was non-combustible. It was an electrical insulator and thermally stable; as a material, it had a working range from near absolute zero up to some 250°C. It became known as Teflon.

The unreactive PTFE proved invaluable in the Manhattan Project, which led to the first atomic bomb, as the intensively reactive fluorine was essential in making UF_6, used in uranium isotope separation. PTFE was used to make fluorine-proof valves and gaskets in the plant. Nowadays, we tend to associate it with applications like non-stick frying pans, where the biggest problem isn't stopping your omelette from sticking to the pan, but getting the PTFE coating to adhere to the pan in the manufacturing stage! A very public application of it was in the 'Millennium Dome' in London; here, PTFE provided an unreactive coating for the roof, which resisted attack by water, heat or UV light, as well as any 'chemical warfare' attack.

Teflon is very unreactive for several reasons. The electron-rich C=C bond is removed in the polymerisation process, so it is not attacked by electrophiles in the addition reactions typical of alkenes. The C–F bonds are very strong, so they are not substituted by the nucleophiles that attack other carbon-X bonds (X=Cl, Br, I). Additionally, the very electronegative fluorine atoms help to protect the carbon backbone from attack.

OTHER PROBLEMATIC ORGANOFLUORINE COMPOUNDS

With the phasing out of CFCs, environmental concerns have shifted to organofluorine chemicals in general, classified as PFAS or PFAs (poly- and perfluoroalkyls), which have become known as 'Forever Chemicals'. This is exemplified by two compounds, PFOA ((8.21), perfluorooctanoic acid) and PFOS ((8.22), perfluorooctane sulfonic acid). These were discovered to be outstanding surfactants back in the 1930s, leading to their use in a wide range of applications, including raincoats, fire-fighting foams and repellent coatings of all kinds, since the fluoroalkyl chains were unparalleled at repelling both water and grease. PFOS was the key ingredient in the fabric protector Scotchgard.

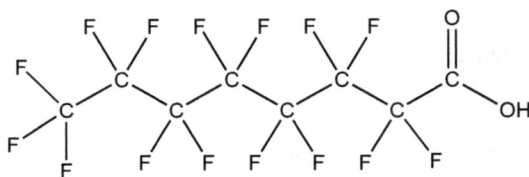

(8.21)

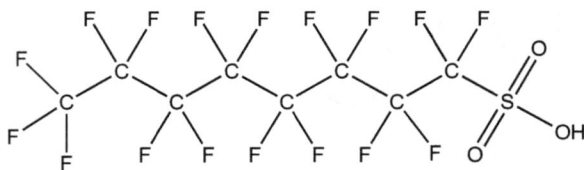

(8.22)

PFAs are now considered as threats to the environment. The strong C–F bonds mean that PFAs do not degrade under normal environmental conditions, and by the mid-1970s, tests showed that around 5,000 of these substances were present in peoples' blood. Furthermore than that, analysis has shown that they accumulate in plants, fish, birds and mammals, throughout the food chain. Their presence has been linked with a greater incidence of conditions including kidney and testicular cancers, liver and kidney disease, pregnancy-induced hypertension, ulcerative colitis and high cholesterol; as a result, more biodegradable substitutes are being introduced.

Visible-light induced degradation of PFAs, employing photocatalysts, is being proposed as one way to eliminate them.

A NATURAL PROBLEM

One natural fluorine compound – and a rarity at that – that presents problems is sodium fluoroacetate (8.23). During World War II, it was discovered to be a rodenticide. It occurs naturally in certain plants of the genus *Gastrolobium*, which take in fluoride from the soil and incorporate it into the fluoroacetate ion. These plants have been studied particularly in Western Australia, where native species like brush-tailed possums and western grey kangaroos are immune to it, but species introduced from elsewhere are not. It also occurs in plants found in parts of Africa and Brazil.

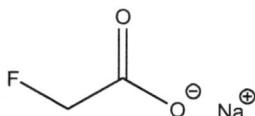

(8.23)

The toxic nature of fluoroacetate is due to the similarity in size between fluorine and hydrogen, which is leveraged in fluorine-containing pharmaceuticals (see the following section).

The enzyme citrate synthase mistakes fluoroacetate for its normal substrate, ace-tate, forming 2-fluorocitrate, rather than citrate; this binds to the enzyme aconitase, stopping the citric acid (Krebs) cycle, a most important metabolic pathway in energy production, leading to heart and brain failure.

FLUORINATED PHARMACEUTICALS

One area in which fluorine-containing organic compounds have been successful is medicinal chemistry, where they have assumed an increasingly important role. In terms of size, fluorine atoms are the nearest substitute for hydrogen atoms. The Van der Waals radius of fluorine is 1.47 Å, so that they are slightly larger than hydrogen atoms, with a Van der Waals radius of 1.20 Å. In contrast, fluorine is much more electronegative than every other non-metal. F has an electronegativity value of 3.98 on the Pauling scale. Corresponding values for H, C and O are 2.20, 2.55 and 3.44, respectively. This means that bonds involving fluorine are very polar.

As noted at the beginning of this chapter, C–F bonds are the strongest known, so that despite their high polarity, C–F bonds are hard to break, meaning that orga-nofluorine compounds undergo slower metabolism, thus having a longer lifetime in the body and being more effective drugs. Other properties are affected, including lipophilicity and the ability to cross membranes.

ANAESTHETICS

Although ethoxyethane (aka diethyl ether) was first synthesised in the 16th cen-tury, it did not come into use as an anaesthetic until the 1840s, around the same time as nitrous oxide (N_2O) and chloroform ($CHCl_3$). All of these suffered from some disadvantages, but it was another century before fluorinated ethers came into use, obviating the evident problem of ether's flammability. Halothane (8.24) and methoxyflurane (8.25) were among the first, their use dating from the 1950s; today, methoxyflurane is no longer used as an anaesthetic but still has a role in pain relief, whilst halothane is not used as an anaesthetic in the Western world. Among their suc-cessors are isoflurane (8.26), desflurane (8.27) and sevoflurane (8.28), the latter being regarded as the best currently in use. Desirable properties of an anaesthetic include high stability, low flammability, non-pungent odour, minimal respiratory and cardio-vascular effects, plus reversible central nervous system effects and lack of reactivity with other medicines. These compounds tend to have atmospheric lifetimes in the area of 1–10 years, rather less than those of CFCs, and some, like halothane, are ozone layer depleters.

(8.24)

(8.25)

(8.26)

(8.27)

(8.28)

Fluorinated pharmaceuticals have come into use to treat a wide range of com-plaints. The one that has received the most publicity is **fluoxetine** (8.29), otherwise known as Prozac, which is an important antidepressant. It is classed as a selective serotonin reuptake inhibitor. It increases the levels of serotonin in the brain, a com-pound believed to improve mood.

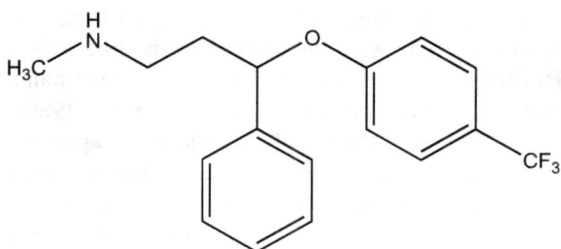

(8.29)

Another fluorine-containing antidepressant, which also works as a selective serotonin reuptake inhibitor, was first marketed as **citalopram**. Citalopram contained two isomeric molecules, as it had a chiral carbon atom. Subsequently, it was marketed as the single (*S*)-isomer (8.30) and became known as **escitalopram** (i.e. (*S*)-citalopram)

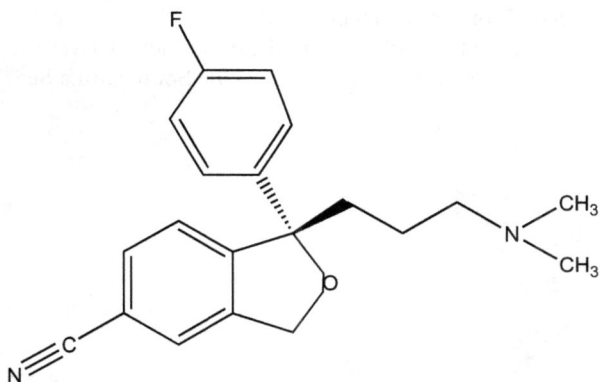

(8.30)

The most controversial organofluorine medication is probably **mefloquine** (8.31), also known as Lariam, a widely used anti-malarial medication.

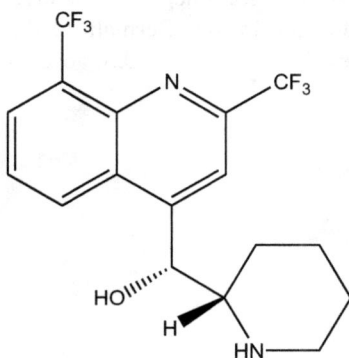

(8.31)

During the Vietnam War, when many U.S. troops were serving in areas where malaria was rife, it became clear that chloroquine was no longer effective as an anti-malarial. Hundreds of potential alternatives were investigated, with one,

WR-142490 (later named Mefloquine), affording 100% cures against chloro-quine-resistant *Plasmodium falciparum*. Trials were undertaken in the mid-1970s, and the drug was marketed by Hoffmann-La Roche in 1979 under the trade name Lariam, going into widespread use. Resistance began to be reported in the 1980s, and problems were recognised by the 1990s with the use of Mefloquine as a prophy-lactic – reports arose of neuropsychiatric episodes, paranoia, depression and even attempts at suicide. Today, with tourism increasing, Mefloquine remains an impor-tant, first-line anti-malarial drug, with travellers to high-risk *Plasmodium falciparum* endemic areas requiring an effective prophylactic. There are indications that women are more at risk, and careful prescribing is important. At present, there is no replace-ment for Mefloquine available or, indeed, in prospect.

5-Fluorouracil is a medication for a number of cancers, including breast, pancre-atic, cervical and stomach cancers; discovered in the 1950s, it came into medical use in 1962. It works by targeting the enzyme thymidylate synthetase, stopping the pro-duction of thymidylate that is needed for the replication of the DNA of cancer cells. 5-Fluorouracil is not a natural drug, but in 2003, some 5-fluorouracil derivatives were isolated from the marine sponge *Phakellia fusca* found in the South China Sea.

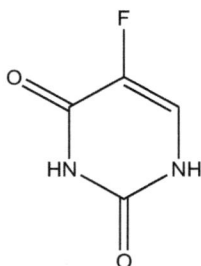

(8.32)

Ciprofloxacin ((8.33), Cipro) is a member of the class of fluoroquinolones. Developed in the 1970s, it came into medical use in 1987. It is an important broad-spectrum antibiotic against both gram-positive and gram-negative bacteria, which came to prominence as an anti-anthrax medication in the aftermath of the September 2001 attacks upon the Twin Towers in New York, when the U.S. govern-ment purchased 100 million tablets of Cipro.

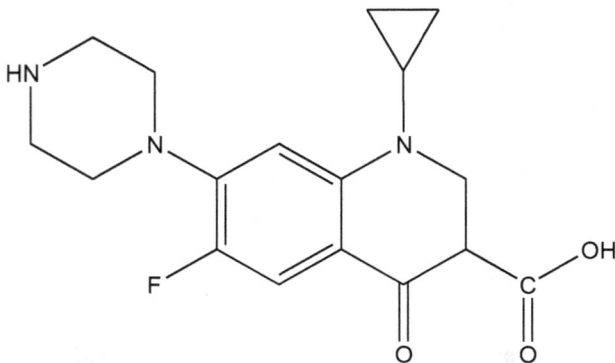

(8.33)

Cipro is used against a wide range of infections, including urinary tract infections. The mode of action of ciprofloxacin is to inhibit the activity of DNA gyrase and topoisomerase, quite unlike that of β-lactam antibiotics like penicillin, which prevent the synthesis of bacterial cell walls.

Cipro is used on a large scale worldwide, not just because of its efficacy, but also due to its few side effects and its low cost. Its downside is that since the early 1990s, there have been reports of bacterial resistance. There are several reasons for this, including the significant amounts used for food-producing animals to improve yields, as well as the use of poor-quality ciprofloxacin formulations, especially in developing countries. Obviously, proper choice of antibiotics in a particular situation is important, as well as ensuring that sufficient doses are taken to eliminate bacteria.

CELECOXIB

The discovery in 1971 that aspirin exerted its effects by inhibiting cyclooxygenase (COX) enzymes, both COX-1 and COX-2, led to the development of synthetic drugs that selectively inhibited the COX-2 enzyme responsible for the synthesis of prostaglandins, responsible for pain. The best known of these are Celebrex, or **celecoxib** (8.34) (Pfizer) and Vioxx (Merck), which came onto the U.S. market in 1998–1999 and were widely promoted as medications providing pain relief against rheumatoid arthritis, osteoarthritis and menstrual pain, without the risk of stomach ulcers.

However, in 2001–2004, evidence emerged linking Vioxx to an increased risk of heart attacks and strokes, leading to its withdrawal from the market in 2004. Celebrex remains on the market as a selective COX-2 inhibitor.

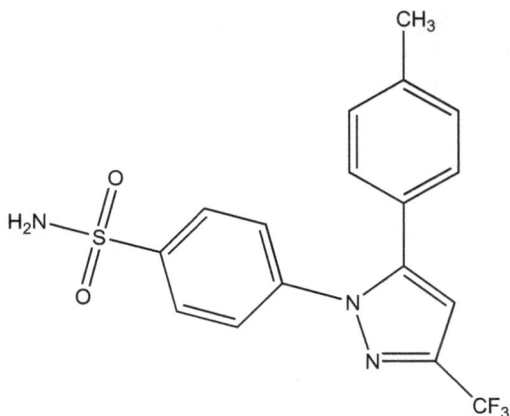

(8.34)

Atorvastatin ((8.35), commercial name Lipitor) is a widely used medication, of the family commonly called statins, which are taken annually by over 200 million people, and used to lower the risk of cardiovascular disease. It was patented (by the company Parke-Davis) in 1986; 10 years later, in December 1996, it was approved for use in the USA.

Statins work by inhibiting the enzyme HMG-CoA reductase, which plays a central role in the synthesis of cholesterol in the liver. High cholesterol levels have been implicated in cardiovascular disease, and we obtain most cholesterol from our bodies, rather than our diet.

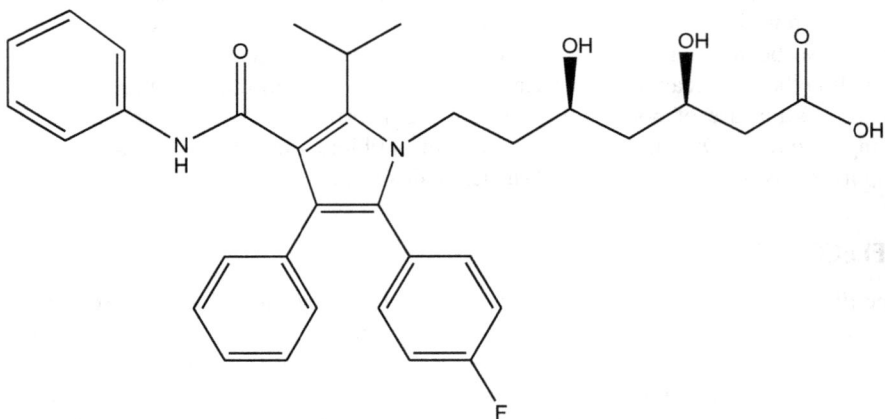

(8.35)

BLOOD SUBSTITUTES

Fluorocarbons can often dissolve large quantities of oxygen, which has led to their use in blood substitutes, notably perfluorodecalin (8.36) and perfluorooctylbromide (8.37). Their development was driven by separate events, such as the demand for transfusions during World War II and the need for safe blood substitutes at during the mad cow disease crisis, as well as during transfusion scandals at the time when HIV emerged.

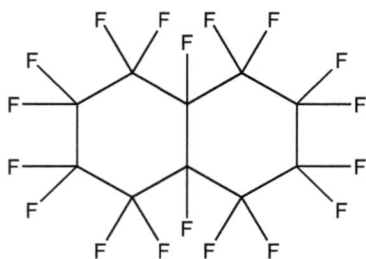

(8.36)

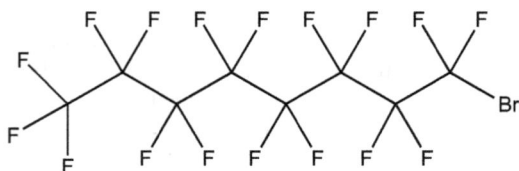

(8.37)

AND THE LATEST NEWS?

New organofluorine compounds continue to be developed, the most remarkable being octafluorocubane, also known as perfluorocubane (8.38), which has a regular cubic structure. Its synthesis was reported in the literature in August 2022. It accepts an electron, which it stores inside the internal cavity – an 'electron in a cube', as the journal *Science* put it – forming the $[C_8F_8]^-$ ion. Readers of the American journal *Chemical and Engineering News* voted octafluorocubane the 'Compound of The Year' for 2022.

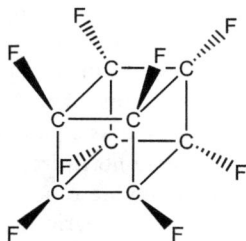

(8.38)

9 Smoking and Vaping

SMOKING

Of over sixty *Nicotiana* species, it is *Nicotiana rustica* and *Nicotiana tabacum* that have been consumed by humans, particularly the latter. They are native to the Americas, where it is thought that humans first encountered them some 18,000 years ago and began to cultivate tobacco between 3000 and 5000 BC, in the Andes. Tobacco was consumed in various ways – sniffing (as in snuff), chewing, drinking (tobacco tea) and smoking, for example. Its insecticidal properties were discovered, and tobacco juice was used on the skin to kill parasites; it was also used as a pain-killer and antiseptic. Shamans used it in religious rites. Tobacco spread north from the Andes, and it was smoked by the Mayans, Aztecs and Caribs, among others, possibly as early as the first century BC. The first Europeans to arrive in America at the end of the 15th century became acquainted with the Indian custom of smoking tobacco (usually as cigars, though pipes were known) and taking snuff. Pipe smoking became the norm in North America among Native Americans.

Tobacco seeds were brought back to Spain and Portugal in the 1550s. Nicotine takes its name from Jean Nicot, the French ambassador to Portugal, who introduced tobacco and smoking to the French court around 1560 for its possible medical applications.

Meanwhile, back in Virginia, settlers who were dissatisfied with the taste of the bitter native *Nicotiana rustica* imported seeds of their preferred *Nicotiana tabacum* (said to be the idea of John Rolfe, who married Pocahontas) and started their own plantations around 1610, leading to an active export trade.

Spanish and Portuguese traders brought tobacco to China and Japan by 1600, and smoking spread there. In China, tobacco was regarded as a protection against malaria, as its smoke was believed to repel mosquitoes. Tobacco's popularity there suffered interruptions; for example, the last Ming Emperor of China introduced the death sentence for smoking in 1640, but he was succeeded in 1644 by the Qing dynasty, who were smokers. Tobacco was banned several times in Japan, with the last ban being repealed in 1625. England embraced tobacco, with advocates including Walter Raleigh, but there were opponents, notably King James I, who wrote a book entitled *A Counterblaste to Tobacco* (1604), and legislated against its cultivation and consumption, with little success. His son and successor, King Charles I, was also opposed. In the end, governments opted to tax it, not just in England. In France, Cardinal Richelieu introduced a customs duty on tobacco in 1629, and taxation became widespread.

Pipe smoking was initially the norm in northern Europe, certainly in England, though cigars were very popular in Spain, including a flourishing export industry. Cigars were adopted as an early method of smoking rolls of tobacco; the practice goes back at least a thousand years in pre-Colonial America, subsequently being

DOI: 10.1201/9781003658115-9

taken back to Europe. Cigar smoking was much more popular than cigarette smoking throughout the 18th century into the early 19th century, especially among British soldiers returning from the Peninsular War, where they had adopted the practice from their Spanish and Portuguese allies. Tennyson, the poet, and William Morris were both pipe smokers, while Karl Marx smoked cheap cigars. Queen Victoria's husband, Prince Albert, was also a cigar smoker, with his own smoking room at Osborne House on the Isle of Wight. Cigarettes were first devised in the Old World in Spain and Turkey (*papeletes* and *cigaritos*) and were then brought back to England in the 1850s, this time in the wake of the Crimean War. However, the practice of cigarette smoking did not catch on immediately, until the American invention (Virginia, again) of the cigarette-making machine in 1880, which chopped cigarettes from a tube of paper-wrapped tobacco, and could produce up to 212 cigarettes per minute. Machine-made cigarettes were pioneered by the American Buchanan Duke, but in 1883, the British firm of W.D. & H. O. Wills installed two of these machines and launched a number of brands in the next decade, including *Woodbines*, which cost 1d. for a pack of five. In 1883, the same year as Wills, a Mr John Player of Nottingham commenced the manufacture of cigarettes, with another well-known British brand being launched.

At one time, cigarette smoking was seen as a somewhat effeminate habit (in contrast, pipes were considered 'manly'), but it surged in popularity during World War I, not least for its convenience for troops in the trenches of the Western Front. Cigarettes were widely consumed in Hollywood films, and World War II again boosted consumption, with tax-free cigarettes shipped to the US Armed Forces. And of course, massive advertising campaigns extolled the product. Cigarette cards, decorated with pictures of sporting stars (or buxom young ladies), were introduced very early, as far back as the 1870s. From the 1960s, menthol-flavoured cigarettes (like *Kool* brand) were targeted in the USA at Black smokers. Since British men were away on the battlefields of World War I, their women worked, notably in the munitions industry; and British women increasingly took up smoking, as did their sisters across the Atlantic.

NICOTINE

The chemical mainly responsible for the increased popularity of smoking was nicotine, which causes both pleasurable and addictive properties. The tobacco plant *Nicotiana tabacum* produces nicotine as a secondary metabolite (a compound produced by an organism that is not directly involved in the growth, development or reproduction of the organism) as a defence against herbivores, its enemies (in the same way that cocaine, caffeine and morphine are produced by the tea plant, *Camellia sinensis*; the coca plant, *Erythroxylum coca*; and the opium poppy, *Papaver somniferum*, respectively). Nicotine is toxic to most insects, with the exceptions being the tobacco hornworm, *Manduca sexta*, and the cigarette beetle, *Lasioderma serricorne*.

Humans have taken advantage of that – at one time, nicotine was widely used as an insecticide. It was in seeking synthetic alternatives to nicotine that Gerhard Schrader synthesised tabun, the first nerve agent, in 1936.

In a celebrated case, some insecticide that contained nicotine was spilled on a chair in a Florida florist's shop one day in 1932. The unobservant florist sat on the chair, whereupon the insecticide soaked through his trousers, leading to nicotine being absorbed by his skin. The very sick florist was soon admitted to hospital, put into pyjamas and observed for a few days; of course, his body duly metabolised the nicotine, so his condition improved rapidly as it was eliminated, resulting in the puzzled doctors discharging him. He was handed back his trousers, which had been kept in a paper bag while he was in bed; the florist put his clothes on again and within an hour suffered a second dose of nicotine poisoning.

Some people are vulnerable to nicotine poisoning as part of their employment, notably tobacco harvesters. With symptoms first recognised three centuries ago, particularly headaches, dizziness and gastrointestinal disorders, 'Green tobacco sickness' is the name given to their 'occupational disease', due to absorbing nicotine through their skin from the wet green tobacco plant. A very recent case of occupational exposure was reported in 2023, in which a man making e-cigarette liquids spilled 300 mL of concentrated nicotine solution onto his leg; despite immediate washing, he experienced headaches and abdominal pain, as well as continuous vomiting. He recovered and was discharged from hospital within 24 hours. It should be noted that the worker was not wearing any protective clothing (nor headwear).

Nicotine contains a chiral carbon atom so can exist as two optical isomers. As usual, Nature produces just the one form, (S)-nicotine (9.1). The (R)-isomer (9.2) is not made by the tobacco plant but can be synthesised in the laboratory. Small amounts are formed in the smoking process, evidently by racemisation. Most of the nicotine taken in by the body is oxidised in the liver to cotinine (9.3) (which exists in (R)- and (S)- isomers, like nicotine). Nicotine has a half-life in the body of about 2 hours. Of course, when nicotine levels drop, an addicted smoker experiences withdrawal symptoms and lights up another cigarette.

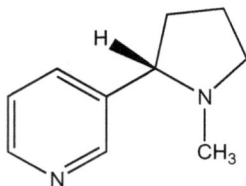

(S)-nicotine   (9.1)

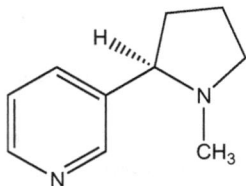

(R)-nicotine   (9.2)

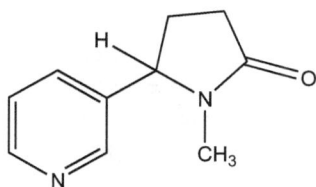

cotinine   (9.3)

The fatal dose of nicotine has been debated. For many years, it was generally stated to be 60 mg or less, linked with warnings that ingestion of five cigarettes could be fatal. It has been pointed out that this would mean that nicotine is as toxic as cyanide. This would correspond to an oral LD50 of around 0.8 mg/kg for nicotine, which is much lower than laboratory values determined for mice, at 3.3 mg/kg, or rats, at over 50 mg/kg. It has more recently been suggested that 500–1,000 mg of ingested nicotine would represent the lower limit for a lethal dose.

Smokers would say that the nicotine they take from tobacco gives them pleasure, as well as reducing stress. It is seen as improving their concentration and focus. When they stop smoking, they feel irritated and anxious, as well as craving tobacco. 'Withdrawal' symptoms, of course.

But how does nicotine work, and why, worldwide, do around a billion people smoke?

The process of smoking oxidises most of the molecules of nicotine present in the tobacco in a cigarette; the fraction (around 10%) that survives is vaporised and transported via the lungs and blood to the brain very rapidly, within 10 seconds. On arrival at the brain, it stimulates the release of neurotransmitters like dopamine, which is responsible for pleasurable sensations. This short gap between the act of taking a drag on the cigarette and the brain's response links the act of smoking with pleasure, which is why smoking can be highly addictive.

Besides dopamine release, nicotine also causes the release of endorphins (such as beta-endorphin), which are natural painkillers involved in relieving stress and pain; promoting calmness and sedation; and improving mood; glutamate, linked to learning and memory; and serotonin, implicated in mood and appetite. There are also indications that nicotine has benefits for memory and cognition in Alzheimer's disease and Parkinson's disease.

Nicotine is a weak base with a pKa of 8.0. In its protonated form, in acidic environments such as most flue-cured tobacco, nicotine is not readily transported across membranes, resulting in little buccal absorption. Smoke from pipe and cigar tobacco, as well as some European cigarettes, has a pH above 6.5, meaning more molecular nicotine is present. Because protonated nicotine (9.4) has a structural similarity to acetylcholine (9.5), it can bind to nicotinic cholinergic receptors (ligand-gated ion channels) in the brain, where dopamine is released, particularly in the mesolimbic area, the corpus striatum and the frontal cortex.

Acetylcholine works as a neurotransmitter, passing messages (nerve impulses) from one nerve to another through receptors at the ends of nerves. The neurotransmitter is released by one nerve cell and then taken up by specific receptors on a neighbouring nerve cell, which then fires, continuing the chain. The junction between nerve cells where the neurotransmitter is released is called a synapse. Because (protonated) nicotine has a similar shape and charge distribution to acetylcholine, it also binds to acetylcholine receptors, leading to a response from these 'nicotinic receptors'.

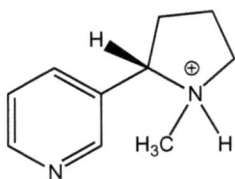

(9.4)

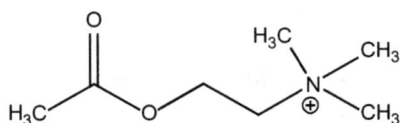

(9.5)

HEALTH EFFECTS OF SMOKING

Nicotine sustains tobacco addiction, which is a major cause of disability and premature death. Although most of the toxicity of smoking is related to other components of cigarette smoke, it is primarily the pharmacologic effects of nicotine that produce addiction to tobacco. Smoking cigarettes remains the leading preventable cause of death worldwide; killing more than eight million people each year, according to the WHO.

Up until the early 20th century, lung cancer was a very rare disease. It became more common after World War I, and the cause was debated, with suggested reasons (along with smoking) including the asphalt from road surfaces ('tarred' roads were a relatively new innovation), air pollution from industry, the influenza pandemic of 1918–1919 and the long-term effects of World War I poison gases. As the rise in cigarette consumption coincided with increased numbers of lung cancer cases, medical statisticians increasingly looked to smoking as the cause.

Before World War II, there were periods of serious anti-smoking movements in Soviet Russia and Nazi Germany. For example, Lenin was a non-smoker who wanted to ban it but did not succeed in getting his way (Stalin was a heavy smoker). Hitler gave up smoking by the time he came to power and was opposed to tobacco. Anti-tobacco propaganda was produced, albeit unsuccessfully. Among the Nazi leaders, Hermann Göring, Heinrich Himmler and Joseph Goebbels were all smokers, as was Eva Braun. The first scientific evidence linking smoking with cancer was published in the USA in 1931, and two German studies from 1939 and 1943 linked smoking with lung cancer, but it was not until 1950 that considerable evidence was published – four papers in the USA and one in the UK – that clearly showed an association between smoking and disease, particularly lung cancer, with increased risk for heavy smokers. Further studies provided evidence in support of this throughout the 1950s, notably by Richard Doll in the UK, so that by the end of the decade, the association between smoking and lung cancer was accepted in the medical profession, with associations between smoking and other diseases established as time went on.

A milestone in 1964 was the report 'Smoking and Health' from the U.S. Surgeon General, which warned the public that smoking causes disease and death.

TOXIC CHEMICALS IN TOBACCO

A smoker is exposed to many toxic chemicals, nearly 10,000 of which are known, although nicotine is not in itself *directly* the most dangerous. Some are gases; some are volatile organic molecules contained in the volatile portion of the aerosol; some are present in the solid particles suspended in the smoke. The most toxic gaseous molecules present are carbon monoxide (CO) and hydrogen cyanide (HCN), as well as oxides of nitrogen. The volatiles include benzene, a known carcinogen, and certain aldehydes including methanal (formaldehyde), ethanal (acetaldehyde) and acrolein, as well as some N-nitrosamines. The most dangerous compounds in the third group are polycyclic aromatic hydrocarbons and certain tobacco-specific nitrosamines (TSNAs). There are compounds of a few metals that are believed to be carcinogens, notably arsenic, cadmium, chromium and nickel.

Tobacco smoke contains over 60 compounds known to be carcinogens, notably many polycyclic aromatic hydrocarbons and TSNAs, many of these are found in the 'tar', which is borne in the smoke to the lungs. The first polycyclic aromatic hydrocarbon present in cigarette smoke to be identified (1954) was 3,4-benzopyrene (9.6); others include dibenzoanthracene (9.7) and 5-methylchrysene (9.8).

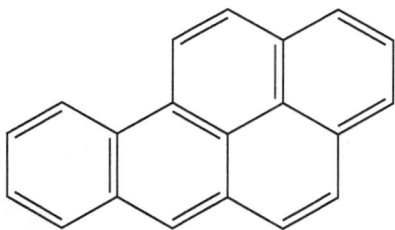

(9.6)

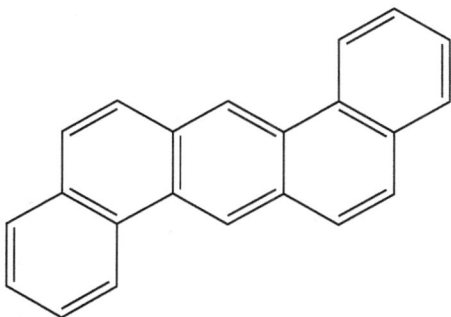

(9.7)

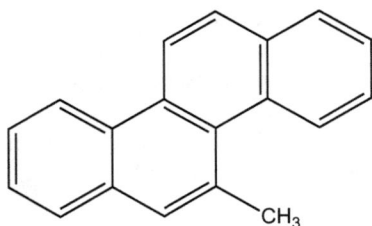

$$(9.8)$$

After being inhaled, polycyclic aromatic hydrocarbons are oxidised in the body by cytochrome P450 enzymes to epoxides; in turn, these are rapidly hydrolysed to diols. The diols are subsequently converted to diol epoxides. Using benzopyrene as an example (9.9):

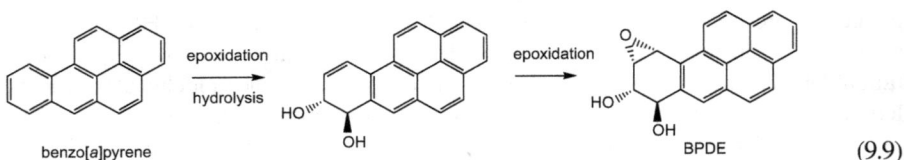

$$(9.9)$$

Benzopyrene diol epoxide reacts readily with the body's DNA, principally at the base guanine (9.10), forming adducts that lead to mutations which deactivate tumour suppressor genes, resulting in aberrant cells and initiating the carcinogenic process.

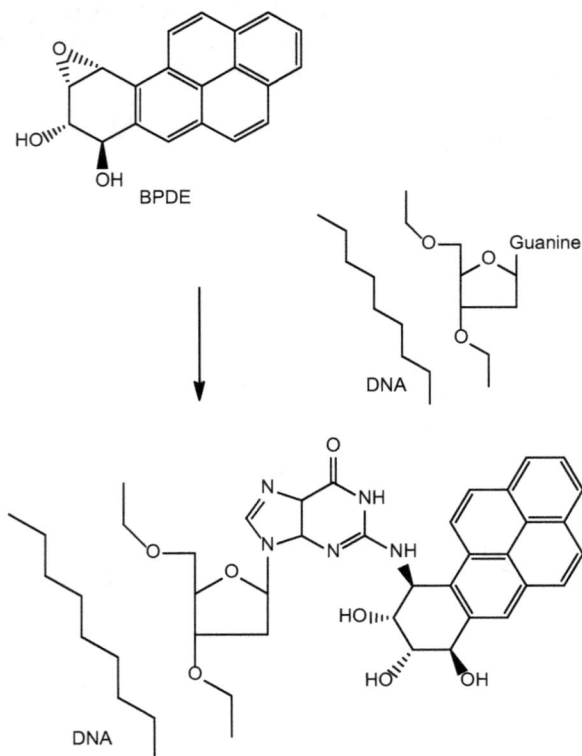

$$(9.10)$$

((9.10) - based on Figure 1 of: K. Alexandrov, M. Rojas and C. Rolando, *Cancer Res.*, 2006, **66** (24), 11938–11945)

TSNAs, notably 4-(methylnitrosamino)-1-(3-pyridyl)-1-butanone (9.11: NNK) and *N'*-nitrosonornicotine (9.12: NNN), are formed substantially during the tobacco curing process and are thus present in unsmoked tobacco.

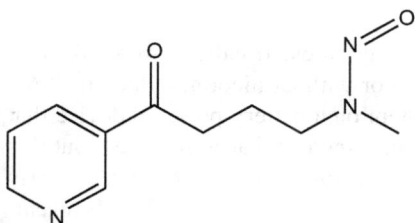

(9.11)

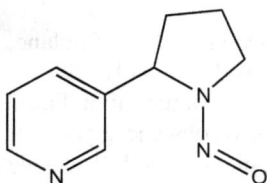

(9.12)

NNN has also been detected in e-cigarette users.

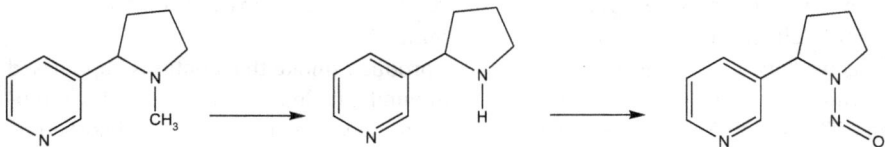

(9.13)

Nicotine is demethylated, forming nornicotine, through the action of the enzyme nicotine N-demethylase; nornicotine can then be nitrosylated by nitrous acid (HNO_2) to form NNN (9.13). NNN (and NNK) are bio-activated through hydroxylation reactions. Taking NNN as an example again, a cytochrome P450 enzyme hydroxylates the five-membered ring in *N'*-nitrosonornicotine (9.14). In human beings, this usually occurs at carbon 2', whilst in non-primate animals, it generally occurs at carbon 5'. This is followed by ring-opening, enabling reaction with the nucleotide base.

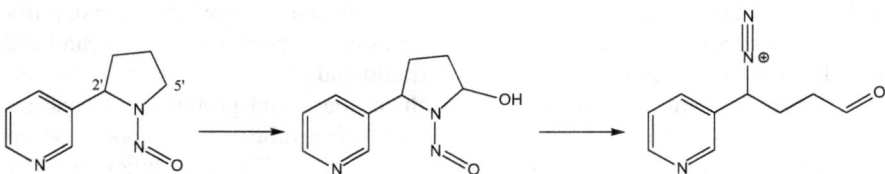

(9.14)

The World Health Organization states that smoking-related diseases kill approximately 8 million people a year (with circa. 600,000 of these from 'passive smoking'). Around 100,000 of these deaths occur in the UK and half a million in the USA. Nicotine does not cause these deaths *directly*, but *addiction* to nicotine does.

VAPING

For over half a century, people have been working on electrically powered devices to provide flavoured hot air for inhalation (with or without nicotine). Back in 1965, the American Herbert Gilbert applied for a patent on a battery-powered device that essentially produced flavoured steam, with no burning or tobacco involved, but this did not get off the ground. E-cigarettes have only become widespread since the early 21st century. A Chinese pharmacist named Hon Lik is often credited with inventing the precursor of modern devices in 2003 as an aid to stopping smoking after his father died of lung cancer (Spoiler: Hon Lik still smokes).

The use of e-cigarettes has mushroomed worldwide since around 2010. In England, for example, much of this increase has occurred since the arrival of COVID-19.

Current e-cigarettes have three main components: (1) a source of a liquid ('e-liquid' composed of a solvent such as propane-1,2-diol (propylene glycol) or propane-1,2,3-triol (aka glycerol or glycerine(e)) containing nicotine and flavouring chemicals, (2) a source of electrical power, such as a rechargeable battery and (3) an 'atomiser'. The atomiser is a heating element that converts the liquid into an aerosol mist of the solvent, flavourings and nicotine, which is inhaled via a mouthpiece, the process usually known as 'vaping'. Therefore, they are tobacco- and combustion-free but still deliver nicotine, as well as other chemicals.

As we have seen, conventional cigarettes produce smoke that contains vapourised nicotine, as well as thousands of other compounds, including well over 50 carcinogens. The nicotine and many other substances are taken up by the lungs, absorbed by the blood and passed to the brain and other parts of the circulatory system. Nicotine is the substance most often common to both smoking tobacco and vaping, though both processes sometimes use cannabis-related substances instead. The vapourised solution from the cartridge uses the solvent to provide the aerosol mist of tiny liquid drops that mimics tobacco smoke.

ISSUES WITH NICOTINE

Most e-liquids contain nicotine, the quantity of which varies from one product to another. The amount of nicotine delivered to the smoker also depends upon the electrical power of the e-cigarette as well as, obviously, the number and duration of puffs that the smoker takes. There is evidence that habitual users of e-cigarettes can have an intake of nicotine comparable to those of traditional cigarettes.

As a base, nicotine accepts protons. Both the base and protonated forms are present in tobacco. The protonated form is said to feel milder and smoother when inhaling tobacco smoke, but the base is absorbed more readily in the lungs. Heating ammonium salts causes decomposition, generating ammonia; tobacco firms mix

ammonium salts with tobacco so that, when the tobacco is burned, the ammonia generated deprotonates the protonated form of nicotine, generating the free base (9.15).

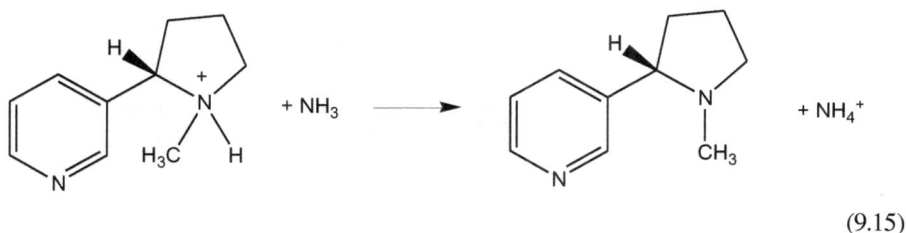

$$(9.15)$$

Analysis of different brands of e-liquids indicates that those of the popular JUUL
brand contain higher concentrations of nicotine, mainly in the 'smoother' protonated
form, than those from its competitors, most of whom contain at least 50% of the less
smooth base form. This may explain the commercial success of JUUL. Nicotine
levels vary; JUUL e-liquids in the UK have around one-third of the nicotine level of
the US ones; and a similar ratio applies to the aerosols.

Concern has been widely expressed about the use of e-cigarettes by adolescents.
Nicotine causes the release of the 'pleasure molecule' dopamine in the brain, so that
when dopamine levels fall again, the user will reach for the cigarette to repeat their
experience. Nicotine exposure may cause changes in the brain with long-term effects,
such as priming users to abuse other substances; in other words, nicotine may be a
'gateway' drug.

Nitrosamines are another chemical of concern, notably TSNAs (TSNAs), which
are generated during the curing of tobacco (see page 197) and are carcinogens. They
have been found at low levels in e-liquids, presumably due to their presence in the
nicotine used. Evidence suggests that carcinogenic emissions from e-cigarettes are
much lower than from tobacco smoke.

METAL HEATING ELEMENTS AND METAL PIECES

Because of the use of metals in e-cigarettes, especially in heating elements, they can
be detected in e-liquids, which can lead to contamination of the aerosols.

Various substances are used to make the heating elements – which are heated
to temperatures of approximately 375°–525° – including nichrome (Ni and Cr),
tungsten, stainless steel and Kanthal (a ferritic iron chromium aluminium alloy).
Analyses have shown the presence of metals including Al, Co, Cr, Mn, Ni, Pb and
Zn in e-cigarette emissions, whilst one particular study has compared the metal concentrations in e-liquid before heating with those in the aerosol after being heated by
the coil, with increases of greater than 2,000% for Pb and Zn and over 600% for Cr,
Ni and Sn.

In one case, giant cell interstitial pneumonia has been linked to cobalt exposure
from the use of marijuana e-cigarettes. A parallel discovery has been that vaping
with aged coils, which increases the emission of toxic aldehydes such as methanal
and ethanal (see pages 195 and 200), causes acute lung injury in mice.

SOLVENTS

The alcohols propane-1,2-diol (propylene glycol or PG) and propane-1,2,3-triol (glycerol or glycerine(e)) are widely used as solvents in e-liquids. They also serve another function, as they help produce aerosols that resemble tobacco smoke, mimicking the smoking experience. In the context of their use in foods, they are regarded as having low toxicity, and of course, glycerol occurs naturally in the human body.

However, these alcohols are not stable under all conditions; thus, they decompose upon heating. Three substances, all aldehydes, formed during the thermal decomposition of glycerol (9.16) and propylene glycol (9.17) have attracted particular attention – acrolein, methanal and ethanal.

You've probably encountered acrolein (propenal) as the substance formed when cooking oil is heated until it begins to smoke ('burnt fat'); it is toxic and also very irritating to the eyes and nasal passages. Indeed, in 1839, Berzelius discovered that acrolein was formed by the thermal decomposition of glycerol. Ethanal (acetaldehyde) and methanal (formaldehyde) are also toxic, with methanal in particular being a well-known Group I carcinogen. Moreover, one study indicates that glycerol- and propylene glycol-derived aerosols contain pulmonary irritants and that methanal formed by solvent decomposition contributes to endothelial dysfunction. It has previously been believed that propylene glycol and glycerol decompose around 350°C–400°C, but it is now known that appreciable amounts of methanal and ethanal are formed below 200°C.

One question is still to be answered: Are these aldehydes produced at dangerous concentrations in e-cigarettes?

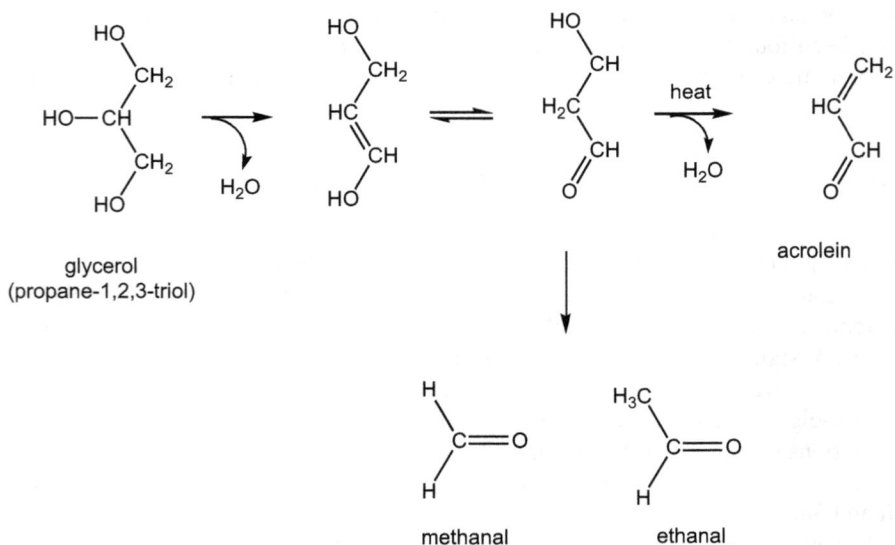

glycerol
(propane-1,2,3-triol)

acrolein

methanal   ethanal

(9.16)

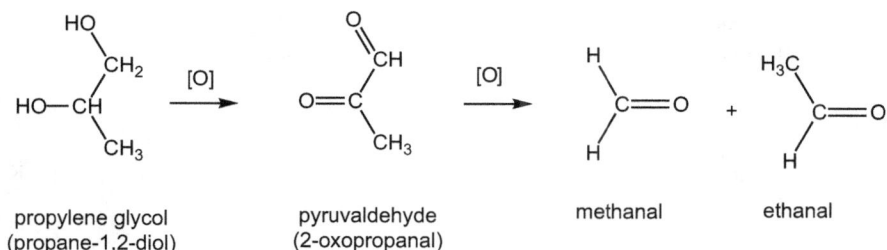

propylene glycol
(propane-1,2-diol)

pyruvaldehyde
(2-oxopropanal)

methanal   ethanal

(9.17)

Another area of concern lies in the fact that a number of the aldehydes used in flavourings in e-liquids, such as vanillin, ethyl vanillin and benzaldehyde, form adducts
(acetals) with propylene glycol and glycerol (9.18).

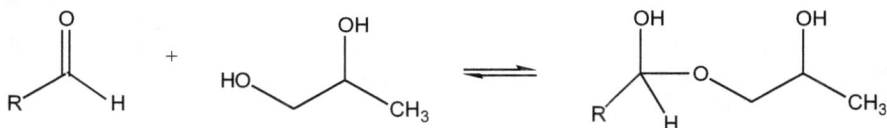

(9.18)

These are cytotoxic and inhibit mitochondrial function in respiratory epithelial cells. The majority of the acetal gets carried into the e-cigarette vapour. They
have been shown to activate aldehyde-sensitive TRPA1 irritant receptors and alde-
hyde-insensitive TRPV1 irritant receptors.

FLAVOURINGS

Compared to conventional cigarettes, flavourings are much more prevalent in e-cigarettes. It is thought that added flavours, especially fruity ones, have been widely used
in e-cigarettes, especially to appeal to young consumers.

Aldehydes are prominent flavouring materials in e-cigarette devices and have
therefore come under scrutiny, as they are recognised to be 'primary irritants' of the
respiratory tract. Vanillin and ethyl vanillin were particularly common in a sample
of 30 e-liquids analysed.

Short-chain aldehydes have been identified as the most toxic products in tobacco
smoke, especially acrolein, methanal and ethanal, and, as already noted, these are
also produced by the thermal decomposition of solvents in e-liquids. Toxic aldehydes
have been suggested to result from the decomposition of sugars in e-liquids, but the
evidence for that is mixed.

Aldehyde concentrations in the aerosol are affected by puffing conditions; e-cigarette users also alter their puffing regime according to nicotine levels in the e-liquid,
which in turn affects the levels of aldehydes like methanal and ethanal. It has been
found that more modern e-cigarette devices produce more aldehydes than the original devices, due to the greater power output from their batteries.

Researchers have suggested that because of the sensitivity of the cardiovascular system to aldehydes, it could suffer significant damage even at low concentrations.

DIACETYL

One component of some e-liquids that has been a cause for particular concern is diacetyl (9.19).

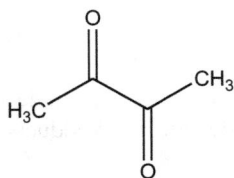

(9.19)

A significant flavouring ingredient in butter, it has been used as a flavouring in foods, notably popcorn, as well as in e-liquids, particularly in the USA. It was banned in the EU as an ingredient in e-liquids in 2016.

It first came to attention in 2002, when eight former workers at a microwave popcorn plant in Missouri, USA, were reported to have bronchiolitis obliterans, a condition in which the alveoli in the lungs are irritated and inflamed, resulting in scarring and causing their narrowing, which leads to respiratory difficulties, or even respiratory failure and the need for lung transplantation. Researchers at Harvard University concluded that there was a strong relationship between the estimated cumulative exposure to diacetyl and the frequency and extent of airway obstruction. A subsequent study of several such plants found that in a number of them, mixers of butter flavouring and packaging-area employees working near tanks of heated oil, who were both exposed to diacetyl, had fixed airway obstruction consistent with bronchiolitis obliterans.

A paper published in 2015 by American researchers reported analyses on 51 types of flavoured e-cigarettes deemed to be 'appealing to youth' (their names included *Cupcake, Fruit Squirts, Waikiki Watermelon, Cotton Candy, Tutti Frutti, Double Apple Hookah, Blue Water Punch, Oatmeal Cookie, Pina Colada and Alien Blood*). Diacetyl was detected in 39 of them; 2,3-pentanedione (9.20: acetylpropionyl) in 23; and acetoin (9.21: 3-hydroxybutanone) in 46.

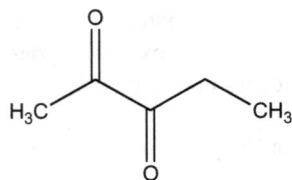

(9.20)

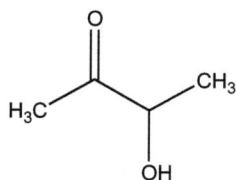

(9.21)

BENZALDEHYDE, CINNAMALDEHYDE AND VANILLIN

Significant numbers of the flavour chemicals contain carbonyl groups, particularly aldehydes, which are often irritants to the mucosal tissue in the respiratory tract when inhaled. Typical examples include benzaldehyde ((9.22): cherry), cinnamaldehyde ((9.23): cinnamon) and the two vanilla odorants, vanillin ((9.24); from the vanilla bean) and ethylvanillin ((9.25); totally synthetic).

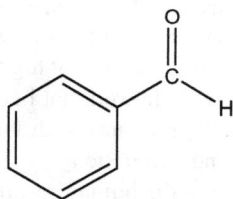

(9.22)

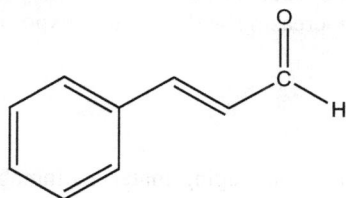

(9.23)

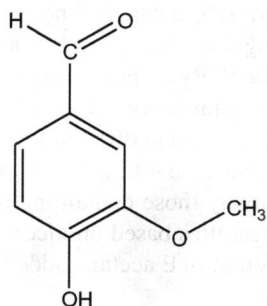

(9.24)

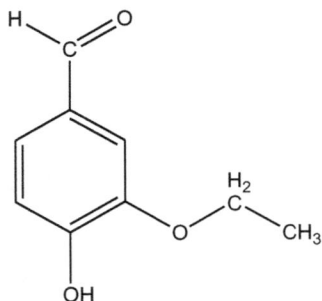

(9.25)

Around half of a sample (2016) of US e-cigarette refill fluids contained cinnamaldehyde at concentrations that are cytotoxic to human embryonic and lung cells. In another study examining the effect of cinnamaldehyde in e-liquids upon human bronchial epithelial cells, it was found to impair mitochondrial function and reduce ciliary beat frequency, thus affecting respiratory defence; it was suggested that e-cigarette users would be at greater risk of respiratory infections. Benzaldehyde was detected in aerosols from over 70% of a sample of 145 different e-cigarettes, at higher levels than those found in conventional cigarettes. A 2024 report showed that benzaldehyde (and its propylene glycol acetal derivative, see p. 201) interacts with lung surfactant monolayers, potentially causing alveolar collapse and cytotoxicity.

There is concern that, in addition to the negative effects upon the human respiratory system, vaping may adversely affect the liver. Vanillin and ethyl vanillin are cytotoxic to human liver cells, with cytotoxicity increasing with repeated exposure to these chemicals.

VITAMIN E ACETATE

Vitamin E acetate has been used as an additive in some vaping materials (notably in illicit cannabis-related products). It came to prominence in the context of e-cigarette or vaping product use-associated lung injury (EVALI), formerly known as vaping-associated pulmonary illness. This was first recognised in April 2019 as a severe vaping-associated pulmonary injury that could be fatal. By 28 February 2020, 68 deaths had been reported in the USA, out of 2,807 hospitalisations. Cases have continued to occur, but because of the rise of COVID-19, it became difficult to distinguish between cases of EVALI and COVID. It soon became clear that the disease was associated with e-cigarettes and vaping products, largely those containing tetrahydrocannabinol oil, as well as some conventional e-cigarettes based on nicotine. A particular chemical responsible has been identified as Vitamin E acetate, added to the e-liquid as a cutting agent.

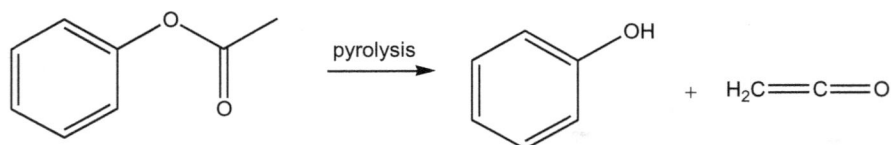

(9.26)

It was already known that the pyrolysis of phenyl acetate (9.26) affords the toxic and highly reactive unsaturated ketone, ethenone (often known as ketene), and it was realised that Vitamin E acetate contained a similar acetate group bound to a benzene ring. Experiments showed that heating Vitamin E acetate similarly gave rise to ethenone; moreover, Δ^9-THC acetate, an unregulated, possibly psychoactive cannabinoid used in cannabis vapes that accumulates in the alveoli, causing lipoid pneumonia and disrupting the pulmonary surfactant, similarly forms ethenone under vaping conditions (9.27). Again, Δ^9-THC acetate contains an acetate group attached to a benzene ring.

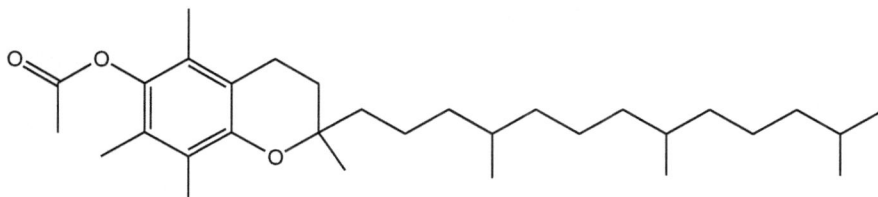

Vitamin E acetate

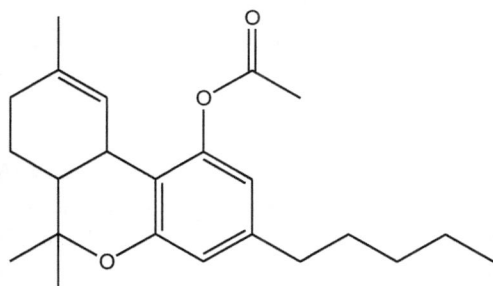

$$H_2C = C = O$$

etheneone
(ketene)

Δ^9-THC acetate

(9.27)

NICOTINE ANALOGUES

There is concern about compounds structurally related to nicotine that are not tobacco-related being used in vaping products. In 2023–2024, two compounds, nicotinamide (9.28) and 6-methyl nicotine (9.29), have appeared on the market, the first in pods and the second in e-liquids. As American legislation is framed, compounds that are not nicotine- or tobacco-based cannot be regulated by the US Tobacco Control Act.

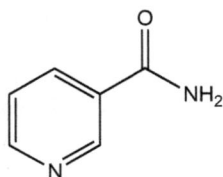

(9.28)

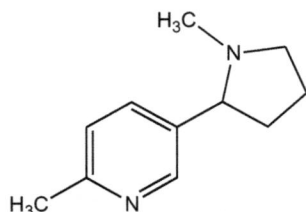

(9.29)

6-Methyl nicotine, marketed under the trade name Metatine, is more potent than nicotine but can be labelled in the USA as 'nicotine free'. Nicotinamide is Vitamin B3. It does not affect the acetylcholine receptors that are targeted by nicotine; its use in vaping products appears to be based on its name so that it can be marketed as some kind of 'safe vitamin vape'.

These compounds appear to be controlled under EU legislation.

STUDENT VAPING

As an example of how smoking habits have changed, surveys in the USA have shown a decrease in the number of teenagers trying 'conventional' cigarettes from 70% in 1991 to 58.4% in 2003 and then to 28.9% in 2017. In contrast, between 2011 and 2018, the percentage of adolescents vaping increased from near zero to over 20%. Another survey reports that there has been a decrease in e-cigarette use among high school students from 14.1% to 10.0%, comparing 2022 with 2023, but it remains to be seen what long-term trends will emerge.

As an example of how vaping, like smoking, predisposes young people to disease, adolescent and young adult e-cigarette users have been found to be five times more likely to be diagnosed with COVID-19 compared with people who did not smoke e-cigarettes. Users of both e-cigarettes and conventional cigarettes were seven times more likely to have a COVID-19 diagnosis.

CONCLUSIONS AND SUMMARY

The health risks of conventional tobacco smoking are well established, particularly with links to lung cancer. In contrast, it is only recently that vaping has emerged, so there has been much less time to assess its safety, and numerous questions remain to be answered.

Vaping has both expanded and developed enormously during the past 15 years or so, with different kinds of electronic devices being followed by 'heated tobacco products'. All are touted as 'safer than smoking'; potential consumers should ask themselves, to use a classical Latin phrase: *'Cui bono?'*, or, in 21st-century parlance: *'What's in it for them?'*

E-cigarettes do not involve combusting tobacco, which means that the carcinogens associated with tobacco smoking are absent. Therefore, it has been assumed that vaping *must* be safer than smoking, without knowing the effects of the inhaled substances upon health. These substances have been largely unregulated.

Vaping is being promoted as a safe way to give up smoking. It may be that e-cigarettes do not cause the health issues associated with cigarette smoking (e.g. lung cancer, pulmonary disease or cardiovascular problems), but the evidence either way is not definite; though on current evidence, e-cigarettes appear less toxic than tobacco smoking.

As already mentioned, vaping, like smoking, can predispose young people to disease. Compared with those who did not smoke e-cigarettes, adolescent and young adult e-cigarette users were five times more likely to be diagnosed positive for COVID-19. Users of both e-cigarettes and conventional cigarettes were seven times more likely to have a COVID-19 diagnosis.

In any case, some people have begun to use e-cigarettes without previously having been tobacco smokers and are finding them addictive, doubtless due to nicotine. We do not understand the implications of the number of young people, both youths and young adults, who have embraced e-cigarettes, with possible issues including the normalisation of smoking behaviour among those who would never have started to smoke combustible tobacco products. These people have been particularly targeted with the use of certain flavours. Until 2013, tobacco and menthol were the most popular flavourings in e-cigarettes, but in recent years, fruit flavours, appealing to the young market, have become the most popular. A number of the flavourings used in e-cigarettes have already been used as food flavourings, in which use they are safe for ingestion. However, inhaling a hot vapour of a substance is a very different route of ingestion compared to consuming the cold liquid, so the inhalation route for these chemicals requires proper assessment. As already mentioned, a warning was sounded by diacetyl, a flavouring naturally present in butter but which caused disease in people working with it in a popcorn manufacturing factory; it is banned from use as an e-flavouring in Europe.

Government policies vary, with some countries (e.g. Turkey, Brazil, India, Singapore and Uruguay) having banned e-cigarettes.

10 Isotopes at Work

ISOTOPES

John Dalton, the proponent of atomic theory (1808), was the first person to determine atomic weights for what we would regard today as a restricted list of elements – not least because many elements had yet to be discovered at that time. Among the key points made by Dalton were that elements are made of extremely small particles called atoms and that atoms of a given element are identical in size, mass and other properties. As the number of known elements increased, so did the accuracy of the determination of atomic masses.

About a century later, in 1912, J. J. Thomson and F. W. Aston found that a magnetic field deflected a beam of gaseous Ne^+ ions onto two different paths because the beam contained two particles of different mass, with differing mass-to-charge ratios. After the interruption of World War I, Aston returned to the problem in 1919 using his 'mass spectrograph' (a primitive mass spectrometer), showing that the two different neon ions had masses of 20 and 22; he later discovered a less abundant particle with a mass of 21 that gave rise to a weaker peak in the spectrum. We now know that the three neon isotopes of masses 20, 21 and 22 have abundances of 90.51%, 0.27% and 9.22%, respectively.

It was Frederick Soddy who first suggested the existence of isotopes (1913), for which he was awarded the Nobel Prize for Chemistry in 1921 (awarded retrospectively in 1922). In his Nobel lecture, Soddy described isotopes in these words: "Put colloquially, their atoms have identical outsides but different insides." Aston was awarded the 1922 Nobel Prize for Chemistry for demonstrating the existence of isotopes with his mass spectrograph. Isotopes were identified at much the same time as radioactivity. Isotopes may be stable or radioactive. Of the unstable radioactive ones, they may occur in nature or be sufficiently short-lived that any present at the formation of the Earth have long since decayed.

We see the existence of isotopes reflected in the relative atomic masses given in today's Periodic Tables. Many of these are close to integral values, notably hydrogen (1.008) and helium (4.003). In some cases, this is simply due to the element in question having only one type of atom (as we now say, an isotope); there are 19 elements found in nature that have just one stable isotope: ^9Be, ^{19}F, ^{23}Na, ^{27}Al, ^{31}P, ^{45}Sc, ^{55}Mn, ^{59}Co, ^{75}As, ^{89}Y, ^{93}Nb, ^{103}Rh, ^{127}I, ^{133}Cs, ^{141}Pr, ^{145}Ho, ^{159}Tb, ^{169}Tm, ^{197}Au and ^{209}Bi. However, in 2003, it was discovered that ^{209}Bi has an incredibly long half-life of 10^{19} years. In addition, there are seven more elements that occur with their one stable isotope accompanied by a radioactive isotope (^{51}V, ^{85}Rb, ^{113}In, ^{139}La, ^{153}Eu, ^{175}Lu and ^{185}Re).

Some other elements have predominantly one stable isotope. In nature, hydrogen is mainly composed of ^1H (99.985%) along with ^2H (0.015%). There is a mere trace

DOI: 10.1201/9781003658115-10

of the radioactive ^{3}H, which has a half-life of 12.262 years. Thus, hydrogen has a relative atomic mass of 1.00794. Similarly, the vast majority of carbon atoms are ^{12}C (98.90%), with just over 1% being ^{13}C (1.10%), and a trace of the radioactive ^{14}C (half-life 5,730 years). Therefore, carbon has a relative atomic mass of 12.011.

The discovery of isotopes solved a particular problem. As more atomic masses were determined and those values became more accurate, it became clear that some had non-integral values. For example, the atomic masses used by Berzelius (1827, 1828, 1835) included 35.5 for chlorine. This value was also adopted by others, including Stanislao Cannizzaro (1860) and by Mendeleev, in his first Periodic Table of 1869. Aston included chlorine in his mass spectrographic studies and showed that it was composed of two stable isotopes, with masses of 35 and 37. We now know that the natural abundances are 75.77% for ^{35}Cl and 24.23% for ^{37}Cl. The element boron consists of 80% ^{10}B and 20% ^{11}B, so the relative atomic mass of boron is 10.8; the composition of bromine is 50.69% ^{79}Br and 49.31% ^{81}Br, hence the relative atomic mass of bromine is 79.90.

Unlike other elements, the heavier isotopes of hydrogen are often referred to by their own symbols, D for ^{2}H (deuterium) and T for the radioactive ^{3}H (tritium). The proportional difference in mass between ^{1}H and ^{2}H is, of course, greater than for any other pair of elements, and they have been an important pair of isotopes for comparison purposes.

In the summer of 1931, Harold Urey at Columbia University, USA, examined the atomic spectrum of gaseous hydrogen and found some weak lines that he thought might be due to a heavier isotope of mass 2. On Thanksgiving Day 1931, Urey led a research team that found that when liquid hydrogen (H_2) was evaporated (its boiling point is 20K, −253°C), the heavy isotope accumulated in the remaining liquid – the 'suspicious' spectral lines became more intense. In 1932, Urey's team found that when water, H_2O, was electrolysed – split up by electrolysis – H_2O molecules were preferentially split up so that D_2O molecules accumulated in the residual water.

At much the same time, the distinguished chemist G.N. Lewis (who had been Urey's PhD supervisor) at the University of California started to examine the deuterium problem. He employed fractional distillation – with one fractionation column 72 feet high – to enrich water samples in D_2O in bulk, then used electrolysis to further enrich the product. He was able to produce over a gram a week of D_2O at over 99% purity.

According to Jacob Bigeleisen, Lewis was generous in supplying 'deuterium samples' (possibly deuterium oxide?) to other researchers – E. O. Lawrence at Berkeley, C. C. Lauritsen at Cal. Tech. and Ernest Rutherford at Cambridge (UK) – for their studies in 1933.

Lewis and his research assistant R. T. Macdonald embarked on a concentrated programme to study the chemical and physical properties of deuterium oxide, resulting in an incredible 26 publications in the space of 16 months between February 1933 and July 1934.

In 1934, the Nobel Prize for Chemistry was awarded for work on deuterium. It was awarded solely to Harold Urey. Gilbert N. Lewis, who was responsible for a fistful of important ideas in chemistry – such as the 'Lewis pair', covalent bonding, acid–base theory, thermodynamics for chemists – was never to win a Nobel Prize, despite being nominated 35 times.

WHAT'S THE DIFFERENCE BETWEEN H_2O AND D_2O?

The structures of the individual molecules are very similar, V-shaped, with slightly different molecular dimensions (10.1). They are not linear molecules because the oxygen atom has eight electrons in its outer shell; four of these are in two two-electron non-bonding pairs ('lone pairs'), so that minimising inter-electronic repulsions leads to an approximately tetrahedral arrangement of the four pairs. O–H is 0.9724 Å in free H_2O molecules versus O–D of 0.9687 Å in D_2O; the bond angle is 104.50° in free H_2O molecules versus 104.35° in D_2O.

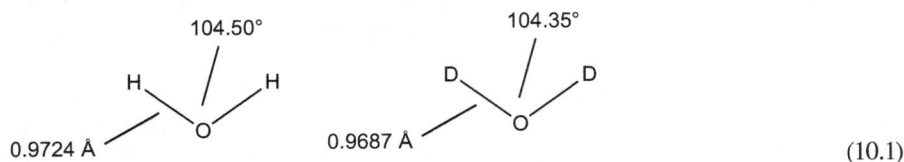

$$(10.1)$$

The most obvious difference between the 'bulk materials' is that D_2O is about 10% denser, at 1.106 g/cm^3, compared to 1.000 g/cm^3, a knock-on effect of deuterium atoms being twice as heavy as 'normal' hydrogen. The freezing point of D_2O is 3.82°C, compared to 0.00°C for H_2O and the boiling point of D_2O is 101.42°, compared to 100.00°C for H_2O; these values reflect the slightly stronger intermolecular hydrogen bonds in D_2O.

Heavy water has low toxicity; naturally, some 0.01% of hydrogen atoms in nature are deuterium. For both plants and animals, fatal effects occur only when approximately 30%–50% of the water content of the organism has been deuterated. Bacteria (and yeast) can function in near-pure deuterium oxide, though with a lower growth rate. Experiments with very pure D_2O have shown that humans perceive it as having a sweeter taste than pure H_2O (mice do not prefer D_2O to H_2O, suggesting that they do not share this perception).

Hydrogen and deuterium are chemically similar but have different spectroscopic and magnetic properties, making deuterium useful to the chemist. Because a deuterium atom has twice the mass of a hydrogen atom, this affects the frequency of vibrations of bonds involving deuterium in the infrared spectra of compounds. Thus, if you replace hydrogen atoms in an O–H group with deuterium, the band due to ν O–H (O–H stretching) at ~3,600 cm^{-1} moves to ~2,600 cm^{-1} (caused by ν O–D). Likewise, C–H stretching frequencies of ~2,900 cm^{-1} shift to ~2.,100 cm^{-1} upon deuteration.

Another application occurs in nuclear magnetic resonance (NMR) spectroscopy. Because of the different nuclear properties of ^1H and ^2H, signals due to deuterium atoms are not observed in a ^1H NMR spectrum. So, if you have an organic compound with a labile O–H (or N–H) group, just add a drop of D_2O to your NMR sample of the compound in a solvent like $CDCl_3$ (whose deuterium is also 'invisible in ^1H NMR). The equilibrium shown in the equation is forced to the right.

$$R{-}O{-}H + D_2O \rightleftharpoons R{-}O{-}D + D{-}O{-}H$$

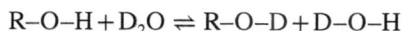

The signal due to the OH group disappears from the ^1H NMR spectrum. This can be seen in the NMR spectrum of a compound such as menthol, where the OH peak at ~1.8 ppm disappears after adding D_2O, showing that the H involved is free to exchange with the solvent (see https://www.chm.bris.ac.uk/motm/menthol/mentholh.htm)

Another classic example of the use of isotopes was reported in 1938, when Irving Roberts and Harold Urey carried out a study on the esterification reaction between benzoic acid and ^{18}O-enriched methanol. They found that none of the ^{18}O ended up in the water produced.

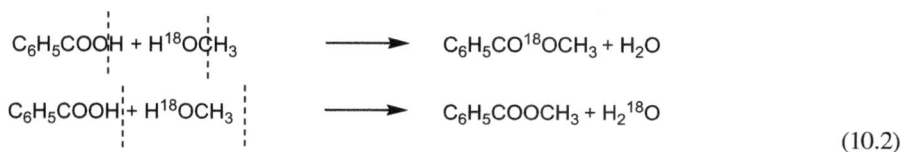

$$C_6H_5COOH + H^{18}OCH_3 \longrightarrow C_6H_5CO^{18}OCH_3 + H_2O$$

$$C_6H_5COOH + H^{18}OCH_3 \longrightarrow C_6H_5COOCH_3 + H_2{}^{18}O$$

$$(10.2)$$

This indicated that the C–OH bond in the benzoic acid broke and that the first of the two pathways shown (10.2) was the one followed. Four years earlier, the hydrolysis of amyl (pentyl) acetate was studied by Polanyi and Szabo at the University of Manchester, who used ^{18}O-enriched water to show that it was the C–O bond next to the C=O that broke, as the reaction produced amyl (pentyl) alcohol containing 'normal' oxygen (10.3).

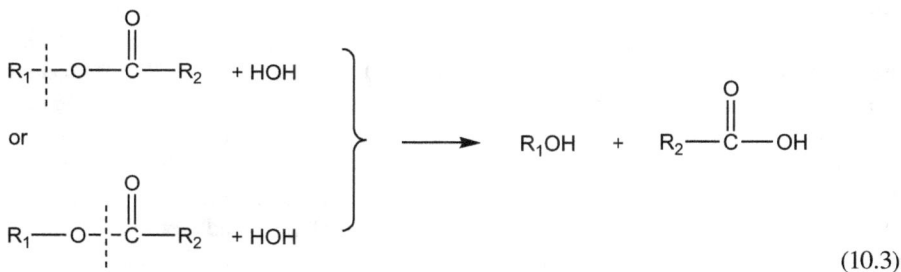

$$R_1{-}O{-}\overset{\overset{\displaystyle O}{\|}}{C}{-}R_2 \;+\; HOH$$

or

$$R_1{-}O{-}\overset{\overset{\displaystyle O}{\|}}{C}{-}R_2 \;+\; HOH \Bigg\} \longrightarrow R_1OH \;+\; R_2{-}\overset{\overset{\displaystyle O}{\|}}{C}{-}OH$$

$$(10.3)$$

LEAD ISOTOPES

Lead, atomic number 82, is the heaviest element to have non-radioactive isotopes. In fact, lead has four stable isotopes. One of these, ^{204}Pb, does not have a radioactive precursor. The other three are all at the end of radioactive decay chains. The decay chain of ^{238}U ends at ^{206}Pb; ^{207}Pb is the final element in the decay chain of ^{232}Th; whilst the decay chain of ^{235}U ends with ^{208}Pb.

Samples of lead in nature have different isotopic makeups because different sources result from different combinations of uranium and thorium ores, leading to differing contributions of various lead isotopes. Scientists have studied the ratios of certain lead isotopes in different mineral deposits containing lead. In general, the $^{206}Pb/^{204}Pb$ and $^{206}Pb/^{207}Pb$ ratios usually found in lead ores worldwide fall within the ranges of 16.0–18.5 and 1.19–1.25, respectively.

One environmental area in which lead isotope ratios have been examined is in connection with the legacy of leaded petrol. Back in 1921, the American engineer Thomas Midgley Jr. discovered that the addition of the compound tetraethyl lead (10.4) to petrol improved the performance of car engines with high compression ratios by preventing the pre-ignition of petrol–air mixtures. Half a century passed before the increased realisation of the damage to health caused by the release of lead into the environment initiated moves towards the adoption of lead-free petrol, as did the installation of catalytic converters, which would be 'poisoned' by lead.

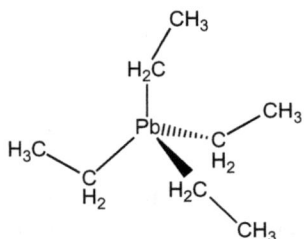

(10.4)

One environmental area in lead isotope ratios that has been examined is in connection with the legacy of leaded petrol. One examination centred upon the region of a former petrol (aka 'gasoline') 'filling station' in the American state of South Carolina, where petrol-contaminated groundwater around three former petrol tanks was found to contain unacceptably high levels of lead. Scientists examined the stable isotope ratios of lead in the water: $^{207}Pb/^{206}Pb$ versus $^{208}Pb/^{206}Pb$, and $^{208}Pb/^{204}Pb$ versus $^{206}Pb/^{204}Pb$. They found an average $^{206}Pb/^{204}Pb$ ratio of 19.038 and an average $^{206}Pb/^{207}Pb$ ratio of 1.213. Tetraethyllead for the European market was mainly made from a lead ore from Broken Hill, Australia, which had a very low $^{206}Pb/^{207}Pb$ ratio (1.03–1.10). Tetraethyllead used in American petrol was made from Mississippi Valley ore deposits, which have a very different lead isotopic composition ($^{206}Pb/^{204}Pb \sim 20.0$; $^{206}Pb/^{207}Pb = 1.31$–1.35), so that American leaded petrol has very different $^{206}Pb/^{207}Pb$ ratios to European petrol. Moreover, the $^{206}Pb/^{207}Pb$ value for local lead-based rocks is ca. 1.18, so it appears that the South Carolina results relate to the local aquifers, rather than any lead contamination from petrol.

Archaeologists have also studied lead isotopes to push their studies into the distant past. For example, the study of Pb isotope ratios in 2000-year-old sediments from the Tiber River and the Trajanic Harbour has shown that "tap water" from ancient Rome had 100 times more lead than local spring waters, reflecting pollution from other mineral sources. Similarly, examination of lead isotopes in sediments in the ancient harbour of Naples has indicated that the eruption of Vesuvius in AD 79 destroyed the existing lead pipe network that supplied water to the city, necessitating the construction of a replacement (which took several decades).

Another example of the use of isotopes by archaeologists was reported in 2023. An ancient iron arrowhead was found in the 19th century at a Bronze Age site at

Mörigen in northwest Switzerland and placed in Bern History Museum. It is largely iron, with a high concentration of nickel (around 8%) typical of iron meteorites. Gamma spectroscopy revealed the presence of the isotope ^{26}Al; natural aluminium is monoisotopic, being 100% the non-radioactive ^{27}Al, whilst ^{26}Al is radioactive, with a half-life of 7.4×10^5 years, providing further evidence for the meteoric origin of the iron. In the days before people learned how to extract iron from natural ores, meteors were used as a source of metallic iron. Three meteorites are known to have a chemical composition similar to that of this arrowhead, found at: Bohumilitz (Czech Republic), Retuerte de Bullaque (Spain) and Kaalijarv (Estonia). It is believed that Estonia was the source in this case, raising questions about long-distance trade links between Switzerland and Estonia.

COINAGE AND ISOTOPES

The Spanish started to look for silver in South America by AD 1500 but only opened mines there in the following century. For some 300 years, around 300 tons of silver were mined annually. About 20% of that remained in America, with around 200 tons reaching Spain (largely through Seville), allowing for around 10% to be used for purchases in Asia and around 15% lost through piracy (etc.). At what point did any of this enormous amount of silver in Spain get used in the mints?

Examination of the composition of lead, copper and silver isotopes present in silver coins shed light on the timing of the incorporation of New World silver. An extensive study looked at coins from Mexico and South America of the 16th–18th centuries, comparing them with Spanish coinage from those centuries, with silver coins from the world of Classical Antiquity (Greece, Rome, etc.) and pre-1492 Spain used for reference. The combined use of Ag, Cu and Pb can distinguish between pre-1492 European silver and Mexican and Andean-sourced silver. As far as the reign of Philip III (1598–1621), Spanish silver coins used European silver – the ancient Spanish silver mines had been reopened – but by the reign of Philip V (1700–1746), these coins incorporated Mexican silver.

URANIUM ISOTOPES AND THEIR SEPARATION

Uranium occurs naturally as a mixture of four long-lived isotopes: ^{234}U (0.005%), ^{235}U (0.720%), ^{236}U a trace and ^{238}U 99.275%. Following experiments carried out by Otto Hahn in December 1938, Lise Meitner and her nephew Otto Frisch realised that uranium atoms could undergo fission with slow neutrons. News of this spread rapidly, and the following month, a team at Columbia University in the USA made the first observation of fission. In order to obtain concentrated amounts of pure ^{235}U, the fissionable isotope, methods had to be devised to separate it from the much more abundant ^{238}U. It was soon realised that gaseous diffusion was the most likely method. This required a volatile compound of uranium, with the hexafluoride UF_6 being the most suitable, as it is volatile, subliming at just over 50°C, at atmospheric pressure. In the diffusion process, gaseous UF_6 is made to diffuse under pressure through porous membranes. The lighter $^{235}UF_6$ molecules diffuse slightly faster than the $^{238}UF_6$ molecules, allowing the two isotopes to be separated, thus enriching the uranium-235.

The separation factor α^* is defined as $\sqrt{M_r(^{238}UF_6)/M_r(^{235}UF_6)} = \sqrt{352/349} = 1.0$ 0429.

Because there is only a small difference in the molecular mass of the two UF_6 molecules, there is only a small amount of separation in one stage, so that in practice around 3,000 cascaded-linked stages have been used to achieve up to 90% enrichment in ^{235}U. The process is very energy-intensive because of the energy needed to push the gas through the membranes. The compound UF_6 is extremely reactive, being hydrolysed readily by even traces of water, one product being the highly toxic hydrogen fluoride. However, one factor favouring the use of UF_6 is that fluorine has only one stable isotope, simplifying the separation.

This process is important for obtaining the enriched uranium used in both nuclear power stations and in nuclear weapons, which derive energy from the fission of uranium-235 atoms. Neutrons can split uranium-235 atoms (fission) into two smaller atoms and release more neutrons. Thus, one reaction that can occur when a neutron (having the right energy) splits a uranium-235 atom is as follows:

$$^1n + {}^{235}U \rightarrow {}^{140}Ba + {}^{93}Kr + 3^1n$$

These three neutrons can potentially go on to split three more uranium atoms, producing nine more neutrons; this can continue to produce a self-sustaining chain reaction.

CARBON ISOTOPES

98.9% of the carbon atoms present in a sample of a carbon compound are the 'normal' isotope, ^{12}C. Aside from a mere trace of the radioactive isotope, ^{14}C, the remaining 1.1% is comprised of the isotope ^{13}C.

Unlike ^{12}C, the isotope of mass 13 has a nuclear spin ($I = \frac{1}{2}$). This makes it amenable to study by NMR spectroscopy. Compared with ^1H, which has an abundance of nearly 100%, ^{13}C suffers from the disadvantage of its low abundance, but with modern spectrometers, that is not a problem; it just requires longer to acquire a spectrum. Separate signals are shown for each carbon environment (Figure 10.1).

There are eight lines, as each carbon atom in the vanillin molecule (10.5) is in a different environment from all the others.

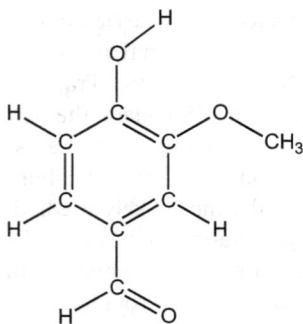

(10.5)

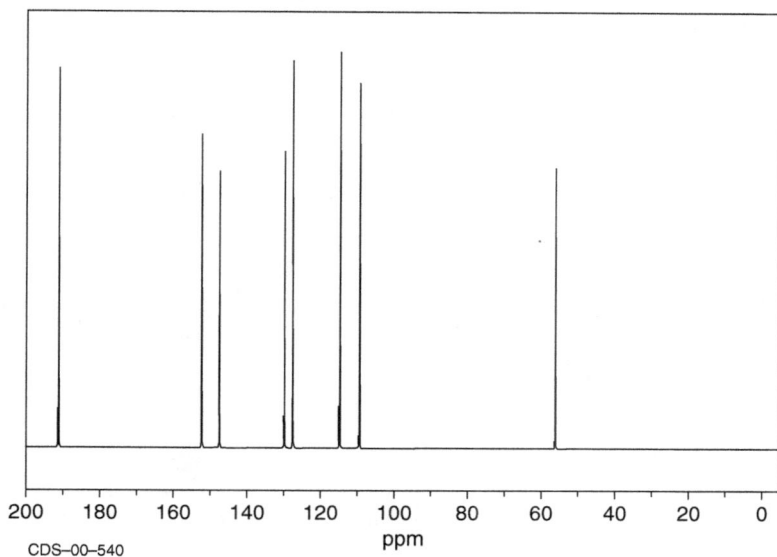

FIGURE 10.1 Proton-decoupled ^{13}C NMR spectrum of vanillin. From sdbs: https://sdbs.db.aist.go.jp/Disclaimer.aspx (National Institute of Advanced Industrial Science and Technology, accessed 18 December 2024).

In the spectrum, the most upfield line at 56.1 pm is caused by the $-O-CH_3$ group, whereas the most downfield, at 191 ppm, is due to the carbon in the aldehyde (–CHO) group. The six aromatic ring carbons give rise to the other six lines, between 109 and 152 ppm.

DETECTING FOOD FRAUD

Vanilla is a very popular flavouring material and is also in demand as a perfume ingredient. Vanilla pods only contain 1%–2% vanillin, its key ingredient, and demand for vanilla has vastly outstripped the supply of natural vanilla. Consequently, vanillin is increasingly obtained from a range of cheaper synthetic routes, such as lignin in wood, glucose (corn, potatoes, rice) or petroleum-derived guaiacol, amongst others. Currently, around 99% of vanillin is synthetic, rather than 'natural'. 'Natural' vanilla bears a substantial premium; the genuine vanilla extract obtained from vanilla pods is up to 200 times more expensive than counterfeit vanilla essence made using synthetic vanillin. Discriminating between genuine vanilla extract and synthetic materials is vital to prevent fraud. Of course, in addition to vanillin, there are hundreds of other molecules present in natural vanilla extract (such as 4-hydroxybenzaldehyde (10.6), 4-hydroxybenzoic acid (10.7) and vanillic acid (10.8)). So some fakes can be detected simply by chromatography, but detection methods have had to keep pace with the cunning of fakers.

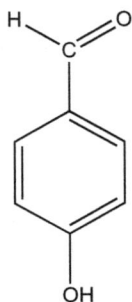

(10.6)

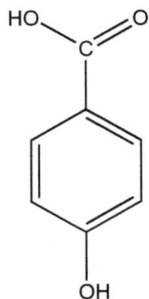

(10.7)

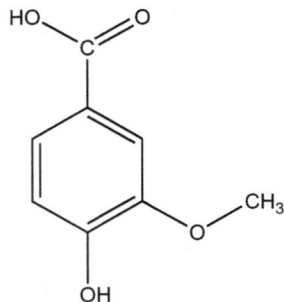

(10.8)

The first method used to detect 'fraudulent' vanillin molecules relied on radio-activity. Vanillin made from petrochemicals, obtained from crude oil that had been formed millions of years ago, had lower levels of ^{14}C (radiocarbon) than vanillin obtained from fresh vanilla pods. However, fakers circumvented this by adding extra ^{14}C to their vanillin.

Plants make vanillin via a biochemical pathway that results in a higher $^{13}C/^{12}C$ ratio than that found in synthetic vanillin. Consequently, analysts examined the $^{13}C/^{12}C$ ratios in the samples using mass spectroscopy. However, counterfeiters evaded detection by incorporating vanillin molecules with extra ^{13}C into their fraudulent "vanilla extract", so that their vanillin samples matched the "natural" ratio.

Carbons outside the aromatic ring are easier to introduce. Because of this chemical inequivalence between the carbon atoms in the aromatic ring and those carbon atoms that are substituent groups, the distribution of the ^{13}C atoms in 'faked' vanillin is non-uniform, with greater numbers found in the aldehyde and methoxy substituent positions (**emboldened** in (10.9)). Therefore, spectroscopists have been able to counteract this by using site-sensitive ^{13}C NMR techniques (SNIF-NMR), and this method has been further developed.

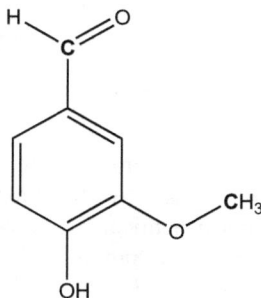

(10.9)

RADIOCARBON

The radioactive isotope ^{14}C has its own particular application. It is generated continuously in the Earth's atmosphere by cosmic ray bombardment. This generates neutrons, which strike nitrogen atoms, the main component of the atmosphere, causing a nuclear reaction that forms ^{14}C:

$$n + {}^{14}_{7}N \rightarrow {}^{14}_{6}C + {}^{1}_{1}H$$

These carbon atoms combine with oxygen, forming carbon dioxide, which is incorporated by plants through photosynthesis. In turn, animals eat plants, thereby acquiring ^{14}C. Consequently, all living things have a certain level of ^{14}C; once they die, the radioactive carbon atoms start to decay. By measuring the amount of ^{14}C in a piece of wood or bone, one can calculate the time since the animal or tree died. The isotope has a half-life of approximately 5,730 years, so it is possible to measure dates as far back as about 50,000 years ago. This is the 'basis of radiocarbon dating', devised at the University of Chicago by Willard Libby, for which he received the 1960 Nobel Prize in Chemistry. In practice, there has not been a consistent level of ^{14}C in the atmosphere throughout time. One recent cause of variation was the testing of nuclear weapons above ground from around 1950 to 1963, which led to a sharp increase in the amount of atmospheric ^{14}C. Since then, levels have been decreasing steadily as carbon is shared with the biosphere on land and in the oceans. Atmospheric levels have also varied further back, so calibration graphs are necessary to relate the dates given by radiocarbon testing to a 'calendar age'.

FRAUDULENT WHITE TRUFFLES

Nowadays, truffles sit at the expensive end of the food spectrum. They possess a particular aroma, and their presence confers distinction upon gourmet products. The best-known truffles are the Italian *Tuber magnatum* Pico ("Alba white truffle") and the French black truffle *Tuber melanosporum* Vittad, a product of Périgord.

Truffles derive their unique aromas from sulfur compounds. In the case of the Italian white truffle, the special compound is bis(methylthio)methane (10.10).

$$(10.10)$$

The food industry uses bis(methylthio)methane to 'improve' the aroma of commercial food products. The compound may come from natural sources or be synthetic bis(methylthio)methane, which owes its identity to the petrochemical industry (and which, in fairness, is a WHO-approved food additive). There is a 'premium' in price of around 50 times in favour of naturally sourced material.

In 2018, a group of eight Italian scientists reported measurements made on 24 samples of fruiting bodies of Italian-grown *Tuber magnatum* Pico and upon 13 commercial food products with truffle aromas available locally, examining the $^{13}C/^{12}C$ ratio for both. The genuine truffles had $\delta^{13}C$ ratios in the range -33.9 to $-42.6‰$. For most of the 'commercial' truffle samples, $\delta^{13}C$ ratios fell between -36.7 and $-43.7‰$, indicating that they had natural origins; however, four of the samples had $\delta^{13}C$ ratios between -53.9 and $-66.5‰$, typical of petrochemical origin.

You can measure the degree of isotope fractionation using the δ (delta) scale, which compares the isotope ratio in the sample material to the isotope ratio in a standard (originally based on Pee Dee Belemnite limestone):

$$\delta^{13}C‰ = \frac{(^{13}C/^{12}C)_{sample} - (^{13}C/^{12}C)_{standard}}{(^{13}C/^{12}C)_{standard}} \times 1,000$$

ISOSCAPES

Studies of isotopes are not just of interest to academics. We have seen that they are employed in other settings, by archaeologists, and they are starting to be used in helping to solve crimes.

One important factor in this has been the discovery of what are called 'isoscapes'. It is now known that the relative abundances of lighter isotopes – such as ^1H and ^2H; ^{12}C and ^{13}C; ^{14}N and ^{15}N; ^{16}O and ^{18}O – vary from one part of the world to another due to "isotopic fractionation"; maps of the isotopic landscapes are called "isoscapes".

Around 96.6% of the world's supply of water is in the oceans, making it the main reservoir of natural hydrogen. Isotopic fractionation is associated with the evaporation, condensation and precipitation of water, so that factors like latitude, altitude, temperature and distance from the ocean where the water evaporated lead to isotopic variations in a particular water sample.

The abundance of ^{13}C and ^{15}N in food can reflect what a person eats (i.e. meat-eater or vegan), just as the abundance of ^2H and ^{18}O can likewise reflect the water we drink or use to make our meals. Thus, if you examine the hydrogen and oxygen isotopic ratios in a bottle of mineral water, the amounts of ^2H and ^{18}O in particular can tell you if that water comes from the part of the world that the label says – the 'genuine article' – or if it is straight from the shopkeeper's tap.

CARBON IN FOODS

Carbon isotope distributions depend upon factors like the different photosynthetic pathways that affect carbon assimilation and fixation. So in sugar beet, typical of cooler climates, carbon fixation proceeds by a C_3 pathway involving 3-phosphoglycerate, whereas in plants grown in hotter climates, like sugar cane, a C_4 dicarboxylic acid is involved. Because of the involvement of plants in the human diet, this leads to a detectable difference in the carbon isotopic makeup of skin and hair in North Americans and South Africans, compared with Central Europeans.

STRONTIUM ISOTOPES

Strontium isotope ratios, more particularly $^{87}Sr/^{86}Sr$, reflect the minerals in contact with – and exchanging with – the water in an area where crops are grown and have thus been used to identify where marijuana has been grown, as well as in mineral waters.

ISOTOPES AND DRUGS

By measuring the ratios for isotopes of N and C, including ^{13}C and ^{15}N, in a sample of cocaine (2000), it is possible to tell from which region of the Andean ridge spanning the growing areas of Peru, Bolivia and Colombia it originates (2000), whether from Bolivia's Chapare valley, Peru's Huallaga/Ucayali and Apurimac valleys (Peru), or Colombia's Guaviare and Putumayo–Caqueta regions, particularly when used in conjunction with the abundance of the minor alkaloids truxilline and trimethoxycocaine (Figure 10.2).

Subsequently (2016), by incorporating additional isotopes (^2H and ^{18}O) into the investigation, the region of study was extended from five to nineteen growing regions, allowing for a more precise identification of the source.

Likewise, forensic scientists have employed isotopic ratios in samples of heroin to determine which of the four principal poppy-growing regions of the world a sample of heroin originates from, similarly using C and H isotope ratios to ascertain whether American marijuana seizures were grown in Canada, Mexico or the USA.

As far as synthetic drugs like MDMA (Ecstasy) are concerned, it has been discovered that carbon and nitrogen isotope ratios in MDMA tablets vary from batch to batch, allowing seized samples to be 'matched' with those from a clandestine lab and presented as evidence in court.

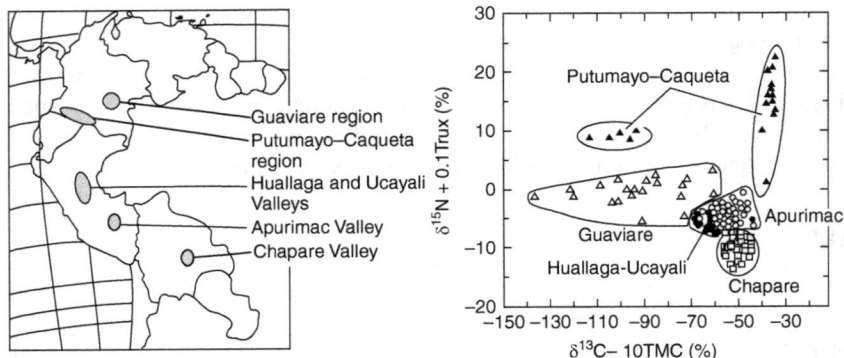

FIGURE 10.2 Identification of geographic regions in South America where coca is commonly grown. (a) Regions producing illicit cocaine; (b) Identification of cocaine-growing regions based on a combined model derived from carbon- and nitrogen-isotope ratios, as well as the abundance of minor alkaloid components truxilline and trimethoxycocaine. Squares represent Bolivia; triangles represent Colombia; and circles represent Peru. Regions within a country are distinguished by black and white symbols. Trux refers to truxilline; TMC refers to trimethoxycocaine. From J. R. Ehleringer, J. F. Casale, M. J. Lott and V. L. Ford, *Nature*, 2000, **408**, 311–312.

KING RICHARD III

On 22 August 1485, two armies met near Market Bosworth in Leicestershire in what was to be the last battle in the Wars of the Roses, deciding the crown of England. The armies were commanded by King Richard III (1452–1485) and Henry Tudor, Earl of Richmond (1457–1509). Richard died fighting bravely in battle; his body was brought to the Grey Friars (Franciscan) church in Leicester for burial. Following the dissolution of the monastery in 1538, this church was demolished, and the site of the burial was lost, not to be rediscovered until excavations in August 2012.

Isotopic analyses were carried out on the skeleton recovered as Richard's, studying ^{13}C, ^{15}N, ^{18}O and $^{87}Sr/^{86}Sr$, to see if the information matched what is known about Richard's life. The analyses examined bioapatite and collagen from two teeth (formed during youth and adolescence, but once formed, having a fixed composition, unlike bones) and from two bones: a femur (which reflects long-term conditions) and a rib (which gives information about the last few years of life).

^{18}O data at the age of around 3 suggested residence in the East of England, consistent with his known birth at Fotheringhay in Northamptonshire, but then shifted, with values around the age of 7 implying that he had moved to western England, in line with Richard being known to live at Ludlow, almost on the border with Wales, in 1459. The $^{87}Sr/^{86}Sr$ ratios are also consistent with this. Subsequently, the oxygen isotope data from his teeth and femur shifted back to values typical of eastern England, in line with known biographical data.

^{13}C and ^{15}N isotope data provided information about Richard's diet. Changes in the nitrogen isotope values towards the end of his life were believed to be caused by consuming more 'luxury items' like game birds and freshwater fish. In his later years, there were also changes in his oxygen isotope values; as he did not move home at that time, this was thought to be due to increased wine consumption, again consistent with his status in life.

ISOTOPES IN SOLVING CRIMES

THE WELSH CASE

One day in November 2006, a badly wounded young man was dumped at the Accident and Emergency (A&E) department of a hospital in South Wales. The hospital staff were unable to speak to him, and he soon died from his stab wounds. Who was he? The police had no documentation to go on; his fingerprints did not match any UK records, and his DNA could not be identified either. UK immigration records were no help, though his facial features were Asian, suggesting he might have been a recent immigrant.

A 14.5 cm length of hair from the man's head, representing about 14.5 months' growth, was sent to Professor Wolfram Meier-Augenstein, who cut it into short sections of 5 mm, each corresponding to 2 weeks of growth, and carried out ^{2}H and ^{18}O isotopic analyses on each, matching their findings with international data.

The data showed that he had only been in the UK for 2–2½ months leading up to his death. Further analysis indicated that in the last 15 months of his life, the deceased had lived in three distinctly different regions: first, for nearly 3 months in Eastern Europe, either Ukraine or Poland, followed by a move to Central Europe, possibly Germany, where he remained for 6–7 months, and finally a staged move over roughly a 2½ month period, ending up on the western side of the UK, where he spent the last couple of months of his life. A match with the fingerprint database of the police in Germany identified the victim as Tran Nyugen, a Vietnamese national, who had been fingerprinted in connection with a crime there. The route he had taken to reach the UK matched the activities of a Vietnamese people-smuggling gang, as other pieces of police evidence fell into place. It transpired that the victim had been smuggled illegally into the UK and had been put to work by the gang as a cannabis farmer to pay off the debt owed to them (around £30 000). However, the cannabis crop was stolen by a rival gang, which led Tran Nyugen's smugglers to accuse him of being complicit in the theft and to torture him for information. In 2008, two gang members were found guilty of the murder of Tran Nyugen and were jailed.

THE 'SCISSOR SISTERS' CASE

One March evening in 2005, some schoolboys playing next to Dublin's Royal Canal discovered the body of a man; it had been dismembered, with the arms and legs separated from the trunk, and the legs had also been cut in half at the knees. His head was never recovered. The body bore 21 stab wounds, which appeared to be the cause of death. On forensic examination, the man was identified as being of African, Afro-Caribbean or Afro-American descent, although this was not obvious at first sight due to the decomposition of the exposed skin pigmentation.

At this point, the Irish police (the Garda) had no further information to rely on, as the fingerprints did not match any on file in either the Irish database or that of Interpol. No 'missing person' report had been filed that matched the victim. It was therefore decided to see if 'stable isotope profiling' could provide clues that would focus the investigation. Isotope analysis examined samples from the victim's hair and fingernails, particularly examining ^{2}H, ^{13}C, ^{15}N and ^{18}O in particular. Because the body obtains nearly all of its hydrogen content from water, either directly or via food, this strongly correlates with the water supply. Carbon and nitrogen are likewise diet-based, so their composition in hair or nails, which are fast-growing, indicates one's location at particular times.

The victim's isotope signatures for ^{2}H, ^{13}C, ^{15}N and ^{18}O in their hair and finger-nails were typical of people residing in Ireland, consistent with them being a Dublin resident. The nitrogen value implied that they were an omnivore, leaning towards a meat-protein-rich diet, while the carbon signature showed no sign of C_4-derived carbon sources, suggesting that the victim had not recently been in North America or Africa.^{18}O data was also obtained from the victim's femur, which 'remodels' approx-imately every 25 years. The data from the 'older' part of the femur matched values obtained for hot, low-altitude coastal regions near the Equator, such as the Horn of Africa. Examining data for the 'newer' part of the femur suggested that the victim had lived in Ireland for 6.3 ± 2.9 years. This information was passed on to the police, who made enquiries among Dublin's African community. With the help of DNA evidence, the victim was identified as a 39-year-old Kenyan named Farah Swaleh Noor, who had immigrated to Ireland 7 years before his death. He was the boyfriend of a Dublin woman, Kathleen Mulhall. In 2006, Mulhall's daughters, Charlotte and Linda (who were dubbed the "Scissor Sisters"), were convicted in court of killing Noor and dismembering his body; one received a life sentence for murder, while the other received a 15-year sentence for manslaughter.

A UNSOLVED CRIME.... FIVE THOUSAND YEARS AGO

A death that will not find a solution in court was reported in February 2024. The skeleton of 'Vittrup Man', a 30–40-year-old Neolithic man, was discovered in a peat bog in Northwest Denmark in 1915. His severely fragmented cranium, the result of at least eight blows, indicated that he died a violent death, quite possibly as a human sac-rifice. ^{14}C dating of his skeleton indicated that he died between 3300 and 3100 years BC. Genetics indicated that he looked different from modern Scandinavians, being rather short with darker skin and hair.

Strontium (^{87}Sr/^{86}Sr) and oxygen isotope (δ^{18}O) analyses were conducted using enamel from an upper premolar and a lower molar 'wisdom tooth'. The former tooth was formed during years 2–7, while the latter tooth was formed during years 9 to 14 of the 13th century. These values are out of sync with the contemporary Funnel Beaker Culture epoch in Denmark; this and other data suggest that he was born on the Scandinavian Peninsula, possibly along the far north Norwegian coast by the Arctic Circle. At the end of his teenage years, there was a sharp drop in the δ^{13}C and δ^{15}N incremental isotope values in his teeth, resembling changes in his bone collagen, which simultaneously dropped, approaching the values seen in his bone collagen and suggesting that he had switched from a fisher-hunter-gatherer diet to a farmer's diet. In other words, during his latter adult years, he lived in a farming society that was several hundred miles from his childhood home. Was he an immigrant? Was he a slave? No one knows.

11 Methane

INTRODUCTION AND HISTORICAL BACKGROUND

This chapter concerns essentially one molecule, methane. One could say that it was a problem solver; but then, the problem changed. Molecules do not come much simpler than the five-atom CH_4, which is the simplest hydrocarbon of all.

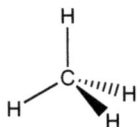

$$(11.1)$$

It has a very regular structure (11.1). The H–C–H bond angles are $109°28'$, as expected for a regular tetrahedron, whilst the bond lengths are identical at $1.086\,\text{Å}$ (108.6 pm).

Methane is ubiquitous. In the universe, for example, it is found in the atmospheres of giant planets and their moons; in comets; and in the ice mantles of interstellar grains. On Earth, methane is formed, along with other gaseous hydrocarbons and petroleum (crude oil), through the anaerobic decomposition of tiny sea animals (e.g. plankton) over a period of millions of years at elevated temperatures, under the pressure of sedimentary strata. The waxy kerogen, the initial decomposition product, subsequently breaks down into smaller molecules, including methane, the smallest of all hydrocarbons, in a natural version of the 'cracking' process. Not only is methane found underground, it also occurs under the seafloor as methane hydrates – these are essentially ice with methane molecules locked up inside 'holes' in the ice structure. These hydrates are also found in permafrost. At one time, there was serious concern that global warming could lead to the decomposition of substantial amounts of methane hydrates, resulting in a catastrophic release of methane, but this view is now discounted. However, methane is generated by other means, with the greatest amounts coming from wetlands (a quantity increased by climate change). Large amounts of methane are released from long-term flooding of rice paddy fields, rice being a major cereal crop for half the earth's population. Animals, particularly ruminants like cattle, not to mention termites, also release methane, and it is also formed in organic waste, whether manure, sewage or landfills.

The Italian scientist Alessandro Volta is usually credited with being the first to collect methane, on Lake Maggiore in November 1776, where he discovered that this gas was flammable. Subsequently, he established that it was formed by the decay of organic matter.

Others were to repeat Volta's collection of methane (known as 'marsh gas'). There is a famous painting of the British chemist John Dalton (1766–1844), a proponent of atomic theory, collecting marsh gas with young people near Manchester. Methane

DOI: 10.1201/9781003658115-11

had long posed an explosive threat in coalmines; it was traditionally known as 'firedamp'. It was present in certain types of coal, absorbed within them and released when the coal was cut into smaller pieces.

This highly flammable gas was ignited by sparks or flames, and the resulting explosions caused many fatalities. Sir Humphry Davy investigated the mine disaster at Felling in County Durham (1812), which resulted in 92 deaths. Davy found that 'firedamp' was largely methane; he went on to develop his significant contribution to mine safety, the Davy lamp (1815). This contained a wick surrounded by a mesh screen, allowing light to escape, enabling methane to enter (and be burned), while preventing the flame of the burning wick from coming into contact with firedamp outside. Others, including the railway pioneer George Stephenson, produced similar devices at that time.

WOOD, COAL AND COAL GAS

Wood is a simple and accessible fuel when you want to keep warm. Just cut the tree down.

It was the fuel used in England and in many other countries until the later Middle Ages, by which time many of the forests had been cut down. Its place was taken by the harder-to-extract coal, which required mining. Coal had been mined in Britain since the Middle Ages. You can see this in place names – Coleorton in Leicestershire, showing the connection with the local coal mines, had its name by 1443, which thus serves as a *terminus ante quem* for coal mining in that area. The great Industrial Revolution in England during the 19th century was made possible by coal. Across the world, coal had been used as a source of heat for thousands of years. It produced more heat and was cleaner than burning wood. One important use of coal was the making of iron, another key material for the development of industry. Originally, charcoal was the essential reducing agent to convert iron oxide into iron, but in 1709, Abraham Darby developed a blast furnace using coke (made by heating coal in the absence of air) that permitted large blast furnaces, leading to the production of large amounts of iron at a lower price. Coal came to have an important role in the propulsion of transport, notably the newly developed steam railways – the pioneering Stockton & Darlington and Liverpool & Manchester lines were opened in the 1820s, leading to a rapid increase in new railway lines in the following decades. Gas works generated domestic gas supplies from coal, as well as chemicals for the new chemical industry (the distillation of coal yielding tar, pitch, coke and other chemicals had been patented in 1781). These chemicals included ammonia (convertible into fertilisers), aromatic hydrocarbons (benzene and xylenes), phenolic compounds, nitrogen-containing bases like pyridines and a wide range of polycyclic aromatic hydrocarbons – such as anthracene and naphthalene. As a source of polycyclic aromatic hydrocarbons, coal is unmatched to this day. So, coal led to a wide range of products (just think of coal tar soap, for a familiar example).

From the beginning of the 19th century, coal gas (also known as 'town gas') developed commercially in the UK, to the extent that by the middle of the century towns had their own 'gas works'. In the process of coal gasification, hot coal reacts with oxygen and steam to produce a mixture of carbon monoxide, carbon dioxide and hydrogen. Besides the obvious benefits to industry, the widespread availability of piped gas by the end of the 19th century greatly benefitted society – gas lamps meant that streets after dark were 'safe'; reading and writing in the home after dark became possible, as did meetings of all kinds. The downside of coal gas was that its carbon monoxide content made it very toxic – would-be suicides could simply stick their head in a gas oven and turn on the gas, perhaps most famously the poet Sylvia Plath in 1963 (she was the wife of Ted Hughes, who later became England's Poet Laureate). You can't do that with natural gas (or a microwave).

COAL, THE ENVIRONMENT AND SMOGS

But all this led to massive environmental effects. The 'blot on the landscape' of all this coal-fired industry (and gas works in particular) was almost taken for granted. A single seismic occurrence did much to drive the move away from coal. London had long been famous for its dense fogs. These fogs were celebrated in the Victorian novels of Dickens and Conan Doyle and have become a stereotype. They were, in fact, not confined to London, being found in other industrial areas, such as the West Midlands and the North, where the same key ingredients were also present. These dense fogs often reduced visibility to a few yards (as this author can attest), making driving a bus or car virtually impossible. These 'smogs' (a combination of 'smoke' and 'fog') were due to smoke produced by industrial and domestic fires, which, apart from particulate matter, also contained toxic gases such as oxides of sulfur and nitrogen, as well as ozone. Smoke also resulted from coal-fired power stations generating electricity, as well as vehicle exhausts, including railway locomotives (at that time largely steam-driven, from coal). Ironically, London had just replaced its electric tram system with diesel-powered buses. From Friday, 5 December 1952, to Tuesday, 9 December 1952, a particularly dense smog settled over a very cold London. People were burning coal to stay warm. The smog infiltrated indoors too – it was reported that on 6 December, a performance of the opera La Traviata was abandoned because the audience could not see the stage. At that time, Britain was exporting its best-quality coal as a means to generate revenue to pay off national debts, resulting in domestic coal being low-quality and high in sulfur content, contributing to elevated levels of toxic SO_2. Total mortality rates for December 1952 were some 80% higher than during the corresponding period in the previous year. Official estimates over the following weeks suggested that as many as 4,000 people had died as a result of the smog, with 100,000 ill due to respiratory infections. More recent analysis suggest that the number of fatalities was an underestimate, with 10,000 to 12,000 being a more likely figure. This event is believed to have been the worst example of air pollution in the history of London and, indeed, Britain. It led to the passing of the Clean Air Act 1956 (4 & 5 Eliz. 2 c.52). This actively promoted smokeless fuels, particularly with the aim of reducing SO_2 levels, as well as levels of particulate matter.

NATURAL GAS AND METHANE

Just when the country was looking for a new fuel, natural gas – which was primarily methane – emerged on the scene. In 1959, a test shipment of liquefied natural gas from the USA was unloaded at Canvey Island in Essex. Canvey Island was subsequently chosen as the location for a pilot scheme in which homes were converted to run on natural gas. Then, in the mid-1960s, regular imports of liquefied natural gas began arriving from Algeria (the terminal at Canvey closed in 1979). In 1965, natural gas was discovered under the North Sea by the drilling rig *Sea Gem*; a terminal at Bacton in Norfolk started to receive gas in 1968 (in place of the Algerian supplies). Major discoveries of new oil and gas fields were made throughout the 1970s and 1980s, and operations received a boost from the substantial increase in prices following the 1973 Arab–Israeli war in the Near East. Following a White Paper in 1967, it was decided to carry out a rapid transition from coal gas to natural gas on a national basis, with domestic gas cookers and heaters being converted to burn natural gas (by fitting new burners) throughout the 1970s, with the last few UK gas works (in Northern Ireland) closing in 1987. Consequently, the 'smogs' became a thing of the past.

This 'dash to gas' was intended to enable the country to benefit from this new 'indigenous energy source', as a resource for homes and industry, as well as for electricity generation. There is a cogent argument that Britain extracted its share of North Sea gas (and oil) too quickly, resulting in a depressed price compared to what could have been obtained with slower extraction – and longer-lived reserves. UK Natural Gas production peaked in the year 2000, at the equivalent of about 1.8 million barrels per day – oil production hit two peaks, in the mid-1980s and late 1990s. Currently, it is around 0.5 million barrels a day.

So, with methane as the new 'clean' fuel, things seemed much better, until growing awareness of global warming was linked to methane's global warming potential and its release from various sources.

METHANE, A GREENHOUSE GAS

Views have changed regarding methane in recent years, largely due to the discovery that it is a 'greenhouse gas'. A greenhouse gas contributes to the 'greenhouse effect', raising the surface temperature of the Earth. These gases absorb infrared radiation emitted by the Earth and trap heat in the atmosphere, causing the Earth's temperature to rise. The main greenhouse gases are water vapour, carbon dioxide, methane, nitrous oxide and ozone. Other greenhouse gases of concern include chlorofluorocarbons (CFCs) and related molecules (see Chapter 8, pp. 174–177 in particular) and sulfur hexafluoride (SF_6).

In fact, the initial observation linking carbon dioxide with global warming was made by Svante Arrhenius in 1896, so the 'greenhouse effect' is not as modern as some might assume. Methane has only recently become an important contributor to this debate. Scientists have measured concentrations of methane using samples of ice from Greenland – these are laid down at a consistent known rate, and by taking core samples, the methane concentrations can be measured at various depths, corresponding to particular times in the past. From 20,000 BC to about 1580 AD, the concentration of methane was essentially constant, but after the latter date, it has risen – particularly fast since around the time of World War I. This can be explained by factors such as increased numbers of herbivores (e.g. cattle) and the cultivation of rice.

Individual gases are assigned a 'Global Warming Potential', which measures their ability to absorb radiation and their atmospheric lifetime. Compared to carbon dioxide, methane has a Global Warming Potential about 25 times greater, but it has a much shorter atmospheric lifetime (which means that restricting emissions of methane could have a more rapid effect on its contribution to global warming). Methane has been said to be responsible for almost a third of the rise in global temperature since the Industrial Revolution. As a result of this, emissions of methane, both actual and potential, have been scrutinised. As already mentioned, this includes emissions from wetlands, including paddy fields; its release from animals such as cattle; and methane hydrates in the oceans. Emissions from permafrost have attracted particular attention. Covering nearly a quarter of the land area in the Northern Hemisphere, permafrost comprises a (frozen) mixture of soil, gravel, sand, ice and organic matter. The fact that it has been frozen for thousands of years means that the organic content – carbon-based, of course – has been protected from warming and decomposing. Until now. It started to thaw around a century ago, releasing carbon as carbon dioxide and methane. Because permafrost is believed to contain the largest reservoir of carbon on the planet (double what is in the atmosphere), its warming is potentially very serious. Studies of a Siberian permafrost site have indicated increased emissions of methane in response to warming. On the other hand, ice core studies of methane from the last deglaciation indicate that methane emissions then were small, suggesting that this could also be the case as the Earth warms today; moreover, it was found that methane emissions from biomass burning in the pre-Industrial Holocene age were comparable to today's values. In another area of emissions, some believe that increased methane release from methane hydrates will be balanced by increasingly efficient oxidation of methane.

So scrutiny of methane continues as the transition to non-carbon-based sources of energy progresses.

This chapter has used methane as an example of molecules that were problem-solvers until the problem changed. One of the best examples of molecules in this category is CFCs (see pp. 172–179). These were ideal for refrigeration, being low-boiling, unreactive and non-toxic, which allowed for the elimination of poisonous refrigerants such as ammonia and sulfur dioxide. At the time that CFCs were introduced, the 1930s, no one could foresee the effect of CFCs upon the ozone layer given the scientific knowledge available. Politicians, often noted for their scientific illiteracy, do not understand this – they are incapable of recognising that scientific knowledge develops.

Bibliography

CHAPTER 1 FOOD

GENERAL

T. P. Coultate, *Food: The Chemistry of Its Components*, 6th edition, Cambridge, Royal Society of Chemistry, 2016 (hereafter 'Coultate')

H. D. Belitz, W. Grosch and P. Schieberle, *Food Chemistry*, 4th edition, Berlin Heidelberg, Springer, 2009

H. McGee, *On Food and Cooking*, Harper Collins 1986; second (enlarged) edition published 2004 (hereafter 'McGee')

CARBOHYDRATES

Coultate, pp. 7–106

S. O'Connell, *Sugar: The Grass that Changed the World*, London, Virgin Books, 2004

R. V. Stock, *Carbohydrates: The Sweet Molecules of Life*, San Diego, Academic Press, 2001

R. L. Myers, *The 100 Most Important Chemical Compounds. A Reference Guide*, Westport, Greenwood Press, 2007, pp. 134–137 (glucose), 268–270 (sucrose)

AMINOACIDS AND PROTEINS

Coultate, pp. 172–230

LIPIDS

Coultate, pp. 107–171

MEAT

R. S. Ramaswamy and J. F. Richards, *Canadian Institute of Food Science and Technology*, 1982, **15**, 7–18 (poultry meat flavour review)

H. Shi and C. T. Ho, The Flavour of Poultry Meat, in: *Flavour of Meat and Meat Products*, F. Shahidi (ed.), Glasgow, Blackie, 1994, pp. 52–69

U. Gasser and W. Grosch, *Zeitschrift für Lebensmittel-Untersuchung und -Forschung*, 1988, **186**, 489–494 (volatiles from cooked beef)

U. Gasser and W. Grosch, *Zeitschrift für Lebensmittel-Untersuchung und -Forschung*, 1990, **190**, 3–8 (primary odorants of chicken broth)

C. Cerny and W. Grosch, *Zeitschrift für Lebensmittel-Untersuchung und -Forschung*, 1992, **194**, 322–325; 1993, **196**, 417–422 (odorants in roasted beef)

H. Guth and W. Grosch, *Lebensmittel-Wissenschaft & Technologie*, 1993, **26**, 171–177 (stewed beef odorants)

K. Specht and W. Baltes, *Journal of Agricultural and Food Chemistry*, 1994, **42**, 2246–2253 (fried beef volatiles)

D. S. Mottram, *Food Chemistry*, 1998, **62**, 415–424 (flavour formation in meat reviewed)

R. Kerscher and W. Grosch, *Zeitschrift für Lebensmittel-Untersuchung und -Forschung A*, 1997, **204**, 3–6 (boiled beef odorants)

N. P. Brunton, D. A. Cronin and F. J. Monahan, *Flavour and Fragrance Journal*, 2002, **17**, 327–334 (turkey volatiles)

S. Rochat and A. Chaintreau, *Journal of Agricultural and Food Chemistry*, 2005, **53**, 9578–9585 (carbonyl compounds in roast beef aroma)

S. Rochat, J-Y. de Saint Laumer and A. Chaintreau, *Journal of Chromatography A*, 2007, **1147**, 85–94 (sulfur compounds in roast beef aroma)

M. Christlbauer and P. Schieberle, *Journal of Agricultural and Food Chemistry*, 2009, **57**, 9114–9122 (beef and pork gravy)

D. D. Jayasena, D. U. Ahn, K. C. Nam and C. Jo, *Asian-Australasian Journal of Animal Sciences*, 2013, **26**, 732–742 (flavour of chicken meat review)

C. R. Kerth and R. K. Miller, *Journal of the Science of Food and Agriculture*, 2015, **95**, 2783–2798 (review, beef flavour)

M. Kozowska, M. A. Majcher and T. Fortuna, *Food Science and Technology, Campinas*, 2017, **37**, 1–7 (meat aroma)

A. Sohail, S. Al-Dalali, J. Wang, J. Xie, A. Shakoor, S. Asimi, H. Shah and P. Patil, *Food Research International*, 2022, **157**, 111385 (aroma compounds in cooked meat reviewed)

J. Bleicher, E. E. Ebner and K. H. Bak, *Molecules*, 2022, **27**, 6703 (odorants in meat reviewed)

H. Yeo, D. P. Balagiannis, J. H. Koek and J. K. Parker, *Molecules*, 2022, **27**, 6712 (odorants in beef and chicken broth)

CHEESE BIBLIOGRAPHY

McGee, 2004, esp. pp. 51–67

Coultate, esp. pp. 196–198

M. H. Tunick, *The Science of Cheese*, Oxford, OUP, 2014

A. Dalby, *Cheese: A Global History*, London, Reaktion Books, 2009

P. M. G. Curioni and J. O. Bosset, *International Dairy Journal*, 2002, **12**, 959–984 (key cheese odorants)

Y. F. Collins, P. L. H. McSweeney and M. G. Wilkinson, *International Dairy Journal*, 2003, **13**, 841–866 (lipolysis and free fatty acid catabolism in cheese)

MOLECULES IN SPECIFIC CHEESES

A. Gallois and D. Langlois, *Lait*, 1990, **70**, 89–106 (Roquefort, bleu d'Auvergne and bleu des causses)

P. Molimard and H. E. Spinnler, *Journal of Dairy Science*, 1996, **79**, 169–184 (Camembert)

L. Moio, P. Piombino and F. Addeo, *Journal of Dairy Research*, 2000, **67**, 273–285 (Gorgonzola)

M. Rychlik and J. O. Bosset, *International Dairy Journal*, 2001, **11**, 895 (Gruyère)

M. Qian, C. Nelson and S. Bloomer, *Journal of the American Oil Chemists' Society*, 2002, **79**, 663–667 (flavour molecules in American blue cheese)

G. Zehentbauer and G. A. Reineccius, *Flavour and Fragrance Journal*, 2002, **17**, 300–305 (Cheddar flavour molecules)

M. Qian and G. A. Reineccius, *Journal of Dairy Science*, 2003, **86**, 770–776 (Parmesan)

T. K. Singh, M. A. Drake and K. R. Cadwallader, *Comprehensive Reviews in Food Science and Food Safety*, 2003, **2**, 139–162 (Cheddar)

J. B. Lawlor, C. M. Delahunty, J. Sheehan and M. G. Wilkinson, *International Dairy Journal*, 2003, **13**, 481–494 (Blue cheeses)

K. Gkatzionis, R. S. T. Linforth and C. E. R. Dodd, *Food Chemistry*, 2009, **113**, 506–512 (Stilton)

F. Sympoura, A. Cornu, P. Tournayre, T. Massouras, J. L. Berdagué and B. Martin, *Journal of Dairy Science*, 2009, **92**, 3040–3048 (Saint-Nectaire)

A. Cornu, N. Rabiau, N. Kondjoyan, I. Verdier-Metz, P. Pradel, P. Tournayre, J. L. Berdagué and B. Martin, *International Dairy Journal*, 2009, **19**, 588–594 (Cantal)

BREAD

Coultate pp. 214–224

McGee, pp. 516–521, 534–550

R. G. Buttery, L. C. Ling, B. O. Juliano and J. G. Turnbaugh, *Journal of Agricultural and Food Chemistry*, 1983, **31**, 823–826 (2-acetyl-1-pyrroline and cooked rice aroma)

P. Schieberle and W. Grosch, *Zeitschrift für Lebensmittel-Untersuchung und -Forschung A*, 1985, **180**, 474–478 (flavour compounds of wheat bread crust)

P. Schieberle, *Journal of Agricultural and Food Chemistry*, 1991, **39**, 1141–1144 (2-acetyl-1-pyrroline, 2-acetyltetrahydropyridine and popcorn odorants)

K. Gassenmeier and P. Schieberle, *Zeitschrift für Lebensmittel-Untersuchung und -Forschung A*, 1995, **201**, 241–248 (aroma compounds in the crumb of French-type wheat bread)

G. Zehentbauer and W. Grosch, *Journal of Cereal Science*, 1998, **28**, 81–92; 1998, **28**, 93–96 (molecules in crust aroma of baguettes)

S. Wongpornchai, T. Sriseadaka and S. Choonvisase, *Journal of Agricultural and Food Chemistry*, 2003, **51**, 457–462 (2-acetyl-1-pyrroline in bread flowers)

A. Hansen and P. Schieberle, *Trends in Food Science & Technology*, 2005, **16**, 85–94 (sourdough volatiles and bread aroma)

B. Reuben and T. Coultate, *Chemistry World*, October 2009, **10**, 54–57 (chemistry of baking)

E. Purlis, *Journal of Food Engineering*, 2010, **99**, 239–249 (browning in bakery products)

I. H. Cho and D. G. Peterson, *Food Science and Biotechnology*, 2010, **19**, 575–582 (chemistry of bread aroma reviewed)

A. N. Birch, M. A. Petersen and Å. S. Hansen, *Food Research International*, 2013, **50**, 480–488 (bread crumb aroma profile)

A. N. Birch, M. A. Petersen, N. Arneborg and Å. S. Hansen, *Food Research International*, 2013, **52**, 160–166 (influence of yeast on bread aroma profiles)

M. N. N. Q. Izzreen, M. A. Petersen and A. S. Hansen, *Cereal Chemistry*, 2015, **93**, 209–216 (volatiles in wheat bread crumb)

J. Pico, J. Bernal and M. Gómez, *Food Research International*, 2015, **75**, 200–215 (wheat bread aroma compounds in crumb and crust reviewed)

C. Helou, P. Jacolot, C. Niquet-Leridon, P. Gadonna-Widehem and F. J. Tessier, *Food Chemistry*, 2016, **190**, 904–911 (Maillard reaction products in bread)

C. Pétel, B. Onno and C. Prost, *Trends in Food Science & Technology*, 2017, **59**, 105–123 (sourdough volatile compounds and bread)

K. Wakte, R. Zanan, V. Hinge, K. Khandagale, A. Nadafa and R. Henry, *Journal of the Science of Food and Agriculture*, 2017, **97**, 384–395 (2-acetyl-1-pyrroline reviewed)

D. Pu, H. Zhang, Y. Zhang, B. Sun, F. Ren and H. Chen, *Journal of Food Processing and Preservation*, 2019, **43**, e13933 (key aroma compounds in white bread)

POTATO

McGee, esp. pp. 302–304

P. Freedman (ed.), *Food: The History of Taste*, London, Thames & Hudson, 2007

H. D. Belitz, W. Grosch and P. Schieberle, *Food Chemistry*, 4th edition, Berlin Heidelberg, Springer, 2009, pp. 788–789

A. F. Smith, *Potato: A Global History*, London, Reaktion Books, 2011

H. Campos and O. Ortiz (eds.), *The Potato Crop. Its Agricultural, Nutritional and Social Contribution to Humankind*, Switzerland, Springer, 2020

J. A. Maga, *Food Reviews International*, 1994, **10**, 1–48 (potato flavour)

R. K. Wagner and W. Grosch, *Lebensmittel-Wissenschaft & Technologie*, 1997, **30**, 164–169

R. K. Wagner and W. Grosch, *Journal of the American Oil Chemists' Society*, 1998, **75**, 1385–1392 ('chips', aka French fries)

W. A. M. Van Loon, J. P. H. Linssen, A. E. M. Boelrijk, M. J. M. Burgering and A. G. J. Voragen, *Journal of Agricultural and Food Chemistry*, 2005, **53**, 6438–6442 ('chips', aka French fries)

J. F. Dresow and H. Böhm, *Landbauforschung - vTI Agriculture and Forestry Research*, 2009, **4**, 309–338 (raw, boiled and baked potatoes)

S. H. Jansky, *American Journal of Potato Research*, 2010, **87**, 209–217 (potato flavour)

TOMATO BIBLIOGRAPHY

R. G. Buttery, R. M. Seifert, D. G. Guadagni and L. C. Ling, *Journal of Agricultural and Food Chemistry*, 1971, **19**, 524–529 (tomato volatiles)

E. A. Baldwin, J. W. Scott, C. K. Shewmaker and W. Schuch, *Horticultural Science*, 2000, **35**, 1013–1022 (aroma compounds in tomato)

E. A. Baldwin, K. Goodner, A. Plotto, K. Pritchett and M. Einstein, *Journal of Food Science*, 2004, **69**, S310–S318 (tomato odorants)

R. Davidovich-Rikanati, Y. Sitrit, Y. Tadmor, Y. Iijima, N. Bilenko, E. Bar, B. Carmona, E. Fallik, N. Dudai, J. E Simon, E. Pichersky and E. Lewinsohn, *Nature Biotechnology*, 2007, **25**, 899–901 (monoterpene accumulation instead of lycopene accumulation improves flavour)

K. Matsui, M. Ishii, M. Sasaki, H. D. Rabinowitch and G. Ben-Oliel, *Journal of Agricultural and Food Chemistry*, 2007, **55**, 4080–4086 (cucumber flavour in ancient wild tomatoes)

E. A. Baldwin, K. Goodner and A. Plotto, *Journal of Food Science*, 2008, **73**, S294–S307 (flavour compounds in tomato)

S. Mathieu, V. Dal Cin, Z. Fei, H. Li, P. Bliss, M. G. Taylor, H. J. Klee and D. M. Tieman, *Journal of Experimental Botany*, 2009, **60**, 325–337 (flavour compounds in tomato)

D. M. Tieman, M. Zeigler, E. Schmelz, M. G. Taylor, S. Rushing, J. B. Jones and H. J. Klee, *The Plant Journal*, 2010, **62**, 113–123 (methyl salicylate and tomato flavour)

D. Tieman, P. Bliss, L. M. McIntyre, A. Blandon-Ubeda, D. Bies, A. Z. Odabasi, G. R. Rodríguez, E. van der Knaap, M. G. Taylor, C. Goulet, M. H. Mageroy, D. J. Snyder, T. Colquhoun, H. Moskowitz, D. G. Clark, C. Sims, L. Bartoshuk and H. J. Klee, *Current Biology*, 2012, **22**, 1035–1039 (optimum tomato flavour)

P. M. Bleeker, R. Mirabella, P. J. Diergaarde, A. VanDoorn, A. Tissier, M. R. Kant, M. Prins, M. de Vos, M. A. Haring and R. C. Schuurink, *Proceedings of the National Academy of Sciences of the United States of America*, 2012, **109**, 20124–20129 (epizingiberene and improved herbivore resistance in tomatoes)

The Tomato Genome Consortium, *Nature*, 2012, **485**, 635–641 (tomato genome)

H. J. Klee and D. M. Tieman, *Trends in Genetics*, 2013, **29**, 257–262 (genetic improvement in tomato flavour)

J. L. Rambla, Y. M. Tikunov, A. J. Monforte, A. G. Bovy and A. Granell, *Journal of Experimental Botany*, 2014, **65**, 4613–4623 (tomato fruit volatile landscape)

B. Zhang, D. M. Tieman, C. Jiao, Y. Xuc, K. Chen, Z. Fec, J. J. Giovannoni and H. J. Klee, *Proceedings of the National Academy of Sciences of the United States of America*, 2016, **113**, 12580–12585 (chilling tomatoes and flavour loss)

M. P. López-Gresa, C. Payá, M. Ozáez, I. Rodrigo, V. Conejero, H. Klee, J. M. Bellés and P. Lisón, *Frontiers in Plant Science*, 2018, **9**, 1855 (volatile esters with a defensive role)

M. Distefano, R. P. Mauro, D. Page, F. Giuffrida, N. Bertin and C. Leonardi, *Agronomy*, 2022, **12**, 376 (aroma volatiles in tomato fruits reviewed)

T. Van Andel, R. A. Vos, E. Michels and A. Stefanaki, *PeerJ*, 2022, **10**, e1279 (Sixteenth-century tomatoes in Europe)

T. Van Andel and A. Stefanaki, *Bauhinia*, 2023, **29**, 139–140 (16th-century tomatoes)

MUSHROOMS

A. J. MacLeod and S. D. Panchasara, *Phytochemistry*, 1983, **22**, 705–709 (volatile aroma components of cooked mushrooms)

E. Combet, J. Henderson, D. C. Eastwood and K. S. Burton, *Mycoscience*, 2006, **47**, 317–326 (C_8 volatiles in mushrooms and fungi)

R. M. Hallock, *McIlvainea*, 2007, **17**(1), 33–41 ('the taste of mushrooms')

S. Grosshauser and P. Schieberle, *Journal of Agricultural and Food Chemistry*, 2013, **61**, 3804–3813 (odorants in fried *Agaricus bisporus*)

C. D. Bertelsen, *Mushroom. A Global History*, London, Reaktion Books, 2013

P. C. Schmidberger and P. Schieberle, *Journal of Agricultural and Food Chemistry*, 2020, **68**, 4493–4506 (aroma compounds in fried shiitake mushrooms)

S-L. Wang, S-Y. Lin, H-T. Du, L. Qin, L-M. Lei and D. Chen, *Foods*, 2021, **10**, 622 (key odorants in shiitake mushrooms)

OTHER MUSHROOM VARIETIES

P. G. De Pinho, B. Ribeiro, R. F. Gonçalves, P. Baptista, P. Valentão, R. M. Seabra and P. B. Andrade, *Journal of Agricultural and Food Chemistry*, 2008, **56**, 1704–1712 (odorants of wild edible mushrooms)

H. Aisala, J. Sola, A. Hopia, K. M. Linderborg and M. Sandell, *Food Chemistry*, 2019, **283**, 566–578 (volatiles of Nordic mushrooms)

ONIONS

C. J. Cavallito and H. J. Bailey, *Journal of the American Chemical Society*, 1944, **66**, 1950–1951

C. J. Cavallito, J. S. Buck and C. M. Suter, *Journal of the American Chemical Society*, 1944, **66**, 1952–1954 (allicin and its antibacterial properties)

E. Block, The chemistry of garlic and onions, *Scientific American*, April 1985, **252**, 114–119

E. Block, S. Naganathan, D. Putman and S-H. Zhao, *Pure and Applied Chemistry*, 1993, **65**, 625–632 (onion and garlic chemistry)

Y. Tokitomo, *Nippon Shokuhin Kagaku Kogaku Kaishi*, 1995, **42**, 279–287 (volatile components of cooked onions)

S. Widder, C. S. Luntzel, T. Dittner and W. Pickenhagen, *Journal of Agricultural and Food Chemistry*, 2000, **48**, 418–423 (discovery of 3-mercapto-2-methylpentan-1-ol in onions)

S. Imai, N. Tsuge, M. Tomotake, Y. Nagatome, H. Sawada, T. Nagata and H. Kumagai, *Nature*, 2002, **419**, 685 (lachrymatory factor synthase)

M. Granvogl, M. Christlbauer and P. Schieberle, *Journal of Agricultural and Food Chemistry*, 2004, **52**, 2797–2802 (3-mercapto-2-methylpentan-1-ol in raw and processed onions)

M. G. Jones, J. Hughes, A. Tregova, J. Milne, A. B. Tomsett and H. A. Collin, *Journal of Experimental Botany*, 2004, **55**, 1903–1918 (biosynthesis of the flavour precursors of onion and garlic)

S. Widder, C. S. Lüntzel, T. Dittner and W. Pickenhagen, *Journal of Agricultural and Food Chemistry*, 2004, **52**, 2797–2802 (3-mercapto-2-methylpentan-1-ol and cooked onion flavour)

L. J. Macpherson, B. H. Geierstanger, V. Viswanath, M. Bandell, S. R. Eid, S. W. Hwang and A. Patapoutian, *Current Biology*, 2005, **15**, 929–934 (activation of TRPA1 and TRPV1 in response to allicin)

V. Lanzotti, *Journal of Chromatography A*, 2006, **1112**, 3–22 (analysis of onion and garlic)

H. Salazar, I. Llorente, A. Jara-Oseguera , R. García-Villegas, M. Munari, S. E. Gordon, L. D Islas and T. Rosenbaum, *Nature Neuroscience*, 2008, **11**, 255–261 (activation of TRPV1 by onion and garlic)

E. Block, *Garlic and Other Alliums: The Lore and the Science*, Cambridge, Royal Society of Chemistry, 2010

M. M. Løkke, M. Edelenbos, E. Larsen and A. Feilberg, *Sensors*, 2012, **12**, 16060–16076 (volatiles from freshly cut onions)

A. Villière, S. Le Roy, C. Fillonneau, F. Guillet, H. Falquerho, S. Boussely and C. Prost, *Flavour*, 2015, **4**, 24 (odour profiles of sué, sautéed, and pan-fried onions)

M. Jay, *Onions and Garlic. A Global History*, London, Reaktion Books, 2016

F. Noe, J. Polster, C. Geithe, M. Kotthoff, P. Schieberle and D. Krautwurst, *Chemical Senses*, 2017, **42**, 195–210 (OR2M3 a human odorant receptor to detect onion odorant)

S. Haasnoot, G. Koopmanschap, H. Colstee, C. Niedeveld, C. Wermes, R. Cannon and N. C. Da Costa, in: *Flavour Science: Proceedings of the XV Weurman Flavour Research Symposium*, B. Siegmund and E. Leitner (eds.), Graz, Verlag der Technischen Universität Graz, 2018, 483–486 (volatile sulfur compounds in onion oil and fresh juice)

J. Borlinghaus, J. Foerster, U. Kappler, H. Antelmann, U. Noll, M. C. H. Gruhlke and A. J. Slusarenko, *Molecules*, 2021, **26**, 1505 (allicin, the odour of freshly crushed garlic)

P. M. Wise and A. Ledyard, *Chemosensory Perception*, 2022, **15**, 70–86 (sensitivity to and taste modulation by 3-mercapto-2-methylpentan-1-ol)

H. G. Mengers, C. Schier, M. Zimmermann, M. C. H. Gruhlke, E. Block, I. M. Blank and A. J. Slusarenko, *Food Chemistry*, 2022, **397**, e54501 (gas phase volatiles from garlic and onion)

N. Alok Sagar, S. Pareek, N. Benkeblia and J. Xiao, *Food Frontiers*, 2022, **3**, 380–412 (onion bioactives reviewed)

STRAWBERRY

G. M. Darrow, *The Strawberry: History, Breeding, and Physiology*, New York, Holt, Rinehart and Winston, 1966

T. Pyysalo, E. Honkanen and T. Hirvi, *Journal of Agricultural and Food Chemistry*, 1979, **27**, 19–22 (comparison of Finnish *Fragaria vesca* with cultivated *Fragaria* x *ananassa*)

P. Schieberle and T. Hofmann, *Journal of Agricultural and Food Chemistry*, 1997, **45**, 227–232 (character impact molecules in strawberry juice)

W. Wozniak, B. Rodajewska, I. Deywori and A. Reselska-Sieciechowiz, *Acta Horticulturae*, 1997, **439**, 333–336 (cultivar and harvest date effects on flavour and other quality attributes of California strawberries)

J. F. Hancock, *Strawberries (Crop Production in Horticulture)*, New York, CABI Publishing, 1999

K. G. Bood and I. Zabetakis, *Journal of Food Science*, 2002, **67**, 2–8 (biosynthesis of strawberry flavour)

A. Aharoni, A. P. Giri, F. W. Verstappen, C. M. Bertea, R. Sevenier, Z. Sun, et al., *Plant Cell*, 2004, **16**, 3110–3131 (genetics of terpenoid production in wild and cultivated strawberries)

F. Lopes da Silva, M. T. Escribano-Bailón, J. J. P. Alonso, J. C. Rivas-Gonzalo and C. Santos-Buelga, *LWT - Food Science and Technology*, 2007, **40**, 374–382 (anthocyanin pigments in strawberry)

W. Schwab et al., *The Plant Journal*, 2008, **54**, 712–732 (biosynthesis of strawberry flavour compounds)

V. Shulaev et al., *Nature Genetics*, 2010, **43**, 109–116 (the genome of woodland strawberry, *Fragaria vesca*)

J. Pet'ka, E. Leitner and B. Parameswarana, *Flavour and Fragrance Journal*, 2012, **27**, 273–279 (musk strawberry flavour)

D. Ulrich and K. Olbricht, *Journal of Applied Botany and Food Quality*, 2013, **86**, 37–46 (*F. vesca* and *Fragaria × ananassa* compared)

G. Bianchi, A. Lovazzano, A. Lanubile and A. Marocco, *Journal of Horticultural Research*, 2014, **22**, 77–84 (aromas of wild and cultivated strawberries)

M. L. Schwieterman et al., *PLoS One*, 2014, **9**, e88446 (chemical compositions of strawberry flavour)

D. Ulrich and K. Olbricht, *Journal of Berry Research*, 2014, **4**, 11–17 (diversity of metabolite patterns and sensory characters in wild and cultivated strawberries)

A. S. Negri, D. Allegra, L. Simoni, F. Rusconi, C. Tonelli, L. Espen and M. Galbiati, *Frontiers in Plant Science*, 2015, **6**, 56 (flavour patterns in *F. moschata* and *F. vesca* compared)

K. M. Folta and H. J. Klee, *Horticulture Research*, 2016, **3**, 16032 (sensory sacrifices when we mass-produce strawberries)

Y. Song, Y-J. Zhang, N. Liu, D-Q. Ye, X. Gong, Y. Qin and Y-L. Liu, *International Journal of Food Properties*, 2017, **20**, S399–S415 (volatiles in Chinese *F. Vesca*)

D. Abouelenein, L. Acquaticci, L. Alessandroni, G. Borsetta, G. Caprioli, C. Mannozzi, R. Marconi, D. Piatti, A. Santanatoglia, G. Sagratini, S. Vittori and A. M. Mustafa, *Molecules*, 2023, **28**, 5810 (strawberry volatiles reviewed)

S. Zheng, J. Cai, P. Huang, Y. Wang, Z. Yang and Y. Yu, *Journal of the Science of Food and Agriculture*, 2023, **103**, 7455–7468 (volatiles in Chinese *F. vesca*)

ORANGES AND LEMONS

P. Laszlo, *Citrus: A History*, Chicago, IL, University of Chicago Press, 2007

T. Sonneman, *Lemon: A Global History*, London, Reaktion Books, 2012

C. Hyman, *Oranges: A Global History*, London, Reaktion Books, 2013

LIMONENE ISOMER SMELLS

L. Friedman and J. G. Miller, *Science*, 1971, **172**, 1044–1046 (likely origin of incorrect statement)

J. Clayden, N. Greeves, S. Warren and P. Wothers, *Organic Chemistry*, 1st edition, Oxford, OUP, 2001, p. 1220

P. Atkins, *Atkins' Molecules*, Cambridge, CUP, 2003, p.14 (examples of incorrect statement)

C. Sell, Scent through the looking glass, *Chemistry & Biodiversity*, 2004, **1**, 1899–1920 (correct statement on isomers of limonene)

L. Kvittingen, B. J. Sjursnes and R. Schmid, *Journal of Chemical Education*, 2021, **98**, 3600–3607 ('Limonene in Citrus: A String of Unchecked Literature Citings?')

R. M. Ikeda, L. A. Rolle, S. H. Vannier and W. L. Stanley, *Journal of Agricultural and Food Chemistry*, 1962, **10**, 98–102 (aldehydes in lemon oil)

A. Hinterholzer and P. Schieberle, *Flavour and Fragrance Journal*, 1998, **13**, 49–55 (odour-active volatiles in fresh, hand-extracted juice of Valencia oranges)

M. H. Boelens, *Perfumer & Flavorist*, 1991, **16**(2), 17–34 (composition of citrus oils, review)

R. Bazemore, K. Goodner and R. Rouseff, *Journal of Food Science*, 1999, **64**, 800–803 (volatiles from orange juice)

A. Buettner and P. Schieberle, *Journal of Agricultural and Food Chemistry*, 2001, **49**, 2387–2394 (aroma differences between hand-squeezed juices from Valencia and Navel oranges)

A. Högnadóttir and R. L. Rouseff, *Journal of Chromatography A*, 2003, **998**, 201–211 (aroma active compounds in orange essence oil)

A. Verzera, A. Trozzi, G. Dugo, G. Di Bella and A. Cotroneo, *Flavour and Fragrance Journal*, 2004, **19**, 544–548 (lemon and sweet orange essential oil composition)

A. Plotto, C. A. Margaría, K. L. Goodner and E. A. Baldwin, *Flavour and Fragrance Journal*, 2008, **23**, 398–406 (key aroma components in an orange juice matrix)

P. R. Perez-Cacho and R. L. Rouseff, *Critical Reviews in Food Science and Nutrition*, 2008, **48**, 681–695 (fresh squeezed orange juice odour reviewed)

M. Averbeck and P. H. Schieberle, *European Food Research and Technology*, 2009, **229**, 611–622 (key aroma compounds in freshly reconstituted orange juice from concentrate)

S. Deterre, B. Rega, J. Delarue, M. Decloux, M. Lebrun and P. Giampaoli, *Flavour and Fragrance Journal*, 2012, **27**, 77–88 (key aroma compounds in bitter orange)

A. Ben Hsouna, N. Ben Halima, S. Smaoui and N. Hamdi, *Lipids in Health and Disease*, 2017, **16**, 146 (citrus lemon essential oil)

S. Feng, J. H. Suh, F. G. Gmitter and Y. Wang, *Journal of Agricultural and Food Chemistry*, 2018, **66**, 203–211 (difference between flavours of sweet orange and mandarin)

M. C. González-Mas, J. L. Rambla, M. P. López-Gresa, M. A. Blázquez and A. Granell, *Frontiers in Plant Science*, 2019, **10**, 12 (volatiles in citrus essential oils, review)

Z. Fan, K. A. Jeffries, X. Su, G. Olmedo, W. Zhao, M. R. Mattia, E. Stover, J. A. Manthey, E. A. Baldwin, S. Lee, F. G. Gmitter, A. Plotto and J. Bai, *Science Advances*, 2024, **10**, eadk2051 (chemical and genetic basis of orange flavour)

CHAPTER 2 VITAMINS

GENERAL

C. Funk, *Journal of Physiology*, 1911, **43**, 395–402

C. Funk, *Journal of State Medicine*, 1913, **20**, 341–368 (the term 'vitamin' first used)

J. Schwarcz, *An Apple A Day*, Toronto, Harper Collins, 2007, pp. 129–132

H. D. Belitz, W. Grosch and P. Schieberle, *Food Chemistry*, 4th edition, Berlin Heidelberg, Springer, 2009,pp. 403–420

M. Eggersdorfer, D. Laudert, U. Létinois, T. McClymont, J. Medlock, T. Netscher and W. Bonrath, *Angewandte Chemie International Edition*, 2012, **51**, 12960–12990 ('One Hundred Years of Vitamins'—Vitamins A, B_1, B_2, B_5, B_6, C, E and H)

T. P. Coultate, *Food: The Chemistry of Its Components*, 6th edition, Cambridge, Royal Society of Chemistry, 2016, pp. 356–407

VITAMIN A

K. L. Penniston and S. A. Tanumihardjo, *American Journal of Clinical Nutrition*, 2006, **83**, 191–201 (acute and chronic toxic effects of vitamin A)

R. E. Black, L. H. Allen, Z. A. Bhutta et al, *The Lancet*, 2008, **371**, 243–260 (worldwide Vitamin A deficiency)

R. D. Semba, *Annals of Nutrition and Metabolism*, 2012, **61**, 192–198 (discovery of Vitamin A)

S. Akhtar, A. Ahmed, M. A. Randhawa, S. Atukorala, N. Arlappa, T. Ismail and Z. Ali, *Journal of Health*, Population and Nutrition, 2013, **31**, 413–423 (Vitamin A deficiency in S Asia)

R. D. Klemm, K. P. West, Jr, A. C. Palmer, Q. Johnson, P. Randall, P. Ranum and C. Northrop-Clewes, *Food and Nutrition Bulletin*, 2010, **31**(No. 1 Suppl), S47–S61 (Vitamin A fortification of wheat flour)

U. H. Lerner, *Frontiers in Endocrinology*, 2024, **15**, 1298851 (Vitamin A - discovery, metabolism etc.)

GOLDEN RICE

X. Ye, S. Al-Babili, A. Klöti, J. Zhang, P. Lucca, P. Beyer and I. Potrykus, *Science*, 2000, **287**, 303–305 (engineering beta-carotene gene into rice)

G. Tang, Y. Hu, S-A. Yin, Y. Wang, G. E. Dallal, M. A. Grusak and R. M. Russell, *American Journal of Clinical Nutrition*, 2012, **96**, 658–664 (comparing beta-carotene sources)

A. J. Kettenburg, J. Hanspach, D. J. Abson and J. Fischer, *Sustainability Science*, 2018, **13**, 1469–1482 (Golden Rice debate)

F. Wu, J Wesseler, D. Zilberman, R. M. Russell, C. Chen and A. C. Dubock, *Proceedings of the National Academy of Sciences of the United States of America*, 2021, **118**, e2120901118 (opinion on Golden Rice)

VITAMIN B_1, THIAMIN

R. R. Williams and J. K. Cline, *Journal of the American Chemical Society*, 1936, **58**, 1504–1505 (Vitamin B_1 synthesis)

Y. Itokawa, *Journal of Nutrition*, 1976, **106**, 581–588 (biography of Kanehiro Takaki)

K. J. Carpenter, *Beriberi, White Rice and Vitamin B*, Berkeley, University of California Press, 2000

M. Eggersdorfer, D. Laudert, U. Létinois, T. McClymont, J. Medlock, T. Netscher and W. Bonrath, *Angewandte Chemie International Edition*, 2012, **51**, 12960–12990

H. Hayashi, B vitamins and folate in context, in: *B Vitamins and Folate: Chemistry, Analysis, Function and Effects (Food and Nutritional Components in Focus No. 4)*, V. R. Preedy (ed.), Cambridge, The Royal Society of Chemistry, 2013

T. J. Smith, C. R. Johnson, R. Koshy, S. Y. Hess, U. A. Qureshi, M. L. Mynak and P. R. Fischer, *Annals of the New York Academy of Sciences*, 2021, **1498**, 9–28 (Vitamin B_1 deficiency disorders)

M. Hrubša, T. Siatka, I. Nejmanová, M. Vopršalová, L. K. Krčmová, K. Matoušová, L. Javorská, K. Macáková, L. Mercolini, F. Remião, M. Máťuš and P. Mladěnka, *Nutrients*, 2022, **14**, 484 (properties of vitamins B_1-B_5)

VITAMIN B_2, RIBOFLAVIN

R. Kuhn, P. György and T. Wagner-Jauregg, *Berichte der Deutschen Chemischen Gesellschaft*, 1933, **66**, 576–580 (isolation from egg yolk)

R. Kuhn, K. Reinemund, F. Weygand and R. Ströbele, *Berichte der Deutschen Chemischen Gesellschaft*, 1935, **68**, 1765–1774 (synthesis of vitamin B_2)

P. György, *Biochemical Journal*, 1935, **29**, 741–759 (differentiation of lactoflavin and the 'rat antipellagra' factor)

C. J. Bates, *World Review of Nutrition and Dietetics*, 1987, **50**, 215–265 (riboflavin in diet)

C. A. Northrop-Clewes and D. I. Thurnham, *Annals of Nutrition and Metabolism*, 2012, **61**, 224–230 (discovery and characterization of riboflavin)

K. Thakur, S. K. Tomar, A. K. Singh, S. Mandal and S. Arora, *Critical Reviews in Food Science and Nutrition*, 2016, **57**, 3650–3660 (riboflavin and health)

N. Suwannasom, I. Kao, A. Pruß, R. Georgieva and H. Bäumler, *International Journal of Molecular Sciences*, 2020, **21**, 950 (health benefits of riboflavin)

N. Olfat, M. Ashoori and A. Saedisomeolia, *British Journal of Nutrition*, 2022, **128**, 1887–1895 (riboflavin as an antioxidant)

M. Hrubša, T. Siatka, I. Nejmanová, M. Vopršalová, L. K. Krčmová, K. Matoušová, L. Javorská, K. Macáková, L. Mercolini, F. Remião, M. Mát'uš and P. Mladěnka, *Nutrients*, 2022, **14**, 484 (properties of vitamins B$_1$-B$_5$)

VITAMIN B$_3$, NIACIN

Coultate, pp. 368–370 (vitamin B$_3$ and its history)

J. Goldberg and G. A. Wheeler, *Public Health Reports*, 1915, **30**, 3336–3339

J. Goldberger, G. A. Wheeler and E. Sydenstricker, *Public Health Reports*, 1920, **35**, 2673–2714 (pellagra and diet)

C. A. Elvehjem, R. J. Madden, F. M. Strong and D. W. Woolley, *Journal of Biological Chemistry*, 1938, **123**, 137–149 (nicotinic acid cures pellagra in dogs)

C. A. Elvehjem, R. J. Madden, F. M. Strong and D. W. Woolley, *JAMA*, 1942, **118**, 823 (naming niacin)

J. D. Stratigos, A. Katsambas, *British Journal of Dermatology*, 1977, **96**, 99–106 (pellagra in 3rd-world countries and war-zones)

M. H. Gómez, C. M. McDonough, L. W. Rooney and R. D. Waniska, *Journal of Food Science*, 1989, **54**, 330–336

J. L. Fernández-Muñoz, M. E. Rodríguez, R. C. Pless, H. E. Martínez-Flores and A. J. Bollet, *Yale Journal of Biology and Medicine*, 1992, **65**, 211–221 (politics and pellagra in the U.S. Southern states)

M. Leal, J. L. Martinez and L. Baños, *Cereal Chemistry*, 2002, **79**, 162–166 (nixtalamisation)

V. S. Kamanna and M. L. Kashyap, *American Journal of Cardiology*, 2008, **101**, 20B–26B (how niacin works)

D. Prabhu, R. S. Dawe and K. Mponda, *Photodermatology, Photoimmunology & Photomedicine*, 2021, **37**, 99–104 (pellagra, a review)

M. Hrubša, T. Siatka, I. Nejmanová, M. Vopršalová, L. K. Krčmová, K. Matoušová, L. Javorská, K. Macáková, L. Mercolini, F. Remião, M. Mát'uš and P. Mladěnka, *Nutrients*, 2022, **14**, 484 (properties of vitamins B$_1$–B$_5$)

VITAMIN B$_5$, PANTOTHENIC ACID

R. J. Williams, C. M. Lyman, G. H. Goodyear, J. H. Truesdail and D. Holaday, *Journal of the American Chemical Society*, 1933, **66**, 2912–2927 (discovery)

R. J. Williams, J. H. Truesdail, H. H. Weinstock, E. Rohrmann, C. M. Lyman and C. H. McBurney, *Journal of the American Chemical Society*, 1938, **60**, 11, 2719–2723 (isolation from liver)

R. J. Williams and R. T. Major, *Science*, 1940, **91**, 246 (structure of pantothenic acid)

R. Leonardi and S. Jackowski, *EcoSal Plus*, 2013, **2**(2), 1–18 (biosynthesis of pantothenic acid and Coenzyme A)

M. Hrubša, T. Siatka, I. Nejmanová, M. Vopršalová, L. K. Krčmová, K. Matoušová, L. Javorská, K. Macáková, L. Mercolini, F. Remião, M. Mát'uš and P. Mladěnka, *Nutrients*, 2022, **14**, 484 (properties of vitamins B_1-B_5)

VITAMIN B_6, PYRIDOXIN

P. György, *Nature*, 1934, **133**, 498–499 (identification)
P. György, *Journal of the American Chemical Society*, 1938, **60**, 983–984
R. Kuhn and G. Wendt, *Berichte der Deutschen Chemischen Gesellschaft*, 1938, **71**, 780–782
S. Lepkovsky, *Science*, 1938, **87**, 169–170 (isolation)
S. A. Harris and K. Folkers, *Science*, 1939, **89**, 347
S. A. Harris and K. Folkers, *Journal of the American Chemical Society*, 1939, **61**, 1245–1247 (synthesis and structure)
S. Mooney, J-E. Leuendorf, C. Hendrickson and H. Hellmann, *Molecules*, 2009, **14**, 329–351 (vitamin B_6 reviewed)
I. H. Rosenberg, *Annals of Nutrition and Metabolism*, 2012, **61**, 236–238 (isolation and identification of vitamin B_6)

VITAMIN B_7, BIOTIN

F. Kogl and B. Tonnis, *Zeitschrift für Physikalische Chemie*, 1932, **242**, 43–73 (isolation of biotin)
V. du Vigneaud, D. B. Melville, K. Folkers, D. E. Wolf, R. Mozingo, J. C. Keresztesy and S. A. Harris, *Journal of Biological Chemistry*, 1942, **146**, 475–485 (structure of biotin)
S. A. Harris, D. E. Wolf, R. Mozingo and K. Folkers, *Science*, 1943, **97**, 447–448 (synthesis of biotin)
O. Livnah, E. A. Bayer, M. Wilchek and J. L. Sussman, *Proceedings of the National Academy of Sciences of the United States of America*, 1993, **90**, 5076–5080 (structure of avidin bound to biotin)
J. Zempleni, S. S. Wijeratne and Y. I. Hassan, *Biofactors*, 2009, **35**, 36–46 (review)
C. E. Chivers, A. L. Koner, E. D. Lowe and M. Howarth, *Biochemical Journal*, 2011, **435**, 55–63 (biotin binding to streptavidin)
D. P. Patel, S. M. Swink and L. Castelo-Soccio, *Skin Appendage Disorders*, 2017, **3**, 166–169 (biotin and hair loss)

VITAMIN B_9, FOLIC ACID

L. Wills, *British Medical Journal*, 1931, **1**, 1059–1064 (discovery)
A. V. Hoffbrand and D. G. Weir, *British Society for Haematology*, 2001, **113**, 579–589 (history of folic acid)
E. P. Quinlivan and J. F Gregory, *American Journal of Clinical Nutrition*, 2003, **77**, 221–225 (folic acid food fortification)
B. Shane, Folate Chemistry and Metabolism, in: *Folate in Health and Disease*, L. B. Bailey (ed.), Boca Raton, CRC Press, 2010, pp. 1–24
K. S. Crider, L. B. Bailey and R. J. Berry, *Nutrients*, 2011, **3**, 370–384 (folic acid food fortification)
A. E. Czeizel, I. Dudás, A. Vereczkey and F. Bánhidy, *Nutrients*, 2013, **5**, 4760–4775 (folate and folic acid: preventing neural-tube defects and congenital heart defects)

Y. Shulpekova, V. Nechaev, S. Kardasheva, A. Sedova, A. Kurbatova, E. Bueverova, A. Kopylov, K. Malsagova, J. C. Dlamini and V. Ivashkin, *Molecules*, 2021, **26**, 3731 (folic acid in health and disease)

VITAMIN B$_{12}$, COBALAMIN

E. L. Rickes, N. G. Brink, F. R. Koniusky, T. R. Wood and K. Folters, *Science*, 1948, **107**, 396–397 (isolation of crystalline vitamin B$_{12}$)

D. C. Hodgkin, J. Kamper, M. Mackay, J. Pickworth, K. N. Trueblood and J. G. White, *Nature*, 1956, **178**, 64–66 (structure of vitamin B$_{12}$)

G. Ferry, *Dorothy Hodgkin: A Life*, London, Granta Books, 1998, pp. 231–285

G. Dodson, *Biographical Memoirs of Fellows of the Royal Society*, 2002, **48**, 179–219 (memoir of Dorothy Hodgkin)

A. D. Smith and H. Refsum, *American Journal of Clinical Nutrition*, 2009, **89**(suppl.), 707S–711S (vitamin B$_{12}$ and cognition in the elderly)

J. M. Scott and A. M. Molloy, *Annals of Nutrition and Metabolism*, 2012, **61**, 239–245 (review, discovery of vitamin B$_{12}$)

A. M. Kimura, R. Kinnno, M. Tsuji and K. Ono, *Alzheimer's & Dementia Journal*, 2021, **17**(Suppl. 9), e055458 (vitamin B12 and cognitive function in the elderly)

D. Osman, A. Cooke, T. R. Young, E. Deery, N. J. Robinson and M. J. Warren, *BBA - Molecular Cell Research*, 2021, **1868**, 118896 (cobalt in B$_{12}$)

A. Sobczyńska-Malefora, E. Delvin, A. McCaddon, K. R. Ahmadi and D. J. Harrington, *Critical Reviews in Clinical Laboratory Sciences*, 2021, **58**, 399–429 (review, B$_{12}$ in health and disease)

E. Azzini, A. Raguzzini and A. Polito, *International Journal of Molecular Sciences*, 2021, **22**, 9694 (review, B$_{12}$ deficiency)

A. A. Lauer, H. S. Grimm, B. Apel, N. Golobrodska, L. Kruse, E. Ratanski, N. Schulten, L. Schwarze, T. Slawik, S. Sperlich, A. Vohla and M. O. W. Grimm, *Biomolecules*, 2022, **12**, 129 (review, B$_{12}$ and Alzheimer's)

B. H. R. Wolffenbuttel, P. J. Owen, M. Ward and R. Green, *BMJ*, 2023, **383**, e071725 (vitamin B$_{12}$ therapetics)

VITAMIN C, ASCORBIC ACID

J. Lind, *A Treatise on the Scurvy*, London, G. Pearch and W. Woodfall, 1772

M. B. Davies, J. Austin and D. A. Partridge, *Vitamin C: Its Chemistry and Biochemistry*, Cambridge, RSC, 1991

G. L. Wheeler, M. A. Jones and N. Smirnoff, *Nature*, 1998, **393**, 365–369 (biosynthesis of Vitamin C in higher plants)

J. Emsley, *Vanity, Vitality and Virility*, Oxford, OUP, 2004, pp. 53–67

R. L. Myers, *The 100 Most Important Chemical Compounds. A Reference Guide*, Westport, Greenwood Press, 2007, pp. 30–32

VITAMIN D, CALCIFEROL

D. D. Bikle, *Chemistry & Biology*, 2014, **21**, 319–329 (vitamin D metabolism, mechanism of action and clinical applications)

G. Jones, *Endocrine Connections*, 2022, **11**, e210594 (historical aspects of vitamin D)

VITAMIN E

R. Brigelius-Flohé, F. J. Kelly, J. T. Salonen, J. Neuzil and J. M. Zingg, *American Journal of Clinical Nutrition*, 2002, **76**, 703–16 (review)

C. Schneider, *Molecular Nutrition & Food Research*, 2005, **49**, 7–30 (chemistry and biology of vitamin E)

E. Reboul, M. Richelle, E. Perrot, C. Desmoulins-Malezet, V. Pirisi and P. Borel, *Journal of Agricultural and Food Chemistry*, 2006, **54**, 8749–8755 (vitamin E in cereals)

E. Niki and M. G. Traber, *Annals of Nutrition and Metabolism*, 2012, **61**, 207–212 (history)

B. C. Blount et al., *New England Journal of Medicine*, 2020, **382**, 697–705 (Vitamin E acetate and vaping–associated lung injury)

D. Wu and D. F. O'Shea, *Proceedings of the National Academy of Sciences of the United States of America*, 2020, **117**, 6349–6355 (benzene, ketene and alkenes formed on vapourising Vitamin E acetate)

R. Feldman, J. Meiman, M. Stanton and D. D. Gummin, *Archives of Toxicology*, 2020, **94**, 2249–2254 (vaping and Vitamin E acetate)

B. Soto, L. Costanzo, A. Puskoor, N. Akkari and P. Geraghty, *Annals of Thoracic Medicine*, 2023, **18**, 1–9 (Vitamin E acetate and vaping reviewed)

VITAMIN K

H. Dam, *Nature*, 1935, **135**, 652–653

H. Dam, *Biochemical Journal*, 1935, **29**, 1273–1285 (discovery)

M. J. Shearer, *Lancet*, 1995, **345**, 229–234

B. Furie, B. A. Bouchard and B. C. Furie, *Blood*, 1999, **93**, 1798–1808 (vitamin K-dependent biosynthesis of gamma-carboxyglutamic acid and clotting)

P. Weber, *Nutrition*, 2001, **17**, 880–887 (vitamin K and bone health)

D. W. Stafford, *Journal of Thrombosis and Haemostasis*, 2005, **3**, 1873–1878 (vitamin K cycle)

J. Adams and J. Pepping, *American Journal of Health-System Pharmacy*, 2005, **62**, 1574–1581 (prevention of osteoporosis)

J. Danziger, *Clinical Journal of the American Society of Nephrology*, 2008, **3**, 1504–1510 (clotting)

CHAPTER 3 SPICES, 'HOT' AND 'COLD'

SPICES

R. Dunn and M. Sanchez, *Delicious: The Evolution of Flavour and How It Made Us Human*, Princeton and Oxford, Princeton University Press, 2021

CHILLIS AND SPICES IN GENERAL

G. Charalambous (ed.), *Spices, Herbs and Edible Fungi*, Amsterdam, Elsevier, 1994

V. A. Parthasarathy, B. Chempakam and T. J. Zachariah (eds.), *Chemistry of Spices*, Wallingford, CABI Publishing, 2008

CURRIES AND BRITISH CURRY-HOUSES

S. Basu, *Curry: The Story of the Nation's Favourite Dish*, Stroud, Sutton, 2003

J. Monroe, *Star of India: The Spicy Adventures of Curry*, Chichester, John Wiley, 2004

L. Collingham, *Curry: A Biography of A Dish*, London, Chatto and Windus, 2005
C. T. Sen, *Curry: A Global History*, London, Reaktion Books, 2009

SPICE HISTORY AND THE SPICE ROUTE

C. Corn, *The Scents of Eden: A History of the Spice Trade*, New York, Kodansha America, 1998
G. Milton, *Nathaniel's Nutmeg: Or, the True and Incredible Adventures of the Spice Trader Who Changed the Course of History*, London, Hodder & Stoughton, 1999
A. Dalby, *Dangerous Tastes: The Story of Spices*, London, British Museum, 2000
C. Caldicott and C. Caldicott, *The Spice Routes: More Recipes from the World Food Café*, London, Frances Lincoln, 2001
P. Le Couteur and J. Burreson, *Napoleon's Buttons: How 17 Molecules Changed History*, New York, Tacher Puttnam, 2003, pp.19–35
J. Turner, *Spice: The History of a Temptation*, London, Harper Collins, 2004
J. Lawton, *Silk, Scents and Spice*, Paris, UNESCO Publishing, 2004
J. Keay, *The Spice Route: A History*, London, John Murray, 2005
P. Freedman, *Out of the East. Spices and the Mediaeval Imagination*, New Haven and London, Yale University Press, 2008
G. P. Nabhan, *Cumin, Camels, and Caravans: A Spice Odyssey*, Berkeley, University of California Press, 2014
H. A. Anderson, *Chillies: A Global History*, London, Reaktion Books, 2016
W. Wang, K. T. K. Nguyen, C. Zhao and H-C. Hung, *Science Advances*, 2023, **9**, eadh5517 (curry spices in Vietnam 2000 years ago)

OTHER MATERIAL ON SPICES

A. Naj, *Peppers: A Story of Hot Pursuits*, London, Knopf, 1991
P. Willard, *Secrets of Saffron: The Vagabond Life of The World's Most Seductive Spice*, Boston, The Beacon Press, 2001
C. McFadden, *Pepper: The Spice that Changed the World*, Bath, Absolute Press, 2008
F. Czarra, *Spices: A Global History*, London, Reaktion Books, 2009
K. M. Friese, K. Kraft and G. P. Nabhan, *Chasing Chiles: Hot Spots Along the Pepper Trail*, Vermont, Chelsea Green, 2011
M. Shaffer, *Pepper: A History of the World's Most Influential Spice*, New York, St Martin's Press, 2013
J. O'Connell, *The Book of Spice. From Anise to Zeodary*, London, Profile Books, 2015

INDIVIDUAL SPICE MOLECULES

F. N. McNamara, A. Randall and M. J. Gunthorpe, *British Journal of Pharmacology*, 2005, **144**, 781–790 (piperine as an agonist at the human human vanilloid receptor (TRPV1))
V. N. Dedov, V. H. Tran, C. C. Duke, M. Connor, M. J. Christie, S. Mandani and B. D. Roufogalis, *British Journal of Pharmacology*, 2002, **137**, 793–798 (gingerols as TRPV1 agonists)
T. Watanabe and Y. Terada, *Journal of Nutritional Science and Vitaminology*, 2015, **61**, S86–S88 (gingerols and shogaols agonists for TRPV1)
Y. Yin, Y. Dong, S. Vu, F. Yang, V. Yarov-Yarovoy, Y. Tian and J. Zheng, *British Journal of Pharmacology*, 2019, **176**, 3364–3377 (6-shogaol, 6-gingerol, and zingerone activate the TRPV1 channel)

M. Bandell, G. M. Story, S. W. Hwang, V. Viswanath, S. R. Eid, M. J. Petrus, T. J. Earley and A. Patapoutian, *Neuron*, 2004, **41**, 849–857

L. J. Macpherson, A. E. Dubin, M. J. Evans, F. Marr, P. G. Schultz, B. F. Cravatt and A. Patapoutian, *Nature*, 2007, **445**, 541–545

C. Legrand, J. M. Merlini, C. de Senarclens-Bezencon and S. Michlig, *Scientific Reports*, 2020, **10**, 11238 (cinnamaldehyde and the TRPA1 receptor)

J. Vriens, B. Nilius and R. Vennekens, *Current Neuropharmacology*, 2008, **6**, 79–96 (eugenol and the TRPM8, TRPV1 and TRPA1 channels)

A. H. Klein, C. L. Joe, A. Davoodi, K. Takechi, M. I. Carstens and E. Carstens, *Neuroscience*, 2014, **271**, 45–55 (eugenol activates TRPV3 and the irritant-sensitive TRPA1)

S-M. Hwang, K. Lee, S-T. Im, E. J. Go, Y. H. Kim and C-K. Park, *Biomolecules*, 2020, **10**, 1513 (combination of eugenol and QX-314 as an anaesthetic)

T. Shirai, K. Kumihashi, M. Sakasai, H. Kusuoku, Y. Shibuya and A. Ohuchi, *ACS Medicinal Chemistry Letters*, 2017, **8**, 715−719 (Δ8′-7-ethoxy-4-hydroxy-3,3′,5′-trimethoxy-8-O-4′-neolignan, a TRPM8 agonist from nutmeg, a promising cooling compound)

L. Bernal, P. Sotelo-Hitschfeld, C. König, V. Sinica, A. Wyatt, Z. Winter, A. Hein, F. Touska, S. Reinhardt, A. Tragl, R. Kusuda, P. Wartenberg, A. Sclaroff, J. D. Pfeifer, F. Ectors, A. Dahl, M. Freichel, V. Vlachova, S. Brauchi, C. Roza, U. Boehm, D. E. Clapham, J. K. Lennerz and K. Zimmermann, *Science Advances,* 2021, **7**, eabf5567 (eugenol is an agonist for the TRPC5 cold sensor)

CAPSAICIN

M. J. Caterina, M. A. Schumacher, M. Tominaga, T. A. Rosen, J. D. Levine and D. Julius, *Nature*, 1997, **389**, 816–824 (discovery of TRPV1 receptor)

J. Billing and P. W. Sherman, *Quarterly Review of Biology*, 1998, **73**, 3–49 (antimicrobial functions of spices)

P. W. Sherman and J. Billing, *BioScience*, 1999, **49**, 453–463 (why we use spices)

J. J. Tewksbury and G. P. Nabhan, *Nature*, 2001, **412**, 203–204 (capsaicin as squirrel deterrent)

H. McGee, *McGee on Food and Cooking*, Hodder, 2004, esp. pp. 418–430

G. M. Story and L. Cruz-Orengo, *American Scientist*, 2007, **95**, 326–333 (hot and cold sensors)

N. R. Gavva, A. W. Bannon, S. Surapaneni, D. N. Hovland Jr, S. G. Lehto, A. Gore, T. Juan, H. Deng, B. Han, L. Klionsky, R-Z. Kuang, A. Le, R. Tamir, J. Wang, B. Youngblood, D. Zhu, M. H. Norman, E. Magal, J. J. S. Treanor and J.-C. Louis, *Journal of Neuroscience*, 2007, **27**, 3366–3374 (TRPV1 receptor and body temperature regulation)

A. M. Binshtok, B. P. Bean and C. J. Woolf, *Nature*, 2007, **449**, 607–610 (capsaicin and anaesthetic)

J. J. Tewksbury, K. M. Reagan, N. J. Machnicki, T. A. Carlo, D. C. Haak, A. L. Calderon-Penaloza and D. J. Levey, *Proceedings of the National Academy of Sciences of the United States of America*, 2008, **105**, 11808–11811 (evolutionary ecology of pungency in wild chillis)

S. J. Conway, *Chemical Society Reviews*, 2008, **37**, 1530–1545 (review of capsaicin and TRPV1)

R. T. Kachoosangi, G. G. Wildgoose and R. G. Compton, *Analyst*, 2008, **133**, 888–895 (carbon nanotube-based sensors for capsaicin)

A. M. Patwardhan, P. M. Scotland, A. N. Akopian and K. M. Hargreaves, *Proceedings of the National Academy of Sciences of the United States of America*, 2009, **106**, 18820–18824

A. M. Patwardhan, A. N. Akopian, N. B. Ruparel, A. Diogenes, S. T. Weintraub, C. Uhlson, R. C. Murphy and K. M. Hargreaves, *Journal of Clinical Investigation*, 2010, **120**, 1617–1626 (linoleic acid metabolites that activate TRPV1 and produce inflammation)

M. de L. Reyes-Escogido, E. G. Gonzalez-Mondragon and E. Vazquez-Tzompantzi, *Molecules*, 2011, **16**, 1253–1270 (capsaicin review)

M. Liao, E. Cao, D. Julius and Y. Cheng, *Nature*, 2013, **504**, 107–112

E. Cao, M. Cao, D. Julius and Y. Cheng, *Nature*, 2013, **504**, 113–118 (TRPV1 ion channel structures)

L. Darré and C. Domene, *Molecular Pharmaceutics*, 2015, **12**, 4454–4465 (binding of capsaicin to the TRPV1 ion channel)

K. Ohbuchi, Y. Mori, K. Ogawa, E. Warabi, M. Yamamoto and T. Hirokawa, *PLoS One*, 2016, **11**, e0162543 (how vanilloids including capsaicin bind to TRPV1)

F. Yang and J. Zheng, *Protein & Cell*, 2017, **8**, 169–177 (mechanism of TRPV1 channel activation by capsaicin)

H. Ledford and E. Callaway, *Nature*, 2021, **598**, 246 (capsaicin and a Nobel Prize)

T. Thornton, D. Mills and E. Bliss, *Nutrients*, 2023, **15**, 1537 (capsaicin's possible health benefits)

SZECHUAN PEPPERS

H. McGee, *On Food & Cooking*, London, Hodder & Stoughton, 2004, pp. 428–429

E. Sugai, Y. Morimitsu and K. Kubota, *Bioscience, Biotechnology, and Biochemistry*, 2005, **69**, 1958–1962 (sanshools in *Xanthoxylum piperitum*)

D. M. Bautista, Y. M. Sigal, A. D. Milstein, J. L. Garrison, J. A. Zorn, P. R. Tsuruda, R. A. Nicoll and D. Julius, *Nature Neuroscience*, 2008, **11**, 772–779 (pungent agents from Szechuan peppers excite sensory neurons by inhibiting two-pore potassium channels)

C. E. Riera, C. Menozzi-Smarrito, M., Affolter, S. Michlig, C. Munari, F. Robert, H. Vogel, S. A. Simon and J. le Coutre, *British Journal of Pharmacology*, 2009, **157**, 1398–1409 (compounds from Sichuan and Melegueta peppers activate TRPA1 and TRPV1 channels)

R. C. Lennertz, M. Tsunozaki, D. M. Bautista and C. L. Stucky, *Journal of Neuroscience*, 2010, **30**, 4353–4361 (tingling paresthesia evoked by hydroxy-a-sanshool)

L. Zhang, B. Shi, H. Wang, L. Zhao and Z. Chen, *Chemical Senses*, 2017, **42**, 575–584 (pungency evaluation of hydroxyl-sanshools)

MENTHOL AND OTHER MINTY SUBSTANCES

K. Tani, T. Yamagata, S. Akutagawa, H. Kumobayashi, T. Taketomi, H. Takaya, A. Miyashita, R. Noyori and S. Otsuka, *Journal of the American Chemical Society*, 1984, **106**, 5208–5217 (menthol synthesis)

D. D. McKemy, W. M. Neuhausser and D. Julius, *Nature*, 2002, **416**, 52–58 (TRPM8 receptor for menthol)

A. M. Peier, A. Moqrich, A. C. Hergarden, A. J. Reeve, D. A. Andersson, G. M. Story, T. J. Earley, I. Dragoni, P. McIntyre, S. Bevan and A. Patapoutian, *Cell*, 2002, **108**, 705–715 (TRPM8 receptor senses cold stimuli and menthol)

R. Noyori, *Angewandte Chemie International Edition*, 2002, **41**, 2008–2022 (Nobel lecture)

R. Noyori, *Green Chemistry*, 2003, **5**, G37–G39 (menthol synthesis)

E. M. Davis, K. L. Ringer, M. E. McConkey and R. B. Croteau, *Plant Physiology*, 2005, **137**, 873–881

R. B. Croteau, E. M. Davis, K. L. Ringer and M. R. Wildung, *Naturwissenschaften*, 2005, **92**, 562–577 (biosynthesis)

D. D. McKemy, *Molecular Pain*, 2005, **1**, 16–22 (TRPM8 receptor)

R. Bentley, *Chemical Reviews*, 2006, **106**, 4099–4112 (isomers and smell)

B. M. Lange, S. S Mahmoud, M. R. Wildung, G. W. Turner, E. M. Davis, I. Lange, R. C. Baker, D. M. Bautista, J. Siemens, J. M. Glazer, P. R. Tsuruda, A. I. Basbaum, C. L. Stucky, S.-E. Jordt and D. Julius, *Nature*, 2007, **448**, 204–208 (TRPM8 receptor as the principal detector of environmental cold)

R. A. Boydston and R. B. Croteau, *Proceedings of the National Academy of Sciences of the United States of America*, 2011, **108**, 16944–16949 (improving peppermint essential oil yield and composition by metabolic engineering)

Y. Yin, M. Wu, L. Zubcevic, W. F. Borschel, G. C. Lander and S-Y. Lee, *Science*, 2018, **359**, 237–241 (structure of the TRPM8 ion channel)

M. M. Diver, Y. Cheng and D. Julius, *Science*, 2019, **365**, 1434–1440 (TRPM8 inhibition and desensitization)

L. Xu, Y. Han, X. Chen, A. Aierken, H. Wen, W. Zheng, H. Wang, X. Lu, Z. Zhao, C. Ma, P. Liang, W. Yang, S. Yang and F. Yang, *Nature Communications*, 2020, **11**, 3790 (menthol molecule binding to the TRPM8 ion channel)

X. Chen, L. Xu, H. Zhang, H. Wen and F. Yang, *Frontiers in Pharmacology*, 2022, 13, 898670 (activation of TRPM8 by the stereoisomers of menthol)

P. S. Gardiner, *Nicotine & Tobacco Research*, 2004, **6**, S55–S65 (Supplement 1) ("The African Americanization of menthol cigarette use in the United States")

S. S. Nath, C. Pandey and D. Roy, *Indian Journal of Anaesthesia*, 2012, **56**, 582–584 (near-fatal exposure to peppermint oil)

N. Kabbani, *Frontiers in Pharmacology*, 2013, **4**, 95 (addictive properties of menthol cigarettes)

R. J. Wickham, *Yale Journal of Biology and Medicine*, 2015, **88**, 279–287 (menthol alters tobacco-smoking behaviour)

A. Kumar, U. Baitha, P. Aggarwal and N. Jamshed, *International Journal of Applied and Basic Medical Research*, 2016, **6**, 137–139 (fatal case of menthol poisoning)

L. Biswas, E. Harrison, Y. Gong, R. Avusula, J. Lee, M. Zhang, T. Rousselle, J. Lage and X. Liu, *Psychopharmacology*, 2016, **233**, 3417–3427 (menthol enhances nicotine self-administration in rats)

K. Wailoo, *Pushing Cool: Big Tobacco, Racial Marketing, and the Untold Story of the Menthol Cigarette*, Chicago, IL, University of Chicago Press, 2021

N. Subbaraman, *Nature*, 2021, **598**, 407–408

D. Mendez and T. T. T. Le, *Tobacco Control*, 2022, **31**, 569–571 (menthol smoking and the African American population 1980–2018)

B. Raudenbush, N. Corley and W. Eppich, *Journal of Sport and Exercise Psychology*, 2001, **23**, 156–160 (peppermint and sport)

S. Barker, P. Grayhem, J. Koon, J. Perkins, A. Whalen and B. Raudenbush, *Perceptual and Motor Skills*, 2003, **97**, 1007–1010 (peppermint odour enhances performance on clerical tasks)

M. Moss, S. Hewitt, L. Moss and K. Wesnes, *International Journal of Neuroscience*, 2008, **118**, 59–77 (peppermint enhances accuracy of memory)

A. Meamarbashi, *Avicenna Journal of Phytomedicine*, 2014, **4**, 72–78 (peppermint essential oil enhances exercise performance)

M. Mahachandra, S. T. Yassierli and E. D. Garnaby, *Procedia Manufacturing*, 2015, **4**, 471–477 (peppermint for an in-vehicle fragrance to maintain drivers' alertness.)

T. Eisner, K. D. McCormick, M. Sakaino, M. Eisner, S. R. Smedley, D. J. Aneshansley, M. Deyrup, R. L. Myers and J. Meinwald, *Chemoecology*, 1990, **1**, 30–37 (chemical defence of a rare mint plant and (+)-*trans*-pulegol)

R. Segelken, 'Gift of Mint', *Chem Matters*, December 1991, 15 (discovery of (+)-*trans*-pulegol)

K. D. McCormick, M. A. Deyrup, E. S. Menges, S. R. Wallace, J. Meinwald and T. Eisner, *Proceedings of the National Academy of Sciences of the United States of America*, 1993, **90**, 7701–7705 (volatile leaf components of *Dicerandra* mints)

MUSTARD

P. Hartley, *The Colman's Mustard Cookbook*, Bath, Absolute Press, 2004 (Colman's mustard)

H. McGee, *On Food and Cooking*, 2nd edition, London, Hodder and Stoughton, 2004, pp. 415–417 (mustard)

T. P. Coultate, *Food: The Chemistry of Its Components*, 6th edition, Cambridge, Royal Society of Chemistry, 2015, pp. 325–327 (mustard)

D. Güzey, *Mustard: A Global History*, London, Reaktion Books, 2019

ALLYL ISOTHIOCYANATE

B. Tollens and A. Henninger, *Justus Liebig's Annalen der Chemie*, 1870, **156**, 134–142 (synth.)

W. P. Reeves and J. V. McClusky, *Tetrahedron Letters*, 1983, **24**, 1585–1588 (synth.)

SINIGRIN AND GLUCOSINOLATES

G. R. Fenwick, R. K. Heaney, W. J. Mullin and C. H. VanEtten, *Critical Reviews in Food Science and Nutrition*, 1983, **18**, 123–201 (glucosinolates and their breakdown products reviewed)

I. T. Johnson, *Phytochemistry Reviews*, 2002, **1**, 183–188 (glucosinolates in diet)

M. E. Cartea and P. Velasco, *Phytochemistry Reviews*, 2008, **7**, 213–229 (glucosinolates and health)

A. Mazumder, A. Dwivedi and J. du Plessis, *Molecules*, 2016, **21**, 416 (sinigrin review)

F. J. Barba, N. Nikmaram, S. Roohinejad, A. Khelfa, Z. Zhu and M. Koubaa, *Frontiers in Nutrition*, 2016, **3**, 24 (review of glucosinolates and their breakdown products)

M. A. Prieto, C. J. López and J. Simal-Gandara, *Advances in Food and Nutrition Research*, 2019, **90**, 305–350 (glucosinolates reviewed)

V. P. Thinh Nguyen, J. Stewart, M. Lopez, I. Ioannou and F. Allais, *Molecules*, 2020, **25**, 4537 (glucosinolates reviewed)

E. L. Connolly, M. Sim, N. Travica, W. Marx, G. Beasy, G. S. Lynch, C. P. Bondonno, J. R. Lewis, J. M. Hodgson and L. C. Blekkenhorst, *Frontiers in Pharmacology*, 2021, **12**, 767975 (glucosinolates and chronic disease)

MUSTARD'S EFFECTS UPON TRP CHANNELS

B. Nilius and G. Owsianik, *Genome Biology*, 2011, **12**, 218 (TRP channels reviewed)

S-E. Jordt, D. M. Bautista, H-H. Chuang, D. D. McKemy, P. M. Zygmunt, E. D. Högestätt, I. D. Meng and D. Julius, *Nature*, 2004, **427**, 260–265 (mustard oil excites sensory nerve fibres through the TRP channel ANKTM1)

L. J. Macpherson, A. E. Dubin, M. J. Evans, F. Marr, P. G. Schultz, B. F. Cravatt and A. Patapoutian, *Nature*, 2007, **445**, 541–545 (activation of TRPA1 ion channels by covalent modification of cysteines)

W. Everaerts, M. Gees, Y. A. Alpizar, R. Farre, C. Leten, A. Apetrei, I. Dewachter, F. van Leuven, R. Vennekens, D. De Ridder, B. Nilius, T. Voets and K. Talavera, *Current Biology*, 2011, **21**, 316–321 (TRPV1 mediates the noxious effects of mustard oil)

Y. A. Alpizar, B. Boonen, M. Gees, A. Sanchez, B. Nilius, T. Voets and K. Talavera, *Pflügers ArchivEuropean Journal of Physiology*, 2014, **466**, 507–515 (allyl isothiocyanate sensitizes TRPV1 to heat stimulation)

C. E. Paulsen, J-P. Armache, Y. Gao, Y. Cheng and D. Julius, *Nature*, 2015, **520**, 511–517 (structure of the TRPA1 ion channel)

J. Zhao, J. V. Lin King, C. E. Paulsen, Y. Cheng and D. Julius, *Nature*, 2020, **585**, 141–145 (how electrophiles activate the TRPA1 receptor)

C. Liu, R. Reese, S. Vu, L. Rougé, S. D. Shields, S. Kakiuchi-Kiyota, H. Chen, K. Johnson, Y. P. Shi, T. Chernov-Rogan, D. M. Z. Greiner, P. B. Kohli, D. Hackos, B. Brillantes, C. Tam, T. Li, J. Wang, B. Safina, S. Magnus, M. Volgraf, J. Payandeh, J. Zheng, A. Rohou and J. Chen, *Neuron*, 2021, **109**, 273–284 (biased agonism of the TRPA1 ion channel)

CHAPTER 4 ABUSED PAINKILLERS

OPIUM, MORPHINE AND HEROIN

A. Conan-Doyle, 'The Man with the Twisted Lip', Strand Magazine, 1891; The Adventures of Sherlock Holmes, 1892

J. Kaplan, *The Hardest Drug*, Chicago, IL, The University of Chicago Press, 1983

F. Zackon, *Heroin*, New York, Chelsea House, 1986

L. D. Kapoor, *Opium Poppy: Botany, Chemistry and Pharmacology*, Binghamton, The Haworth Press, 1995

M. Booth, *Opium: A History*, London, Simon and Schuster, 1996

J. Durlacher, *Agenda Heroin*, London, Carlton Books, 2000

B. Hodgson, *Opium: A Portrait of the Heavenly Demon*, London, Souvenir Press, 2000

B. Hodgson, *In the Arms of Morpheus: The Tragic History of Laudanum, Morphine, and Patent Medicines*, Buffalo, NY, Firefly Books, 2001

R. Davenport-Hines, *The Pursuit of Oblivion. A Global History of Narcotics 1500–2000*, London, Weidenfeld and Nicolson, 2001

D. D. Courtwright, *Forces of Habit*, Cambridge (Mass.) and London, Harvard University Press, 2001, esp. pp. 31–38

R. G. Homan, B. Hudson and R. C. Rowe, *Popular Medicines*, London, Pharmaceutical Press, 2008, pp. 53–59 (Dr. Collis Browne's Chlorodyne)

D. Ciccarone, *International Journal of Drug Policy*, 2009, **20**, 277–282 (heroin and America)

P-A. Chouvy, *Opium: Uncovering the Politics of the Poppy*, London, I. B. Tauris & Co., 2009

T. Dormandy, *Opium: Reality's Dark Dream*, New Haven and London, Yale University Press, 2012

L. Inglis, *Milk of Paradise*, London, Macmillan, 2018 (opium)

D. Ciccarone, *International Journal of Drug Policy*, 2019, **71**, 183–188 (the triple wave epidemic and its suppliers)

A. Gottås, E. L. Øiestad, F. Boix, V. Vindenes, Å. Ripel, C. H. Thaulow and J. Mørland, *British Journal of Pharmacology*, 2013, **170**, 546–556

D. Perekopskiy and E. A. Kiyatkin, *ACS Chemical Neuroscience*, 2019, **10**, 8, 3409–3414 (6-monoacetylmorphine as a metabolite of heroin)

FENTANYL

T. H. Stanley, *Journal of Pain and Symptom Management*, 1992, **7**, S3–S7 (history and development of fentanyls)

T. H. Stanley, T. D. Egan and H. Van Aken, *Anesthesia & Analgesia*, 2008, **106**, 451–462 (tribute to Dr. Paul A. J. Janssen)

I. Ojanperä, M. Gergov, M. Liiv, A. Riikoja and E. Vuori, *International Journal of Legal Medicine*, 2008, **122**, 395–400

M. C. Gerald, *The Drug Book*, New York, Sterling, 2013, pp. 352–353

J. J. Li, *Blockbuster Drugs*, New York, OUP, 2014, pp. 146–152 (also Duragesic)

T. H. Stanley, *The Journal of Pain*, 2014, **15**, 1215–1224 ('the fentanyl story')

R. S. Vardanyan and V. J. Hruby, *Future Medicinal Chemistry*, 2014, **6**, 385–412 (fentanyl-related compounds and derivatives)

J. Mounteney, I. Giraudon, G. Denissov and P. Griffiths, *International Journal of Drug Policy*, 2015, **26**, 626–631 (fentanyl deaths rising in Europe)

M. P. Prekupec, P. A. Mansky and M. H. Baumann, *Addiction Medicine*, 2017, **11**, 256–265 (misuse of synthetic opioids, esp. fentanyls)

K. McLaughlin, *Science*, 2017, **355**, 1364–1366 (deadly opioid chemistry)

J. B. Zawilska, *Frontiers in Psychiatry*, 2017, **8**, 110 (abused fentanyls)

S. M. Burns, C. W. Cunningham and S. L. Mercer, *ACS Chemical Neuroscience*, 2018, **9**, 2428–2437 (DARK classics in chemical neuroscience: fentanyl)

K. Megget, *Chemistry & Industry*, 2018, **10**, 19–21 (fentanyl drug crisis)

P. Armenian, K. T. Vo, J. Barr-Walker and K. L. Lynch, *Neuropharmacology*, 2018, **134**, 121–132 (fentanyl, fentanyl analogs and novel synthetic opioids reviewed)

D. Ciccarone, *International Journal of Drug Policy*, 2019, **71**, 183–188 (the triple wave epidemic and its suppliers)

B. Westhoff, *Fentanyl, Inc.*, London, Grove Atlantic, 2019

A. Uusküla, A. Talu, S. Vorbjov, M. Salekešin, J. Rannap, L. Lemsalu and D. Des Jarlais, *International Journal of Drug Policy*, 2020, **81**, 102757 (fentanyl epidemic in Estonia)

Q. N. Vo, P. Mahinthichaichan, J. Shen and C. R. Ellis, *Nature Communications*, 2021, **12**, 984 (how μ-opioid receptor recognizes fentanyl)

P. Brunetti, F. Pirani, J. Carlier, R. Giorgetti, F. P. Busardò and A. F. Lo Faro, *Journal of Analytical Toxicology*, 2021, **45**, 537–554 (acute intoxications and fatalities from illicit fentanyl and analogues 2017–2019)

C. Noble, D. M. Papsun, S. Diaz and B. K. Logan, *Emerging Trends in Drugs, Addictions, and Health*, 2021, **1**, 100022 (detection of carfentanil and 3-methylfentanyl in forensics)

D. Ciccarone, *Current Opinion in Psychiatry*, 2021, **34**, 344–350 (the rise of illicit fentanyls, stimulants and the fourth wave of the opioid overdose crisis)

A. N. Edinoff, D. M. Garza, S. P. Vining, M. E. Vasterling, E. D. Jackson, K. S. Murnane, A. M. Kaye, R. N. Fair, Y. J. L. Torres, A. E. Badr, E. M. Cornett and A. D. Kaye, *Pain and Therapy*, 2023, **12**, 399–421 (dangers of fentanyls and nitazenes)

CARFENTANIL

P. A. J. Janssen and G. H. P. Van Daele, U.S. Patent No. 4179569, published Dec. 18, 1979 (synthesis of carfentanil)

J. L. Galzi, B. Ilien, E. J. Simon, M. Goeloner and C. Hirth, *Tetrahedron Letters*, 1987, **28**, 401–404 (synth.)

L. P. Reiff and P. B. Sollman, U. S. Patent US 5106983 A, published April 21, 1992 (improved synthesis of carfentanil and analogues)

T. Wang and J. T. Berner, *Journal of Analytical Toxicology*, 2006, **30**, 335–341 (analysis of carfentanil and other fentanyls in urine; MS of carfentanil)

M. G. Feasel, A. Wohlfarth, J. M. Nilles, S. Pang, R. L. Kristovich and M. A. Huestis, *AAPS Journal*, 2016, **18**, 1489–1499 (metabolism of carfentanil)

J. F. Casale, J. R. Mallette and E. M. Guest, *Forensic chemistry*, 2017, **3**, 74–80 (IR, MS, NMR of carfentanil; analysis of illicit carfentanil)

J. Suzuki and S. El-Haddada, *Drug and Alcohol Dependence*, 2017, **171**, 107–116 (review of fentanyl and non-pharmaceutical fentanyls)

S. P. Elliott and E. H. Lopez, *Journal of Analytical Toxicology*, 2018, **42**, e41–e45 (UK deaths involving carfentanil)

N. Misailidi, I. Papoutsis, P. Nikolaou, A. Dona, C. Spiliopoulou and S. Athanaselis, *Forensic Toxicology*, 2018, **36**, 12–32 (synth. of carfentanil; review of ocfentanil and carfentanil)

J. L. Leen and D. N. Juurlink, *Canadian Journal of Anaesthesia.*, 2019, **66**, 414–421 (carfentanil concerns)

A. E. Ringuette, M. Spock, C. W. Lindsley and A. M. Bender, *ACS Chemical Neuroscience*, 2020, **11**, 3955–3967 (DARK Classics in Chemical Neuroscience: Carfentanil)

J. B. Zawilska, K. Kuczyńska, W. Kosmal, K. Markiewicz and P. Adamowicz, *Forensic Science International*., 2021, **320**, 110715

CARFENTANIL AS A VETERINARY SEDATIVE AND ITS USE

J. C. Haigh, L. J. Lee and R. E. Schweinsburg, *Journal of Wildlife Diseases*, 1983, **19**, 140–144 (carfentanil to immobilise polar bears)

M. D. Kock and J. Berger, *Journal of Wildlife Diseases*, 1987, **23**, 625–633 (carfentanil to immobilise North American bison)

B. Welsch, E. R. Jacobson, G. V. Kollias, L. Kramer, H. Gardner and C. D. Page, *Journal of Zoo and Wildlife Medicine*, 1989, **20**, 446–453 (tusk extraction in the African elephant using carfentanil)

T. J. Portas, *Australian Veterinary Journal*, 2004, **82**, 542–549 (carfentanil to immobilise rhinoceros)

A. V. George, J. J. Lu, M. V. Pisano, J. Metz and T. B. Erickson, *American Journal of Emergency Medicine*, 2010, **28**, 530–532 (the elk story)

CARFENTANIL AND THE MOSCOW THEATRE SIEGE

P. M. Wax, C. E. Becker and S. C. Curry, *Annals of Emergency Medicine*, 2003, **41**, 700–705 (indication that carfentanil was involved)

J. R. Riches, R. W. Read, R. M. Black, N. J. Cooper and C. M. Timperley, *Journal of Analytical Toxicology*, 2012, **36**, 647–656 (analysis of clothing and urine shows carfentanil used in siege)

OXYCODONE AND OXYCONTIN

J. J. Li, *Blockbuster Drugs*, New York, OUP, 2014, p. 142

M. C. Gerald, *The Drug Book*, New York, Sterling, 2013, pp. 444–445

B. Meier, *Pain Killer: A "Wonder" Drug's Trail of Addiction and Death*, Emmaus, PA, Rodale Press, 2003 (Oxycontin)

J. Abramson, *Overdo$ed America. The Broken Promise of American Medicine*, New York, Harper Collins, 2004, pp. 102–103, 121

R. Waters, *Prescription Painkillers: Oxycontin, Percocet, Vicodin, & Other Addictive Analgesics*, Broomhall, PA, Mason Crest, 2014 (useful introduction and summary on addictive analgesics)

J. Forman, *A Nation in Pain: Healing Our Biggest Health Problem*, New York, OUP, 2014 (pain and painkillers in the USA)

S. Quinones, *Dreamland: The True Tale of America's Opiate Epidemic*, New York, Bloomsbury Press, 2015

B. Macy, *Dopesick*, London, Apollo Books, 2018 (American opioid addiction)

C. McGreal, *American Overdose. The Opioid Tragedy in Three Acts*, New York, PublicAffairs, 2018

P. R. Keefe, *Empire of Pain*, London, Picador, 2021 (Purdue Pharma and Oxycontin)

P. R. Keefe, *The Lancet*, July 24, 2021, **398**, 277 (Lancet Editorial: A time of crisis for the opioid epidemic in the USA)

PREPARATION OF OXYCODONE FROM CODEINE OR THEBAINE

M. Freund and E. Speyer, *Journal für praktische Chemie*, 1916, **94**, 135–178
B. S. Huang, L. Yansong, B. Y. Ji and A. P. Christodoulou, US Patent, 6008355 A, Filed 16 July 1998, published 28 December 1999 (see also B. S. Huang, L. Yansong, B. Y. Ji and A. P. Christodoulou, US Patent 5869669 A, Filed 11 July 1997, published 9 February 1999.)
G. B. Kok and P. J. Scammells, *RSC Advances*, 2012, **2**, 11318–11325 (improved synth.)

HOW CODEINE AND OXYCODONE WORK

F. B. Ross and M. T. Smith, *Pain*, 1997, **73**, 151–157
E. Kalso, *Pain*, 2007, **132**, 227–228
M. T. Smith, *Current Opinion in Anesthesiology*, 2008, **21**, 596–601

OXYCONTIN ABUSE

T. J. Cicero, J. A. Inciardi and A. Muñoz, *Journal of Pain*, 2005, **6**, 662–672 (trends in abuse of Oxycontin and other opioid analgesics in the United States: 2002–2004)
S. S. Jayawant and R. Balkrishnan, *Therapeutics and Clinical Risk Management*, 2005, **1**, 77–82 (OxyContin abuse: issues and solutions)
D. Carise, K. L. Dugosh, A. T. McLellan, A. Camilleri, G. E. Woody and K. G. Lynch, *American Journal of Psychiatry*, 2007, **164**, 1750–1756 (prescription OxyContin abuse)
H. Hedegaard, B. A. Bastian, J. P. Trinidad, M. Spencer and M. Warner, *National Vital Statistics Reports*, 2018, **67**, 1–12 (drugs most frequently Involved in US overdose deaths 2011–2016)
N. D. Volkow, *America's Addiction to Opioids: Heroin and Prescription Drug Abuse*, at https://www.drugabuse.gov/about-nida/legislative-activities/testimony-to-congress/2014/americas-addiction-to-opioids-heroin-prescription-drug-abuse

ABUSE-RESISTANT FORMULATIONS

R. B. Raffa and J. V. Pergolizzi, *Drugs*, 2010, **70**, 1657–1675 (opioid formulations designed to resist/deter abuse)
K. L. Hahn, *American Health & Drug Benefits*, 2011, **4**, 107–114 (strategies to prevent opioid misuse, abuse, and diversion)
T. J. Cicero, M. S. Ellis and H. L. Surratt, *The New England Journal of Medicine*, 2012, **367**, 187–189 (abuse-deterrent formulation of OxyContin)
R. B. Raffa, J. V. Pergolizzi Jr, E. Muñiz, R. Taylor and J. Pergolizzi, *Pain Research and Treatment*, 2012, **8**, 282981 (designing opioids that deter abuse)
J. V. Pergolizzi, R. Taylor and R. B. Raffa, *Advances in Therapy*, 2015, **32**, 485–495 (abuse-resistance)

DESOMORPHINE

E. Mosettig, F. L. Cohen and L. F. Small, *Journal of the American Chemical Society*, 1932, **54**, 793–801
L. F. Small and D. E. Morris, *Journal of the American Chemical Society*, 1933, **55**, 2874–2885
N. B. Eddy and H. A. Howes, *Journal of Pharmacology and Experimental Therapeutics*, 1935, **55**, 257–267 (synthesis of desomorphine)

C. I. Wright and F. A. Barbour, *Journal of Pharmacology and Experimental Therapeutics*, 1937, **61**, 422–439 (respiratory effects)

C. K. Himmelsbach, *Journal of Pharmacology and Experimental Therapeutics*, 1939, **67**, 239–249 (addiction characteristics)

L. E. Lee, *Journal of Pharmacology and Experimental Therapeutics*, 1944, **78**, 375–385 (morphine and desomorphine compared)

P. A. Janssen, *British Journal of Anaesthesia*, 1962, **34**, 260–268

L. J. Sargent and E. L. May, *Journal of Medicinal Chemistry*, 1970, **13**, 1061–1063 (potency of desomorphine compared with morphine)

S. A. Savchuk, S. S. Barsegyan, I. B. Barsegyan and G. M. Kolesov, *Journal of Analytical Chemistry*, 2008, **63**, 361–370 (mass spectra of desomorphine and impurities)

W. A. Banks, *BMC Neurology*, 2009, **9**(Suppl 1), S3 (characteristics of compounds that cross the blood-brain barrier)

S. Srimurugan, C.-J. Su, H-C. Shu, K. Murugan and C. Chen, *Monatshefte für Chemie*, 2012, **143**, 171–174 (improved synthesis of desomorphine, ir, nmr, cmr)

KROKODIL

R. Skowronek, R. Celiński and C. Chowaniec, *Clinical Toxicology*, 2012, **4**, 269–269 ("Crocodile"– new dangerous designer drug of abuse from the East)

M. Gahr, R. W. Freudenmann, C. Hiemke, I. M. Gunst, B. J. Connemann and C Schönfeldt-Lecuona, *Journal of Addictive Diseases*, 2012, **31**, 407–412 (Desomorphine Goes "Crocodile")

R. Heimer, *International Journal of Drug Policy*, 2013, **24**, 275–277 (patterns of new drug emergence: A comment in light of 'krokodil')

R. E. Booth, *International Journal of Drug Policy*, 2013, **24**, 277–280 ('Krokodil' and other home-produced drugs for injection: A perspective from Ukraine)

J.-P. C. Grund, A. Latypov and M. Harris, *International Journal of Drug Policy*, 2013, **24**, 265–274. (Breaking worse: The emergence of krokodil and excessive injuries among people who inject drugs in Eurasia)

D. V. Thekkemuriyi, S. G. John and U. Pillai, *American Journal of Medicine*, 2014, **127**(3), e1–e2 ('Krokodil'—A Designer Drug from Across the Atlantic, with Serious Consequences)

M. Katselou, I. Papoutsis, P. Nikolaou, C. Spiliopoulou and S. Athanaselis, *Life Sciences*, 2014, **102**, 81–87 (a "Krokodil" emerges from the murky waters of addiction. Abuse trends of an old drug)

E. A. Alves J. X. Soares, C. M. Afonso, J.-P. C. Grund, A. S. Agonia, S. M. Cravo, A. D. P. Netto, F. Carvalho and R. J. Dinis-Oliveira, *Forensic Science International*, 2015, **257**, 76–82 (the harmful chemistry behind "krokodil": street-like synthesis and product analysis)

J. F. Neves, E. A. Alves, J. X. Soares, S. M. Cravo, A. M. S. Silva, A. D. P. Netto, F. Carvalho, R. J. Dinis-Oliveira and C. M. Afonso, *Data Analysis*, 2016, **6**, 83–88 (data analysis of "krokodil" samples obtained by street-like synthesis)

A. Haskin, N. Kim and C. Aguh, *JAAD Case Reports*, 2016, **2**, 174–176 (krokodil-induced skin necrosis in an intravenous drug user)

E. A. Alves, P. Brandão, J. F. Neves, S. M. Cravo, J. X. Soares, J.-P. C. Grund, J. A. Duarte, C. M. Afonso, A. D. P. Netto, F. Carvalho and R. J. Dinis-Oliveira, *Human Psychopharmacology: Clinical and Experimental*, 2017, **32**, e2572 (subcutaneous administrations of krokodil causes skin necrosis and internal organs toxicity in Wistar rats)

D. H. Â. Florez, A. M. dos Santos Moreira, P. R. da Silva, R. Brandão, M. M. C. Borges, F. J. M. de Santana and K. B. Borges, *Drug and Alcohol Dependence*, 2017, **173**, 59–68 (overview of its chemistry, pharmacology, metabolism, toxicology and analysis)

J. Mella-Raipán, J. Romero-Parra and G. Recabarren-Gajardo, *ACS Chemical Neuroscience*, 2020, **11**, 3905–3927 (DARK Classics in Chemical Neuroscience: Heroin and Desomorphine)

E. A. Alves, *ACS Chemical Neuroscience*, 2020, **11**, 3968–3978 (DARK Classics in Chemical Neuroscience: Krokodil)

H. Hussain, S. Angly, G. Michel, E. Garcia and A. Fadel, *European Journal of Case Reports in Internal Medicine*, 2023, **10**, 004181 (Krokodil-associated cutaneous ulceration, cardiac arrhythmia and thrombi)

'SPICE'

P. I. Dargan and D. M. Wood (eds.), *Novel Psychoactive Substances: Classification, Pharmacology and Toxicology*, Amsterdam, Academic Press, 2013, esp. pp. 317–343

J. B. Zawilska and J. Wojcieszak, *International Journal of Neuropsychopharmacology*, 2014, **17**, 509–525 (review)

B. Mills, A. Yepes and K. Nugent, *American Journal of the Medical Sciences*, 2015, **350**, 59–62 (review)

B. M. Ford, S. Tai, W. E. Fantegrossi and P. L. Prather, *Trends in Pharmacological Sciences*, 2017, **38**, 257–276 (review)

V. Abbate, M. Schwenk, B. C. Presley and N. Uchiyama, *Pure and Applied Chemistry*, 2018, **90**, 1255–1282

'SPICE' IN PRISONS

R. Ralphs, L. Williams, R. Askew and A. Norton, *International Journal of Drug Policy*, 2017, **40**, 57–69

C. Norman, G. Walker, B. McKirdy, C. McDonald, D. Fletcher, L. H. Antonides, O. B. Sutcliffe, N. Nic Daéid and C. McKenzie, *Drug Testing and Analysis*, 2020, **12**, 538–554 (detecting synthetic cannabinoid infused in paper in prisons)

O. Corazza, S. Coloccini, S. Marrinan, M. Vigar, C. Watkins, C. Zene, A. Negri, A. Aresti, S. Darke, R. Rinaldi, A. Metastasio and G. Bersani, *Frontiers in Psychiatry*, 2020, **11**, 450 ('Spice' users in gaols)

C. Norman, S. Halter, B. Haschimi, D. Acreman, J. Smith, A. J. Krotulski, A. L. A. Mohr, B. K. Logan, N. Nic Daéid, V. Auwärter and C. McKenzie, *Drug Testing and Analysis*, 2021, **13**, 841–852 (synthetic cannabinoids in prison vs. the general population)

T. J. McMann, A. Calac, M. Nali, R. Cuomo, J. Maroulis and T. K Mackey, *JMIR Infodemiology*, 2022, **2**, e37632 (synthetic cannabinoids in prisons)

Urgent action needed to curb rise in prison deaths linked to spice, *The Guardian*, Jan. 31 2023

Five inmates die after taking spice laced with fentanyl, *The Sun*, Aug. 3 2023

Smuggling Spice into Nebraska jails, *Journal Star* 2023

JWH-018

J. W. Huffman, D. Dai, B. R. Martin and D. R. Compton, *Bioorganic & Medicinal Chemistry Letters*, 1994, **4**, 563–566 (cannabimimetic 1-alkyl-2methyl-3-(1-naphthoyl)indoles))

J. A. H. Lainton, J. W. Huffman, B. R. Martin and D. R. Compton, *Tetrahedron Letters*, 1995, **36**, 1401–1404 (1-alkyl-3-(1-naphthoyl)pyrroles)

J. W. Huffman et al, *Bioorganic and Medicinal Chemistry*, 2005, **13**, 89–112 (1-alkyl-3-(1-naph-thoyl)indoles as highly selective CB(2) receptor agonists)

V. Auwärter, S. Dresen, W. Weinmann, M. Müller, M. Pütz and N. Ferreirós, *Journal of Mass Spectrometry*, 2009, **44**, 832–837 (JWH-018, first detection in 'Spice')

R. Lindigkeit, A. Boehme, I. Eiserloh, M. Luebbecke, M. Wiggermann, L. Ernst and T. Beuerle, *Forensic Science International*, 2009, **191**, 58–63 (JWH-018 and JWH-073, NMR and MS)

B. K. Atwood, J. W. Huffman, A. Straiker and K. Mackie, *British Journal of Pharmacology*, 2010, **160**, 585–593. (JWH-018)

T. Sobolevsky, I. Prasolov and G. Rodchenkov, *Forensic Science International*, 2010, **200**, 141–147 (MS, JWH-018 metabolites in urine)

S. Every-Palmer, *Drug and Alcohol Dependence*, 2011, **117**, 152–157 (JWH-018 and psychosis)

J. L. Wiley, J. A. Marusich, J. W. Huffman, R. L. Balster and B. F. Thomas, 'Hijacking of basic research: The case of synthetic cannabinoids'. RTI Press publication No. OP-0007-1111. Research Triangle Park, NC, RTI Press, 2011

AMB-FUBINACA AND RELATED MOLECULES

I. P. Buchler, M. J. Hayes, S. G. Hedge, S. L. Hockerman, D. E. Jones, S. W. Kortum, J. G. Rico, R. E. Tenbrink and K. K. Wu, Indazole derivatives as CB1 receptor modulators and their preparation and use in the treatment of CB1-mediated diseases. Patent WO 2009/106982 (2009)

N. Uchiyama, S. Matsuda, M. Kawamura, R. Kikura-Hanajiri and Y. Goda, *Forensic Toxicology*, 2013, **31**, 223–240 (ADB-FUBINACA in illegal drugs in Japan)

S. D. Banister, M. Longworth, R. Kevin, S. Sachdev, M. Santiago, J. Stuart, J. B. C. Mack, M. Glass, I. S. McGregor, M. Connor and M. Kassiou, *ACS Chemical Neuroscience*, 2016, **7**, 1241–1254 (pharmacology of AMB-FUBINACA and related drugs)

K. G. Shanks, W. Clark and G. Behonick, *Journal of Analytical Toxicology*, 2016, **40**, 236–239 (death associated with ADB-FUBINACA)

J. Carlier, X. Diao, A. Wohlfarth, K. Scheidweiler and M. A. Huestis, *Current Neuropharmacology*, 2017, **15**, 682–691 (ADB-FUBINACA metabolism)

A. J. Adams, S. D. Banister, L. Irizarry, J. Trecki, M. Schwartz and R. Gerona, *New England Journal of Medicine*, 2017, **376**, 235–242 ('zombie' deaths associated with AMB-FUBINACA in 2016 in New York)

T. F. Gamage, C. E. Farquhar, T. W. Lefever, J. A. Marusich, R. C. Kevin, I. S. McGregor, J. L. Wiley and B. F. Thomas, *Journal of Pharmacology and Experimental Therapeutics*, 2018, **365**, 437–446 (MMB- and MDMB-FUBINACA and other synthetic cannabinoids)

N. Brandehoff, A. Adams, K. McDaniel, S. D. Banister, R. Gerona and A. A. Monte, *Clinical Toxicology*, 2018, **56**, 193–198 (ADB-FUBINACA (aka Black Mamba) abuse in Colorado)

P. Adamowicz, et al., *EClinicalMedicine*, 2020, **25**, 100460 (deaths associated with AMB-FUBINACA in Auckland NZ in 2017–2019)

J. A. Hvozdovich, C. W. Chronister, B. K. Logan and B. A. Goldberger, *Journal of Analytical Toxicology*, 2020, **44**, 298–300 (synthetic cannabinoids including AMB-FUBINACA and deaths in Florida prisons)

S. Darke, J. Duflou, M. Farrell, A. Peacock and J. Lappin, *Clinical Toxicology (Phila)*, 2020, **58**, 368–374 (characteristics and circumstances of synthetic cannabinoid-related deaths)

J. C. Hancox, N. J. Kalk and G. Henderson, *Therapeutic Advances in Drug Safety*, 2020, **11**, 1–4 (synthetic cannabinoids and potential cardiac arrhythmia risk)

R. Roque-Bravo, R. S. Silva, R. F. Malheiro, H. Carmo, F. Carvalho, D. D. da Silva and J. P. Silva, *Annual Review of Pharmacology and Toxicology*, 2023, **63**, 187–209 (synthetic cannabinoids overview)

S. Cotton, 'Spice compounds never tested on humans', *The Sunday Times*, Nov 4th 2018 https://www.thetimes.com/world/us-world/article/spice-compounds-never-tested-on-humans-8n8039mjc

NITAZENES

Y. Bao, S. Men, J. Shi and L. Lu, *Lancet Psychiatry*, 2019, **6**, e15 (control of fentanyl-related substances in China.)

P. Blanckaert, A. Cannaert, K. Van Uytfanghe, F. Hulpia, E. Deconinck, S. Van Calenbergh and C. Stove, *Drug Testing and Analysis*, 2020, **12**, 422–430 (emergence of isotonitazene)

C. M. Zagorski, J. M. Myslinski and L. G. Hill, *International Journal of Drug Policy*, 2020, **86**, 102939 (isotonitazene as a contaminant of concern in the illegal opioid supply)

C. L. Shover, T. O. Falasinnu, R. B. Freedman and K. Humphreys, *Journal of Addiction Medicine*, 2021, **15**, 429–431 (characteristics of isotonitazene-Involved overdose deaths)

I. Ujváry, R. Christie, M. Evans-Brown, A. Gallegos, R. Jorge, J. de Morais and R. Sedefov, *ACS Chemical Neuroscience*, 2021, **12**, 1072–1092 (etonitazene and related benzimidazoles)

M. Siczek, M. Zawadzki, M. Siczek, A. Chłopaś-Konowałek and P. Szpot, *Forensic Toxicology*, 2021, **39**, 146–155 (etazene)

F. Mueller, C. Bogdal, B. Pfeiffer, L. Andrello, A. Ceschif, A. Thomas and E. Grata, *Forensic Science International*, 2021, **320**, 110686 (three Swiss fatalities involving isotonitazene)

A. Di Trana, S. Pichini, R. Pacifici, R. Giorgetti and F. P. Busardò, *Frontiers in Psychiatry*, 2022, **13**, 858234 (synthetic benzimidazole opioids)

K. Hasegawa, K. Minakata, M. Suzuki and O. Suzuki, *Forensic Toxicology*, 2022, **40**, 234–243 (nitazenes reviewed)

J. Pergolizzi, R. Raffa, J. A. K. LeQuang, F. Breve and G. Varrassi, *Cureus*, 2023, **15**, e40736 (nitazenes reviewed)

N. J. Malcolm, B. Palkovic, D. J. Sprague, M. M. Calkins, J. K. Lanham, A. L. Halberstadt, A. G. Stucke and J. D. McCorvy, *iScience*, 2023, **26**, 107121 (potency of isotonitazene and its metabolite, N-desethyl isotonitazene)

A. N. Edinoff, D. M. Garza, S. P. Vining, M. E. Vasterling, E. D. Jackson, K. S. Murnane, A. M. Kaye, R. N. Fair, Y. J. L. Torres, A. E. Badr, E. M. Cornett and A. D. Kaye, *Pain and Therapy*, 2023, **12**, 399–421 (dangers of fentanyls and nitazenes)

N. Eastwood and S. Schlossenberg, *BMJ*, 2023, **383**, 2421 (nitazenes on the UK illegal drugs market)

A. Holland, C. S. Copeland, G. W. Shorter, D. J Connolly, A. Wiseman, J. Mooney, K. Fenton and M. Harris, *Lancet Public Health*, 2024, **9**, e71–e72 (nitazenes and the UK drugs crisis)

M. M. Vandeputte, G. C. Glatfelter, D. Walther, N. K. Layle, D. M. St. Germaine, I. Ujváry, D. M. Iula, M. H. Baumann and C. P. Stove, *Pharmacological Research*, 2024, **210**, 107503 (nitazenes: *in vitro* μ-opioid receptor assays and *in vivo* behavioral studies in mice)

CHAPTER 5 NASTY SMELLING MOLECULES

HYDROGEN SULFIDE

J. G. Moore, L. D. Jessop and D. N. Osborne, *Gastroenterology*, 1987, **93**, 1321–1329 (flatus)

B. Tvedt, K. Skyberg, O. Aaserud, A. Hobbesland and T. Mathiesen, *American Journal of Industrial Medicine*, 1991, **20**, 91–101 (H$_2$S and brain damage)

R. J. Reiffenstein, W. C. Hulbert and S. H. Roth, *Annual Review of Pharmacology and Toxicology*, 1992, **32**, 109–134 (toxicology of H_2S)

F. L. Suarez, J. Springfield and M. D. Levitt, *Gut*, 1998, **43**, 100–104 (H_2S in flatus)

H. Sato, T. Hirose, T. Kimura et al., *Journal of Health Sciences*, 2001, **47**, 483–490 (malodorants in faeces and urine)

H. Sato, H. Morimatsu, T. Kimura, Y. Moriyama, T. Yamashita and Y. Nakashima, *Journal of Health Sciences*, 2002, **48**, 179–185 (malodorants in faeces)

R. G. Hendrickson, A. Chang and R. J. Hamilton, *American Journal of Industrial Medicine*, 2004, **45**, 346–350 (H_2S fatalities)

C. E. Garner, S. Smith, B. de L. Costello, P. White, R. Spencer, C. S. J. Probert and N. M. Ratcliffe, *FASEB Journal*, 2007, **21**, 1675–1688 (VOCs in faeces)

D. H. Truong, M. A. Eghbal, W. Hindmarsh, S. H. Roth and P. J. O'Brien, *Drug Metabolism Reviews*, 2006, **38**, 733–744 (molecular mechanisms of H_2S toxicity)

A. Christia-Lotter, C. Bartoli, M. Piercecchi-Marti, D. Demory, A. Pelissier-Alicot, A. Sanvoisin and G. Leonetti, *Forensic Science International*, 2007, **169**, 206–209 (fatal inhalation of H_2S)

C. N. Harvey-Woodworth, *British Dental Journal*, 2013, **214**, E20 (H_2S and Me_2S in halitosis)

V. S. Lina, A. R. Lippert and C. J. Chang, *Proceedings of the National Academy of Sciences of the United States of America*, 2013, **110**, 7131–7135 (monitoring hydrogen sulfide in the body)

E. M. Bos, H. Van Goor, J. A. Joles, M. Whiteman and H. G. Leuvenink, *British Journal of Pharmacology*, 2015, **172**, 1479–1493 (H_2S in the human body)

DIMETHYL SULFIDE

L. Pierce and M. Hayashi, *Journal of Chemical Physics*, 1961, **35**, 479–485

T. Iijima, S. Tsuchiya and M. Kimura, *Bulletin of the Chemical Society of Japan*, 1977, **50**, 2564–2567 (Me_2S structure)

F. L. Suarez, J Springfield and M. D. Levitt, *Gut*, 1998, **43**, 100–104 (Me_2S in flatulence)

A. Tangerman, *International Dental Journal*, 2002, **52**, 201–206

A. Tangerman and E. G. Winkel, *Journal of Clinical Periodontology*, 2007, **34**, 748–755

C. N. Harvey-Woodworth, *British Dental Journal*, 2013, **214**, E20

M. Monfort-Codinach, E. Chimenos-Küstner, R. Alburquerque and J. López-López, *Oral Health and Dental Management*, 2014, **13**, 975–981 (Me_2S and halitosis)

T. Talou, A. Gaset, M. Delmas, M. Kulifaj and C. Montant, *Mycological Research*, 1990, **94**, 277–278 (Me_2S and truffle hunting)

F. Bellesia, A. Pinetti, A. Bianchi and B. Tirillini, *Flavour and Fragrance Journal*, 1998, **13**, 56–58 (volatile organic compounds of Italian black truffle)

G. C. Kite, *Kew Bulletin*, 2000, **55**, 237–240

M. C. Stensmyr, I. Urru, I. Collu, M. Celander, B. S. Hansson and A.-M. Angioy, *Nature*, 2002, **420**, 625–626 (*Helicodiceros muscivorus*)

P. Díaz, E. Ibáñez, F. J. Señoráns and G. Reglero, *Journal of Chromatography A*, 2003, **1017**, 207–214

F. Martin et al., *Nature*, 2010, **464**, 1033–1038 (genome of the Périgord black truffle, *Tuber melanosporum*)

R. Splivallo, S. Ottonello, A. Mello and P. Karlovsky, *New Phytologist*, 2011, **189**, 688–699

P. C. Schmidberger and P. Schieberle, *Journal of Agricultural and Food Chemistry*, 2017, **65**, 9287–9296 (aroma compounds in white Alba truffle and Burgundy truffle (*Tuber uncinatum*))

A. M. Mustafa, S. Angeloni, F. K. Nzekoue, D. Abouelenein, G. Sagratini, G. Caprioli and E. Torregiani, *Molecules*, 2020, **25**, 5948 (truffle volatiles)

R. Epping, L. Bliesener, T. Weiss and M. Koch, *Molecules*, 2022, **27**, 5169 (truffle marker volatiles)

J. Niimi, A. Deveau and R. Splivallo, *New Phytologist*, 2021, **230**, 1623–1638 (variations in white truffle aroma)

G. B. Cunningham, V. Strauss and P. G. Ryan, *Journal of Experimental Biology*, 2008, **211**, 3123–3127 (African penguins detect Me_2S)

G. A. Nevitt, *Journal of Experimental Biology*, 2008, **211**, 1706–1713

M. S. Savoca and G. A. Nevitt, *Proceedings of the National Academy of Sciences of the United States of America*, 2014, **111**, 4157–4161 (Me_2S and seabirds)

J. M. González, A. W. B. Johnston, M. Vila-Costa and A. Buchan, Genetics and Molecular Features of Bacterial Dimethylsulfoniopropionate (DMSP) and Dimethyl Sulfide (DMS) Transformations, in: *Handbook of Hydrocarbon and Lipid Microbiology*, K. N. Timmis (ed.), Berlin Heidelberg, Springer-Verlag, 2010, pp. 2101–1211 (bacterial formation of Me_2S)

H. Schäfer, N. Myronova and R. Boden, *Journal of Experimental Botany*, 2010, **61**, 315–334 (Me_2S in the biosphere)

DIMETHYL DISULFIDE

G. C. Kite and W. L. A. Hetterscheid, *Phytochemistry*, 1997, **46**, 71–75 (Inflorescence odours of *Amorphophallus* and *Pseudodracontium (Araceae)*)

G. C. Kite, *Kew Bulletin*, 2000, **55**, 237–240

M. C. Stensmyr, I. Urru, I. Collu, M. Celander, B. S. Hansson and A.-M. Angioy, *Nature*, 2002, **420**, 625–626 (*Helicodiceros muscivorus*)

M. Shirasu, K. Fujioka, S. Kakashima, S. Nagai, Y. Tomizawa, H. Tsukaya, J. Murata, Y. Manome and K. Touhara, *Bioscience, Biotechnology, and Biochemistry*, 2010, **74**, 2550–2554 (*Amorphophallus titanum*)

W. L. A. Hetterscheid, A. Wistuba, V. B. Amoroso, M. P. Medecilo and C. Claudel, *Botanical Studies*, 2012, **53**, 415–420 (*Amorphophallus natolii*)

H. K. Stutz, G. J. Silverman, P. Angeline and R. E. Levin, *Journal of Food Science*, 1991, **56**, 1147–1153

S. D. Johnson and A. Jürgens, *South African Journal of Botany*, 2010, **76**, 796–807 (decaying meat)

UNSATURATED THIOLS IN SKUNKS AND ELSEWHERE

T. B. Aldrich, *Journal of Experimental Medicine*, 1896, **1**, 323–340 (Common Skunk, *Mephitis mephitica*)

W. F. Wood, *Journal of Chemical Ecology*, 1990, **16**, 2057–2065 (Striped skunk)

W. F. Wood, C. G. Morgan and A. Miller, *Journal of Chemical Ecology*, 1991, **17**, 1415–1420 (Spotted skunk)

W. F. Wood, C. O. Fisher and G. A. Graham, *Journal of Chemical Ecology*, 1993, **19**, 837–841 (Hog-nosed skunk)

W. F. Wood, B. G. Sollers, G. A. Dragoo and J. W. Dragoo, *Journal of Chemical Ecology*, 2002, **28**, 1865–1870 (Hooded skunk)

W. F. Wood, *The Chemical Educator*, 1999, **4**, 44–50 (review of skunk spray research)

D. De Keukeleire, A. Heyerick, K. Huvaere, L. H. Skibsted and M. L. Andersen, *Cerevisia*, 2008, **33**, 133–144 (lightstruck beer)

H. Tamura, A. Fujita, M. Steinhaus, E. Takahisa, H. Watanabe and P. Schieberle, *Journal of Agricultural and Food Chemistry*, 2011, **59**, 10211–10218 (roasted sesame seeds)

F. San-Juan, J. Cacho, V. Ferreira and A. Escudero, *Talanta*, 2012, **99**, 225–231 (3-methyl-2-bu-tene-1-thiol in wine)
I. W. H. Oswald, M. A. Ojeda, R. J. Pobanz, K. A. Koby, A. J. Buchanan, J. Del Rosso, M. A. Guzman and T. J. Martin, *ACS Omega*, 2021, **6**, 31667–31676 (thiols in cannabis)

THIOLS IN WINES

A. Roland, R. Schneider, A. Razungles and F. Cavelier, *Chemical Reviews*, 2011, **111**, 7355–7376 (thiols in wines reviewed)
T. Allen, M. Herbst-Johnstone, M. Girault, P. Butler, G. Logan, S. Jouanneau, L. Nicolau and P. A. Kilmartin, *Journal of Agricultural and Food Chemistry*, 2011, **59**, 10641–10650 (thiols in Sauvignon blanc wines)
M. Mestres, O. Busto and J. Guasch, *Journal of Chromatography A*, 2000, **881**, 569–581 (organic sulfur compounds in wine aroma)
L. Mateo-Vivaracho, J. N. Zapata, J. Cacho and V. Ferreira, *Journal of Agricultural and Food Chemistry*, 2010, **58**, 10184–10194 (thiols in white wine)
D. Rowe, K. Auty, S. Jameson, P. Setchell and G. Carter-Jones, *Perfume Flavor*, 2004, **29**(7), 2–20 (high impact sulfur compounds in perfumes)
P. Fuchsmann, M. T. Stern, Y-A. Brügger and K. Breme, *Journal of Agricultural and Food Chemistry*, 2015, **63**, 7511–7521 (sulfur compounds in Swiss Tilsit cheeses)

SULFUR COMPOUNDS IN FRUITS

D. Lehmann, A. Dietrich, U. Hener and A. Mosandl, *Phytochemical Analysis*, 1995, **6**, 255–257
A. Buettner and P. Schieberle, *Journal of Agricultural and Food Chemistry*, 1999, **47**, 5189–5193 (grapefruit odorants)
R. J. Cannon and C.-T. Ho, *Journal of Food and Drug Analysis*, 2018, **26**, 445–468 (volatile sulfur compounds in tropical fruits)
Y. Y. Voon, N. Sheikh, A. Hamid, G. Rusul, A. Osman and S. Y. Quek, *Food Chemistry*, 2007, **103**, 1217–1227 (durian fruit)
J-X. Li, P. Schieberle and M. Steinhaus, *Journal of Agricultural and Food Chemistry*, 2012, **60**, 11253–11262 (Thai durian)

NATURAL GAS ODORANTS

K. Liszka, M. Łaciak and A. Oliinyk, *AGH Drilling, Oil, Gas*, 2014, **31**(1), 59–71
D. L. Tenkrat, T. Hlincik and O. Prokes, in: P. Potocnik (ed.), *Natural Gas Odorization*, Sciyo, 2010. https://doi.org/10.5772/9825
D. R. Michanowicz, O. M. Leventhal, J. K. Domen, S. R. Williams, E. D. Lebel, L. A. L. Hill, J. J. Buonocore, C. L. Nordgaard, A. S. Bernstein and S. B. C. Shonkoff, *Current Environmental Health Reports*, 2023, **10**, 337–352 (natural gas odorants reviewed)

THIOLS IN BODILY ODOURS

S. Li, L. Ahmed, R. Zhang, Y. Pan, H. Matsunami, J. L. Burger, E. Block, V. S. Batista and H. Zhuang, *Journal of the American Chemical Society*, 2016, **138**, 13281–13288 (human odorant receptor OR2T11 responds to thiols)
F. Noe, J. Polster, C. Geithe, M. Kotthoff, P. Schieberle and D. Krautwurst, *Chemical Senses*, 2017, **42**, 195–210 (OR2M3 human odorant receptor for the sensitive detection of onion key food odorant 3-mercapto-2-methylpentan-1-ol)

A. Tangerman and E. G. Winke, *Phosphorus, Sulfur, and Silicon and the Related Elements*, 2013, 188, 396–402 (volatile sulfur compounds and bad breath reviewed)

Y. Hasegawa, M. Yabuki and K. Takeuchi, *Chemistry & Biodiversity*, 2004, **1**, 2042–2050

K. Takeuchi, M. Yabuki and Y. Hasegawa, *Flavour and Fragrance Journal*, 2013, **28**, 223–230

M. Troccaz, C. Starkenmann, Y. Niclass, M. van de Waal and A. J. Clark, *Chemistry & Biodiversity*, 2004, **1**, 1022–1035

A. Natsch, J. Schmid and F. Flachsmann, *Chemistry & Biodiversity*, 2004, **1**, 1058–1072

M. Troccaz, G. Borchard, C. Vuilleumier, S. Raviot-Derrien, Y. Niclass, S. Beccucci and C. Starkenmann, *Chemical Senses*, 2009, **34**, 203–210 (odorants in human axillary odour)

M. Granvogl, M. Christlbauer and P. Schieberle, *Journal of Agricultural and Food Chemistry*, 2004, **52**, 2797–2802 (3-mercapto-2-methylpentan-1-ol in onions)

M. Christlbauer and P. Schieberle, *Journal of Agricultural and Food Chemistry*, 2011, **59**, 13122–13130 (mercapto-2-methylpentan-1-ol in beef and pork vegetable gravies)

A. M. Sourabié, H.-E. Spinnler, P. Bonnarme, A. Saint-Eve and S. Landaud, *Journal of Agricultural and Food Chemistry*, 2008, **56**, 4674–4680 (ethyl 3-mercaptopropionate in Munster and Camembert)

B. Quintanilla-Casas, N. Dulsat-Serra, N. Cortés-Francisco, J. Caixach and S. Vichi, *LWT - Food Science and Technology*, 2015, **64**, 1085–1090 (thiols in brewed coffee)

M. J. I. Mattina, J. J. Pignatello and R. K. Swihart, *Journal of Chemical Ecology*, 1991, **17**, 451–462 (thiols in bobcat urine)

P. Apps, M. Claase, B. Yexley and J. W. McNutt, *Journal of Ethology*, 2017, **35**, 153–159 (3-mercapto-3-methylbutanol in wildcat and leopard urine)

V. Varlet and X. Fernandez, *Food Science and Technology International*, 2010, **16**, 463–503 (sulfur-containing volatiles in seafood reviewed)

TRIMETHYLAMINE

M. Al-Waiz, R. Ayesh, S. C. Mitchell, J. R. Idle and R. L. Smith, *Lancet*, 1987, **329**, 634–635

R. Ayesh, S. C. Mitchell, A. Q. Zhang and R. L. Smith, *BMJ*, 1993, **307**, 655–657

S. C. Mitchell, *Perspectives in Biology and Medicine*, 1996, **39**, 223–235

S. C. Mitchell and R. L. Smith, *Drug Metabolism and Disposition*, 2001, **29**, 517–521 (trimethylaminuria)

B. A. Seibel and P. J. Walsh, *Journal of Experimental Biology*, 2002, **205**(3), 297–306 (Me$_3$NO in fish)

H. U. Rehman, *Postgraduate Medical Journal*, 1999, **75**, 451–452 (treatment of trimethylaminuria)

J. R. Cashman, K. Camp, S. S. Fakharzadeh, P. V. Fennessey, R. N. Hines, O. A. Mamer, S. C. Mitchell, G. P. Nguyen, D. Schlenk, R. L. Smith, S. S. Tjoa, D. E. Williams and S. Yannicelli, *Current Drug Metabolism*, 2003, **4**, 151–170 (review)

E. A. Fraser-Andrews, N. J. Manning, P. Eldridge, J. McGrath and H. du P. Menage, *Clinical and Experimental Dermatology*, 2003, **28**, 203–205 (diagnosis)

J. Messenger, S. Clark, S. Massick and M. Mechtel, *Journal of Clinical and Aesthetic Dermatology*, 2013, **6**, 45–48 (trimethylaminuria reviewed)

Y. Zhu, E. Jameson, M. Crosatti, H. Schäfer, K. Rajakumar, T. D. H. Bugg and Y. Chen, *Proceedings of the National Academy of Sciences of the United States of America*, 2014, **111**, 4268–4273 (production of Me$_3$N from carnitine)

S. Rath, B. Heidrich, D. H. Pieper and M. Vital, *Microbiome*, 2017, **5**, 54 (Me$_3$N-producing bacteria of the human gut)

D. Roddy, P. McCarthy, D. Nerney, J. Mulligan-Rabbitt, E. Smith and E. P. Treacy, *JIMD Reports*, 2020, **57**, 1–9 (trimethylaminuria and daily psychosocial functioning)

A. C. Schmidt and J-C. Leroux, *Drug Discovery Today*, 2020, **25**, 1710–1717 (trimethylaminuria treatment reviewed)

MOLECULES AND SWEAT

S. Munk, P. Münch, L. Stahnke, J. Adler-Nissen and P. Schieberle, *Journal of Surfactants and Detergents*, 2000, **3**, 505–515 (odorants in soiled laundry odour)

A. Natsch, H. Geller, P. Gygax, J. Schmid and G. Acuna, *Journal of Biological Chemistry*, 2003, **278**, 5718–5727 (axillary odour precursors)

M. Troccaz, C. Starkenmann, Y. Niclass, M. van de Waal and A. J. Clark, *Chemistry & Biodiversity*, 2004, **1**, 1022–1035 (3-methyl-3-sulfanylhexan-1-ol in sweat)

A. Natsch, J. Schmid and F. Flachsmann, *Chemistry & Biodiversity*, 2004, **1**, 1058–1071 (sulfanylalkanols in sweat)

A. G. James, J. Casey, D. Hylands and G. Mycock, *World Journal of Microbiology and Biotechnology*, 2004, **20**, 787–793 (fatty acids and axillary malodour)

Y. Hasegawa, M. Yabuki and M. Matsukane, *Chemistry & Biodiversity*, 2004, **1**, 2042–2050 (3-hydroxy-3-methylhexanoic acid and 3-sulfanylalkan-1-ols in axillary sweat)

C. Starkenmann, Y. Niclass, M. Troccaz and A. J. Clark, *Chemistry & Biodiversity*, 2005, **2**, 705–716 (the precursor to 3-methyl-3-sulfanylhexan-1-ol in sweat)

A. Natsch, S. Derrer, F. Flachsmann and J. Schmid, *Chemistry & Biodiversity*, 2006, **3**, 1–20 (carboxylic acids and axillary secretions)

D. J. Penn, E. Oberzaucher, K. Grammer, G. Fischer, H. A. Soini, D. Wiesler, M. V. Novotny, S. J. Dixon, Y. Xu and R. G. Brereton, *Journal of the Royal Society Interface*, 2007, **4**, 331–340 (B. O. 'fingerprint molecules')

R. C. Smallegange, N. O. Verhulst and W. Takken, *Trends in Parasitology*, 2011, **27**, 143–148 (sweaty skin and vectors for pathogens)

E. Fredrich, H. Barzantny, I. Brune and A. Tauch, *Trends in Microbiology*, 2013, **21**, 305–312 (human axilla reviewed)

C. J. Denawaka, I. A. Fowlis and J. R. Dean, *Journal of Chromatography A*, 2016, **1438**, 216–225 (volatiles identified in clothing malodour)

C. J.-F. Chappuis, Y. Niclass, I. Cayeux and C. Starkenmann, *Flavour and Fragrance Journal*, 2016, **31**, 95–100 (toilet malodour compounds)

M. Rudden, R. Herman, M. Rose, D. Bawdon, D. S. Cox, E. Dodson, M. T. G. Holden, A. J. Wilkinson, A. G. James and G. H. Thomas, *Scientific Reports*, 2020, **10**, 12500 (thiol production in human B.O.)

MOLECULES AND HUMAN REMAINS

A. A. Vass, *Microbiology Today*, 2001, **28**, 190–192 (human decomposition)

M. Statheropoulos, C. Spiliopoulou and A. Agapiou, *Forensic Science International*, 2005, **153**, 147–155 (human decay molecules)

M. Statheropoulos, A. Agapiou, C. Spiliopoulou, G. C. Pallis and E. Sianos, *Science of the Total Environment*, 2007, **385**, 221–227 (VOCs in early stages of human decomposition)

A. A. Vass, R. R. Smith, C. V. Thompson, M. N. Burnett, N. Dulgerian and B. A. Eckenrode, *Journal of Forensic Sciences*, 2008, **53**, 384–391 (odour from human remains)

E. M. Hoffman, A. M. Curran, N. Dulgerian, R. A. Stockham and B. A. Eckenrode, *Forensic Science International*, 2009, **186**, 6–13 (volatiles from human remains)

D. T. Hudson, A. M. Curran and K. G. Furton, *Journal of Forensic Sciences*, 2009, **54**, 1270–1277 (human scent from crime scenes)

L. E. DeGreeff and K. G. Furton, *Analytical and Bioanalytical Chemistry*, 2011, **401**, 1295–1307 (volatiles from human remains)

M. E. Cablk, E. E. Szelagowski and J. C. Sagebiel, *Forensic Science International*, 2012, **220**, 118–125 (VOCs in the headspace above human and animal remains)

A. A. Vass, *Forensic Science International*, 2012, **222**, 234–241 (odour associated with soil in contact with human remains)

A. Hussain, L. R. Saraivaa, D. M. Ferrero, G. Ahuja, V. S. Krishna, S. D. Liberles and S. I. Korsching, *Proceedings of the National Academy of Sciences of the United States of America*, 2013, **110**, 19579–19584 (trace amine associated receptor 13c (TAAR13c) for cadaverine)

I. Riezzo, M. Neri, M. Rendine, A. Bellifemina, S. Cantatore, C. Fiore and E. Turillazzi, *Forensic Science International*, 2014, **244**, 213–221 (do cadaver dogs work?)

E. Rosier, S. Loix, W. Develter, W. Van de Voorde, J. Tytgat and E. Cuypers, *PLoS One*, 2015, **10**, e0137341 (esters associated with death smell)

S. Everts, *Chemical & Engineering News*, 2016, **94**(14), 16–18

MOLECULES AND MOSQUITOES

F. Acree, R. B. Turner, H. K. Gouck, M. Beroza and N. Smith, *Science*, 1968, **161**, 1346–1347 (L-Lactic acid as a mosquito attractant)

B. G. J. Knols, *Lancet*, 1996, **348**, 1322 (Limburger cheese attracts *Anopheles gambiae*)

B. G. J. Knols, J. J. van Loon, A. Cork, R. D. Robinson, W. Adam, J. Meijerink, R de Jong and W. Takken, *Bulletin of Entomological Research*, 1997, **87**, 151–159 (butanoic acid as mosquito attractant)

A. G. James, J. Casey, D. Hyliands and G. Mycock, *World Journal of Microbiology & Biotechnology*, 2004, **20**, 787–793 (fatty acid bacteria)

J. G. Logan, M. A. Birkett, S. J. Clark, S. Powers, N. J. Seal, L. J. Wadhams, A. J. Mordue (Luntz) and J. A. Pickett, *Journal of Chemical Ecology*, 2008, **34**, 308–322 (compounds that reduce mosquito attractiveness)

R. C. Smallegange, Y. T. Qiu, G. Bukovinszkiné-Kiss, J. J. A. Van Loon and W. Takken, *Journal of Chemical Ecology*, 2009, **35**, 933–943 (propanoic acid, butanoic acid, 3-methylbutanoic acid and other carboxylic acids attract mosquitoes)

R. C. Smallegange, N. O. Verhulst and W. Takken, *Trends in Parasitology*, 2011, **27**, 143–148 (review- sweaty skin: an invitation to bite?)

M. E. De Obaldia, T. Morita, L. C. Dedmon, D. J. Boehmler, C. S. Jiang, E. V. Zeledon, J. R. Cross and L. B. Vosshall, *Cell*, 2022, **185**, 4099–4116 (people with differential mosquito attraction produce different amounts of carboxylic acids)

D. Giraldo, S. Rankin-Turner, A. Corver, G. M. Tauxe, A. L. Gao, D. M. Jackson, L. Simubali, C. Book, J. C. Stevenson, P. E. Thuma, R. C. McCoy, A. Gordus, M. M. Mburu, E. Simulundu and C. J. McMeniman, *Current Biology*, 2023, **33**, 2367–2382 (butanoic acid, 2-methylpropanoic acid and 3-methylbutanoic acid attractive to *Anopheles gambiae*)

EUGLOSSINE BEES

R. L. Dressler, *Annual Review of Ecology, Evolution, and Systematics*, 1982, **13**, 373–394 (biology of the orchid bees)

T. Eltz, W. M. Whitten, D. W. Roubik and K. E. Linsenmair, *Journal of Chemical Ecology*, 1999, **25**, 157–176

T. Eltz, Y. Zimmermann, J. Haftmann, R. Twele, W. Francke, J. J. G. Quezada-Euan and K. Lunau, *Proceedings of the Royal Society B*, 2007, **274**, 2843–2848 (orchid bees collecting)

CHAPTER 6 WAR & PEACE IN NATURE

M. D. Gordin, *Journal of Chemical Education*, 1996, **83**, 561–565 (Borodin)

GREEN AROMA

A. Hatanaka, *Phytochemistry*, 1993, **34**, 1201–1218 (biogeneration of green odour)

R. K. Monson, M. T. Lerdau, T. D. Sharkey, D. S. Schimel and R. Fall, *Atmospheric Environment*, 1995, **29**, 2989–3002

R. Fall, T. Karl, A. Hansel, A. Jordan and W. Lindinger, *Journal of Geophysical Research*, 1999, **104**, 15963–15974 (C_6 volatiles emitted after leaf wounding)

H. Akutsu, T. Kikusui, Y. Takeuchi, K. Sano, A. Hatanaka and Y. Mori, *Physiology & Behavior*, 2002, **75**, 355–360 (hexenal/hexenol mixture reduces stress in rats)

T. Sasabe, M. Kobayashi, Y. Kondo, H. Onoe, S. Matsubara, S. Yamamoto, H. Tsukada, K. Onoe, H. Watabe, H. Iida, M. Kogo, K. Sano, A. Hatanaka, T. Sawada and Y. Watanabe, *Chemical Senses*, 2003, **28**, 565–572 ('Green Odour' and reduction of stress)

I. Kubo, K-I. Fujita, A. Kubo, K-I. Nihei and T. Ogura, *Journal of Agricultural and Food Chemistry*, 2004, **52**, 3329–3332 (antibacterial C_6 compounds in coriander, against *Salmonella choleraesuis*)

T. Karl, F. Harren, C. L. Warneke, J. de Gouw, C. Grayless and R. Fall, *Journal of Geophysical Research*, 2005, **110**, D15302 (VOCs from grass crops)

K. Kishimoto, K. Matsui, R. Ozawa and J. Takabayashi, *Plant and Cell Physiology*, 2005, **46**, 1093–1102

K. Kishimoto, K. Matsui, R. Ozawa and J. Takabayashi, *Plant Science*, 2006, **170**, 715–723 (C_6 aldehydes and resistance to a fungal pathogen)

T. Oka, S. Hayashida, Y. Kaneda, M. Takenaga, Y. Tamagawa, S. Tsuji and A. Hatanaka, *BioPsychoSocial Medicine*, 2008, **2**, 2 (anti-stress effect of green odour in healthy humans)

S. Allmann and I. T. Baldwin, *Science*, 2010, **329**, 1075–1078 (when feeding, tobacco horn-worms summon predators)

A. Scala, S. Allmann, R. Mirabella, M. A. Haring and R. C. Schuurink, *International Journal of Molecular Sciences*, 2013, **14**, 17781–17811 (GLVs as weapons against herbivores and pathogens)

M. Naeem ul Hassan, Z. Zainal and I. Ismail, *Plant Biotechnology Journal*, 2015, **13**, 727–739 (GLV review)

M. Ameye, S. Allmann, J. Verwaeren, G. Smagghe, G. Haesaert, R. C. Schuurink and K. Audenaert, *New Phytologist*, 2018, **220**, 666–683 (GLV volatiles produced by plants: a meta-analysis)

W. Kirstine, I. Gabally, Y. Ye and M. Hooper, *Journal of Geophysical Research*, 1998, **103**, 10605–10619 (VOCs emitted from pasture)

M. Wikelski, M. Quetting, Y. Cheng, W. Fiedler, A. Flack, A. Gagliardo, R. Salas, N. Zannoni and J. Williams, *Scientific Reports*, 2021, **11**, 12912 (GLVs attract white storks to freshly cut meadows)

K. Matsui and J. Engelberth, *Plant and Cell Physiology*, 2022, **63**, 1378–1390 (GLVs, plant responses to biotic attack)

PLANT DEFENCE

P. W. Paré and J. H. Tumlinson, *Plant Physiology*, 1999, **121**, 325–331 (plant volatiles as a defence against insect herbivores)

L. L. Walling, *Journal of Plant Growth Regulation*, 2000, **19**, 195–216 (plant responses to herbivores)

J. Engelberth, H. T. Alborn, E. A. Schmelz and J. H. Tumlinson, *Proceedings of the National Academy of Sciences of the United States of America*, 2004, **101**, 1781–1785 (airborne signals prime plants against insect herbivore attack)

K. L Korth, *Genome Biology*, 2003, **4**, 221 (response of plants to herbivorous insects)

T. G. E. Davies, L. M. Field, P. N. R. Usherwood and M. S. Williamson, *Life*, 2007, **59**, 151–162 (DDT, pyrethrins and pyrethroids)

A. R. War, M. G. Paulraj, T. Ahmad, A. A. Buhroo, B. Hussain, S. Ignacimuthu and H. C. Sharma, *Plant Signalling & Behavior*, 2012, **7**, 1306–1320 (mechanisms of plant defence against insect herbivores)

S. Lev-Yadun, *Plant Signaling & Behavior*, 2016, **11**, e1179419 (plants versus herbivores)

R. W. J. Kortbeek, M. van der Gragt and P. M. Bleeker, *European Journal of Plant Pathology*, 2019, **154**, 67–90 (plant metabolites against insects)

A. Steppuhn, K. Gase, B. Krock, R. Halitschke and I. T. Baldwin, *PLoS Biology*, 2004, **2**, e217 (nicotine's defensive function in nature)

M. W. Sabelis and H. E. Vandebaan, *Entomologia Experimentalis et Applicata*, 1983, **33**, 303–314

M. W. Sabelis, J. E. Vermaat and A. Groeneveld, *Physiological Entomology*, 1984, **9**, 437–446 (methyl salicylate (MeSA), predatory mites and spider mites)

V. Shulaev, P. Silverman and I. Raskin, *Nature*, 1997, **385**, 718–721 (airborne signalling by MeSA in plant pathogen resistance)

T. Karl, A. Guenther, A. Turnipseed, E. G. Patton and K. Jardine, *Biogeosciences*, 2008, **5**, 1287–1294 (walnuts and MeSA)

K. Ament, V. Krasikov, S. Allmann, M. Rep, F. L. W. Takken and R. C. Schuurink, *The Plant Journal*, 2010, **62**, 124–134 (MeSA in tomato and spider mites)

D. M. A. Dempsey and D. F. Klessig, *BMC Biology*, 2017, **15**, 23 (salicylic acid combats disease in plants)

E. E. Farmer and C. A. Ryan, *Proceedings of the National Academy of Sciences of the United States of America*, 1990, **87**, 7713–7716 (Methyl jasmonate (MeJa) induces synthesis of defensive proteinase inhibitor proteins)

I. T. Baldwin, *Proceedings of the National Academy of Sciences of the United States of America*, 1998, **95**, 8113–8118 (jasmonate-induced responses in plants)

J. S. Thaler, *Nature*, 1999, **399**, 686 (jasmonate-inducible plant defences)

M. L. Campos, J. H. Kang and G. A. Howe, *Journal of Chemical Ecology*, 2014, **40**, 657–675 (jasmonate-triggered plant immunity)

T. A. Dar, M. Uddin, M. M. Khan, K. R. Hakeem and H. Jaleel, *Environmental and Experimental Botany*, 2015, **115**, 49–57 (jasmonates counter plant stress)

T. Lortzing and A. Steppuhn, *Current Opinion in Insect Science*, 2016, **14**, 32–39 (jasmonate signalling in plants)

WEAPONISED INSECTS

T. Eisner, M. Eisner and M. Siegler, *Secret Weapons: Defenses of Insects, Spiders, Scorpions and Other Many-Legged Creatures*, Cambridge, MA, The Belknap Press, 2005

FIRE ANTS

J. G. MacConnell, M. S. Blum and H. M. Fales, *Science*, 1970, **168**, 840–841

E. G. LeBrun, N. T. Jones and L. E. Gilbert, *Science*, 2014, **343**, 1014–1017

D. J. C. Kronauer, *Current Biology*, 2014, **24**, R372–R374

H. Srisong, S. Sukprasert, S. Klaynongsruang, J. Daduang and S. Daduang, *Journal of Venomous Animals and Toxins including Tropical Diseases*, 2018, **24**, 23

BOMBARDIER BEETLE

D. J. Aneshansley, T. Eisner, J. M. Widom and B. Widom, *Science*, 1969, **165**, 61–63

T. Eisner and J. Dean, *Proceedings of the National Academy of Sciences of the United States of America*, 1976, **73**, 1365–1367

T. Eisner and D. J. Aneshansley, *Proceedings of the National Academy of Sciences of the United States of America*, 1999, **96**, 9705–9709

NECRODES SURINAMENSIS

T. Eisner and J. Meinwald, *Psyche*, 1982, **89**, 357–367

J. Meinwald, B. Roach and T. Eisner, *Journal of Chemical Ecology*, 1986, **13**, 35–38

B. Roach, T. Eisner and J. Meinwald, *Journal of Organic Chemistry*, 1990, **55**, 4047–4951

LUBBER GRASSHOPPERS ROMALEA GUTTATA

C. G. Jones, T. A. Hess, M. S. Blum, D. W. Whitman, P. J. Silk and M. S. Blum, *Journal of Chemical Ecology*, 1986, **12**, 749–761

Eisner, Eisner and Siegler, 'Secret Weapons', pp. 97–101

WALKINGSTICK INSECTS, ANISOMORPHA BUPRESTOIDES

A. T. Dossey, S. S. Walse and A. S. Edison, *Journal of Chemical Ecology*, 2008, **34**, 584–590

Eisner, Eisner and Siegler, 'Secret Weapons', pp. 93–96

AUSTRALIAN TERMITE NASUTITERMES EXITIOSUS

T. Eisner, I. Kriston and D. J. Aneshansley, *Behavioural Ecology and Sociobiology*, 1976, **1**, 83–125

B. P. Moore, *Journal of Insect Physiology*, 1968, **14**, 33–39

Eisner, Eisner and Siegler, 'Secret Weapons', pp. 82–88

BLISTER BEETLES

J. E. Carrel and T. Eisner, *Science*, 1974, **183**, 755–757

Eisner, Eisner and Siegler, 'Secret Weapons', pp. 220–227

BURNET MOTHS

N. B. Jensen, M. Zagrobelny, K. Hjernø, C. E. Olsen, J. Houghton-Larsen, J. Borch, B. L. Møller and S. Bak, *Nature Communications*, 2011, **2**, 273

E. S. Briolat, M. Zagrobelny, C. E. Olsen, J. D. Blount and M. Stevens, *Evolution*, 2018, **72**, 1460–1474

SOIL CENTIPEDES AND POLYDESMID MILLIPEDES

T. H. Jones, W. E. Conner, J. Meinwald, H. E. Eisner and T. Eisner, *Journal of Chemical Ecology*, 1976, **2**, 421–429

S. S. Duffey, M. S. Blum, H. M. Fales, S. L. Evans, R. W. Roncadori, D. L. Tiemann and Y. Nakagawa, *Journal of Chemical Ecology*, 1977, **3**, 101–113
Eisner, Eisner and Siegler, 'Secret Weapons', pp. 33–36 and 43–47
T. Eisner, *For Love of Insects*, Cambridge, MA, The Belknap Press, 2003, pp. 58–67

AFRICAN WEAVER ANT, OECOPHYLLA LONGINODA

O. Roux, J. Billen, J. Orivel and A. Dejean, *PLoS One*, 2010, **5**, e8957

PLANTS ATTRACTING INSECTS

H. E. M. Dobson, 'Relationship between Floral Fragrance Composition and Type of Pollinator', in: *Biology of Floral Scent*, Natalia Dudareva and Eran Pichersky (eds.), Boca Raton, Taylor and Francis, 2006, pp. 147–198
R. A. Raguso, *Annual Review of Ecology and Systematics*, 2008, **39**, 549–569 (ecology and evolution of floral scent)
B. J. Glover, *Current Biology*, 2011, **21**, R307–R308 (pollinator attraction)
F. P. Schiestl and S. D. Johnson, *Trends in Ecology & Evolution*, 2013, **28**, 307–315 (pollinator-mediated evolution of floral signals)
H. Bouwmeester, R. C. Schuurink, P. M. Bleeker and F. Schiestl, *The Plant Journal*, 2019, **100**, 892–907 (volatiles in plant communication)
D. Bisrat and C. Jung, *Journal of Ecology and Environment*, 2022, 46, 03 (roles of flower scent in bee–flower mediations)

EUGLOSSINE BEES

J. D. Ackerman, *Biotropica*, 1989, **21**, 340–347 (choice of odorants for 44 different species of male Euglossine bees)
G. Gerlach and R. Schill, *Botanica Acta*, 1991, **104**, 379 (composition of orchid scents attracting euglossine bees)
F. P. Schiestl and D. W. Roubik, *Journal of Chemical Ecology*, 2003, **29**, 253–257 (electroantennogram response of *Euglossa cybelia* and *Eulaema polychrome* to odorants)
T. Eltz, Y. Zimmermann, J. Haftmann, R. Twele, W. Francke, J. J. G. Quezada-Euan and K. Lunau, *Proceedings of the Royal Society of London. Series B*, 2007, **274**, 2843–2848 (perfume collection in orchid bees)
D. R. Roberts, W. D. Alecrim, J. M. Heller, S. R. Erhardt and J. B. Lima, *Nature*, 1982, **297**, 62–63
V. Walter and D. Roberts, *Science of the Total Environment*, 2007, **377**, 371–377 (a DDT collecting bee in Brazil)

INSECT REPELLENTS

T. M. Katz, J. H. Miller and A. A. Hebert, *Journal of the American Academy of Dermatology*, 2008, **58**, 865–871 (insect repellents reviewed)
M. F. Maia and S. J. Moore, *Malaria Journal*, 2011, **10**(Suppl 1), S11 (plant-based insect repellents: a review)
S. D. Rodriguez, H-N. Chung, K. K. Gonzales, J. Vulcan, Y. Li, J. A. Ahumada, H. M. Romero, M. De La Torre, F. Shu and I. A. Hansen, *Journal of Insect Science*, 2017, **17**, 24 (comparison of DEET, PMD, citronellol candles and other devices)
F. Okumu, *Malaria Journal*, 2020, **19**, 260 (mosquito nets review)

E. T. McCabe, W. F. Barthel, S. I. Gertler and S. A. Hall, *Journal of Organic Chemistry*, 1954, **19**, 493–498 (DEET synthesis)

J. A. Pickett, M. A. Birkett and J. G. Logan, *Proceedings of the National Academy of Sciences of the United States of America*, 2008, **105**, 13195–13196

Z. Syed and W. S. Leal, *Proceedings of the National Academy of Sciences of the United States of America*, 2008, **105**, 13598–13603 (mosquitoes smell and avoid DEET)

D. R. Swale, B. Sun, F. Tong and J. R. Bloomquist, *PLoS One*, 2014, **9**, e103713 (neurotoxicity and mode of action of DEET)

V. Corbel, M. Stankiewicz, C. Pennetier, D. Fournier, J. Stojan, E. Girard, M. Dimitrov, J. Molgó, J-M. Hougard and B. Lapied, *BMC Biology*, 2009, **7**, 47 (DEET inhibits cholinesterases in insect nervous systems)

E. J. Dennis, O. V. Goldman and L. B. Vosshall, *Current Biology*, 2019, **29**, 1551–1556 (*Aedes aegypti* uses legs to sense DEET)

D. R. Swale and J. R Bloomquist, *Pest Management Science*, 2019, **75**, 2068–2070 (review of DEET safety)

D. R. Barnard, U. R. Bernier, K. H. Posey and R-D. Xue, *Journal of Medical Entomology*, 2002, **39**, 895–899 (comparison of DEET and TMD against black salt marsh mosquitoes)

J. Drapeau, M. Rossano, D. Touraud, U. Obermayr, M. Geier, A. Rose and W. Kunz, *Comptes Rendus Chimie*, 2011, **14**, 629–635 (green PMD synthesis)

I. Kurnia, A. Yoshida, N. Chaihad, T. Prakoso, S. Li, X. Du, X. Hao, A. Abudula and G. Guan, *New Journal of Chemistry*, 2020, **44**, 10441 (PMD synthesis)

L. G. Borrego, N. Ramarosandratana, E. Jeanneau, E. Métay, V. V. Ramanandraibe, M. T. Andrianjafy and M. Lemaire, *Journal of Agricultural and Food Chemistry*, 2021, **69**, 11095–11109 (activity of PMD isomers towards *Aedes albopictus*)

CHAPTER 7 ORGANOCHLORINE COMPOUNDS

N. Winterton, *Green Chemistry*, 2000, **2**, 173–225

G. W. Gribble (ed.), Natural Production of Organohalogen Compounds, *The Handbook of Environmental Chemistry*, Volume 3 Part P, Berlin, Springer 2003

R. M. Moore in Gribble, 2003, pp. 85–102 (marine sources of volatile organohalogens)

G. W. Gribble, *Chemosphere*, 2003, **52**, 289–297

G. W. Gribble, *American Scientist*, 2004, **92**, 342–349 ('Amazing Organohalogens')

F. Keppler, D. B. Harper, T. Röckmann, R. M. Moore and J. T. G. Hamilton, *Atmospheric Chemistry and Physics*, 2005, **5**, 2403–2411 (natural CH_3Cl production)

J. W. Dini, *Plating & Surface Finishing*, 2009, **107**, 14–17

G. W. Gribble, *Progress in the Chemistry of Organic Natural Products*, 2023, **121**, 1–546

CHLORINATING WATER

J. Emsley, *Nature's Building Blocks*, Oxford, OUP, 2001, pp. 126–132 (chlorinating water supplies)

C. Anderson, *Nature*, 1991, **354**, 255 (Peruvian cholera outbreak)

E. Gotuzzo, J. Cieza, J. Estremadoyro and C. Seas, *Infectious Disease Clinics of North America*, 1994, **8**, 183–205

M. J. McGuire, '*The Chlorine Revolution: Water Disinfection and the Fight to Save Lives*', Denver, American Water Works Association, 2014; see also S. Andrew, *Journal of Chemical Education*, 2014, **91**, 466–467

CH₃CL

M. Rossberg, W. Lendle, G. Pfleiderer, A. Tögel, T. R. Torkelson and K. K. Beutel, *'Chloromethanes', Ullmann's Encyclopedia of Industrial Chemistry*, 2012, **9**, 15–42

V. P. Yandrapu and N. R. Kanidarapu, *Periodica Polytechnica Chemical Engineering*, 2022, **66**, 341–353 (indust. synth.)

J. E. Lovelock, *Nature*, 1975, **256**, 193–194 (detection of CH₃Cl in air over England)

NATURAL SOURCES OF CH₃CL

R. Watling and D. B. Harper, *Mycological Research*, 1998, **102**, 769–787 (fungi on wood)

H. Anke and R. W. S. Weber, *Mycologist*, 2006, **20**, 83–89 (fungi on wood)

M. O. Andreae and P. Merlet, *Global Biogeochemical Cycles*, 2001, **15**, 955–966 (burning)

M. A. Cochrane, *Nature*, 2003, **421**, 913–919 (burning)

X. Ni and L. P. Hager, *Proceedings of the National Academy of Sciences of the United States of America*, 1998, **95**, 12866–12871 (CH₃Cl from *Batis* mari*tima*, a halophytic plant that grows abundantly in salt marshes)

R. C. Rhew, B. R. Miller and R. F. Weiss, *Nature*, 2000, **403**, 292–295 (salt marshes)

Y. Yokouchi, M. Ikeda, Y. Inuzuka and T. Yukawa, *Nature*, 2002, **416**, 163–165 (vegetation)

R. E. Stoiber, D. C. Leggett, T. F. Jenkins, R. P. Murrmann and W. I. Rose, *Geological Society of America Bulletin*, 1971, **82**, 2299–2302 (volcanic gas from Santiaguito, Guatemala 1969)

E. C. Y. Inn, J. F. Vedder, E. P. Concon and D. O'Hara, *Science*, 1981, **211**, 821–823 (Mount St. Helens)

R. A. Rasmussen, M. A. K. Khalil, R. W. Dalluge, S. A. Penkett and B. Jones, *Science*, 1982, **215**, 665–667 (Mount St. Helens)

F. Tassi, F. Capecchiacci, J. Cabassi, S. Calabrese, O. Vaselli, D. Rouwet, G. Pecoraino and G. Chiodini, *Journal of Geophysical Research*, 2012, 117, D17305 (Mount Etna)

E. C. Fayolle et al., *Nature Astronomy*, 2017, **1**, 703–708 (CH₃Cl in interstellar space and in the coma of comet 67P/Churyumov–Gerasimenko (67P/C-G))

Y. Sailaukhanuly, Z. Sárossy, L. Carlsen and H. Egsgaard, *Chemosphere*, 2014, **111**, 575–579

W. C. McRoberts, F. Keppler, D. B. Harper and J. T. G. Hamilton, *Environmental Chemistry*, 2015, **12**, 426 (pectin as a source of chloromethane)

OTHER HALOMETHANES

D. B. Harper and J. T. G. Hamilton, Natural Production of Organohalogen Compounds, in: *The Handbook of Environmental Chemistry*, G. W. Gribble (ed.), Volume 3 Part P, Berlin, Springer, 2003, pp. 17–41

M. A. K. Khalil and R. A. Rasmussen, *Atmospheric Environment*, 1999, **33**, 1151 (atmospheric chloroform)

DDT

O. Zeidler, *Berichte der Deutschen Chemischen Gesellschaft*, 1874, **7**, 1180 (synthesis of DDT)

T. F. West and G. A. Campbell, *DDT and Newer Persistent Insecticides*, 2nd edition, London, Chapman and Hall, 1950

R. Carson, *Silent Spring*, Boston, Houghton Mifflin, 1962

R. L. Metcalf, *Journal of Agricultural and Food Chemistry*, 1973, **21**, 511–519 ('A Century of DDT')

R. G. Beatty, *The DDT Myth*, New York, John Day Co, 1973

E. M. Whelan, *Toxic Terror*, Buffalo, Prometheus Press, 1993, esp. pp. 91–120

J. Emsley, *Molecules at an Exhibition*, Oxford, OUP, 1995, pp. 163–167

R. Tron and R. Bate, *Malaria and the DDT Story*, London, Institute of Economic Affairs, 2001

C. M. Anelli, C. H. Krupke and R. P. Prasad, *American Entomologist*, 2006, **52**, 224–233

C. H. Krupke, R. P. Prasad and C. M. Anelli, *American Entomologist*, 2007, **53**, 16–26 ('Professional Entomology and the 44 Noisy Years since *Silent Spring*')

R. L. Myers, *The 100 Most Important Chemical Compounds. A Reference Guide*, Westport, Greenwood Press, 2007, pp. 95–97

R. J. Hargrove, DDT, in: *Molecules That Matter*, R. J. Giguere (ed), Philadelphia, Chemical Heritage Foundation, 2008, pp. 129–141

D. Kinkella, *DDT and The American Century*, Chapel Hill, The University of Carolina Press, 2011

E. Conis, *How to Sell Poison. The Rise, Fall and Toxic Return of DDT*, New York, Bold Type Books, 2022

DIELDRIN

J. Robinson, V. K. H. Brown, A. Richardson and M. Roberts, *Life Sciences*, 1967, **6**, 1207–1220 (quail and pigeon killed by dieldrin)

D. M. Jones, D. Bennett and K. E. Elgar, *Nature*, 1978, **272**, 52 (deaths of owls traced to insecticide-treated timber)

J. L. Jorgenson, *Environmental Health Perspectives*, 2001, **109** (supplement 1), 113–139 (Aldrin and dieldrin as persistent pollutants)

V. Zitko, Chlorinated pesticides: Aldrin, DDT, Endrin, Dieldrin, Mirex, in: *The Handbook of Environmental Chemistry, Vol. 3, Part O Persistent Organic Pollutants*, H. Fiedler (ed.), Berlin and Heidelberg, Springer-Verlag, 2003, pp. 47–90

M. Honeycutt and S. Shirley, *Encyclopedia of Toxicology*, Waltham, Elsevier Science, 2014, (Third Edition), pp. 107–110

P. Tsiantas, E. N. Tzanetou, H. Karasali and K. M. Kasiotis, *Agriculture*, 2021, **11**, 314 (dieldrin persistence)

R. L. Lipnick and D. C. G. Muir, Persistent, Bioaccumulative, and Toxic Chemicals I, in: *ACS Symposium Series*, R. L. Lipnick et al., (ed.), Washington, DC, American Chemical Society, 2000, pp. 1–12

SERTRALINE (*ZOLOFT*)

J. J. Li, *Laughing Gas, Viagra and Lipitor*, Oxford, OUP, 2006, p. 149

E. J. Corey, B. Czakó and L. Kürti, *Molecules and Medicine*, New Jersey, John Wiley, 2007, p. 231

J. J. Li, *Top Drugs*, Oxford, OUP, 2015, p. 123

DETTOL AND TCP

M. K. Bruch, 'Chloroxylenol: An Old-New Antimicrobial', in: *Handbook of Disinfectants and Antiseptics*, J. M. Ascenzi (ed.), New York, Marcel Dekker, 1996, p. 265; vide: https://en.wikipedia.org/wiki/Chloroxylenol

Jim Clark in *Chemguide* https://www.chemguide.co.uk/qandc/tcp.html

TETRACYCLINE

L. H. Conover, W. T. Moreland, A. R. English, C. R. Stephens and F. J. Pilgrim, *Journal of the American Chemical Society*, 1953, **75**, 4622–4623 (tetracycline)

C. R. Stephens, L. H. Conover, R. Pasternack, F. A. Hochstein, W. T. Moreland, P. P. Regna, F. J. Pilgrim, K. J. Brunings and R. B. Woodward, *Journal of the American Chemical Society*, 1954, **76**, 3568–3575 (chlorotetracycline structure)

J. Donohue, J. D. Dunitz, K. N. Trueblood and M. S. Webster, *Journal of the American Chemical Society*, 1963, **85**, 851–856 (crystal structure of chlorotetracycline)

T. H. Jukes, *Reviews of Infectious Diseases*, 1985, **7**, 702–707 (chlorotetracycline review)

M. Wainwright, *Miracle Cure. The story of Antibiotics*, Oxford, Blackwell, 1990, pp. 152–153 (chlorotetracycline discovery)

I. Chopra and M. Roberts, *Microbiology and Molecular Biology Reviews*, 2001, **65**, 232–260 (tetracycline antibiotics reviewed)

J. J. Li, *Laughing Gas, Viagra and Lipitor*, Oxford, OUP, 2006, pp. 67–69 (discovery of chlorotetracycline)

M. L. Nelson, A. Dinardo, J. Hochberg and G. J. Armelagos, *American Journal of Physical Anthropology*, 2010, **143**, 151–154 (tetracycline used in Sudanese Nubia 350–550 CE)

M. L. Nelson and S. B. Levy, *Annals of the New York Academy of Sciences*, 2011, **1241**, 17–32 (history of the tetracyclines)

A. O. Legendre, L. R. R. Silva, D. M. Silva, I. M. L. Rosa, L. C. Azarias, P. J. de Abreu, M. B. de Araújo, P. P. Neves, C. Torres, F. T. Martins and A. C. Doriguetto, *CrystEngComm*, 2012, **14**, 2532–2540 (structure of doxycycline monohydrate)

J. S. Grant, C. Stafylis, C. Celum, T. Grennan, B. Haire, J. Kaldor, A. F. Luetkemeyer, J. M. Saunders, J-M. Molina and J. D. Klausner, *Clinical Infectious Diseases*, 2020, **70**, 1247–1253 (doxycycline prophylaxis for bacterial STDs)

VANCOMYCIN

R. C. Moellering, *Clinical Infectious Diseases*, 2006, **42**, S3–S4

D. P. Levine, *Clinical Infectious Diseases*, 2006, **42**, S5–S12 (vancomycin: a history)

E. Rubinstein and Y. Keynan, *Frontiers in Public Health*, 2014, **2**, 217 (vancomycin revisited)

E. Mühlberg, F. Umstätter, C. Kleist, C. Domhan, W. Mier and P. Uhl, *Canadian Journal of Microbiology*, 2020, **66**, 11–16 (vancomycin and breaking antibiotic resistance in multidrug-resistant bacteria)

TEICOPLANIN

F. Parenti, G. Beretta, M. Berti and V. Arioli, *Journal of Antibiotics*, 1978, **31**, 276–283 (discovery)

A. H. Hunt, R. M. Molloy, J. L. Occolowitz, G. G. Marconi and M. Debono, *Journal of the American Chemical Society*, 1984, **106**, 4891–4895

J. C. J. Barna, D. H. Williams, D. J. M. Stone, T. W. C. Leung and D. M. Doddrell, *Journal of the American Chemical Society*, 1984, **106**, 4895–4902 (structure)

V. Vimberg, *Pharmaceuticals*, 2021, **14**, 1227 (review)

CHLORAMPHENICOL

J. Ehrlich, Q. R. Bartz, R. M. Smith, D. A. Joslyn and P. R. Burkholder, *Science*, 1947, **106**, 417

D. Gottlieb, P. K. Bhattacharyya, H. W. Anderson and H. E. Carter, *Journal of Bacteriology*, 1948, **55**, 409–417 (discovery)

J. Controulis, M. C. Rebstock and H. M. Crooks, *Journal of the American Chemical Society*, 1949, **71**, 2463–2468 (synthesis of chloramphenicol)

M. C. Gerald, *The Drug Book*, New York, Sterling, 2013, pp. 250–251

DICHLOROPHENOL

W. C. Agosta, *Chemical Communication*, New York and Oxford, WH Freeman, 1992, pp. 52–56

R. S. Berger, *Science*, 1972, **177**, 704–705 (2,6–Dichlorophenol, sex pheromone of the lone star tick)

Y. S. Chow, C. B. Wang and L. C. Lin, *Annals of the Entomological Society of America*, 1975, **68**, 485–488 (the brown dog tick, *Rhipicephalus sanguineus*)

D. E. Sonenshine, R. M. Silverstein, E. Plummer, J. R. West and T. McCullough, *Journal of Chemical Ecology*, 1976, **2**, 201–209 (the Rocky Mountain tick, *Dermacentor Andersoni* and the American dog tick, *Dermacentor variabilis*)

P. E. Hanson, J. A. Yoder, J. L. Pizzuli and C. I. Sanders, *Journal of Medical Entomology*, 2002, **39**, 945–947 (2,4-dichlorophenol in females of the American dog tick)

C. C. B. Louly, D. N. Silveira, S. F. Soares, P. H. Ferri, A. C. C. Melo and L. M. F. Borges, *Memórias do Instituto Oswaldo Cruz is a peer*, 2008, **103**, 60–65 (the Brazilian ticks *Rhipicephalus sanguineus* and *Amblyomma cajennense*)

K. K. Gachoka, D. P. Costa, C. C. B. Louly, L. L. Ferreira, L. M. F. Borges, L. C. Fariab and P. H. Ferri, *Journal of the Brazilian Chemical Society*, 2010, **21**, 1642–1647 (the South American tick *Amblyomma cajennense*)

HALOGENATED COMPOUNDS FROM MARINE FUNGI

C. Wang, H. Lu, J. Lan, K. H. A. Zaman and S. Cao, *Molecules*, 2021, **26**, 458 (review)

EPIBATIDINE AND ABT-594

J. W. Daly, H. M. Garraffo, T. F. Spande, M. W. Decker, J. P. Sullivan and M. Williams, *Natural Product Reports*, 2000, **17**, 131–135 (epibatidine discovery and other frog alkaloids reviewed)

H. M. Garraffo, T. F. Spande and M. Williams, *Heterocycles*, 2009, **79**, 207–217 (epibatidine review)

I. C. Umana, C. A. Daniele and D. S. McGehee, *Biochemical Pharmacology*, 2013, **86**, 1208–1214 (neuronal nicotinic receptors as analgesic targets and ABT-594)

R. E. de Oliveira Filho and A. T. Omori, *Journal of the Brazilian Chemical Society*, 2015, **26**, 837–850 (epibatidine syntheses reviewed)

R. D. Tarvin, C. M. Borghese, W. Sachs, J. C. Santos, Y. Lu, L. A. O'Connell, D. C. Cannatella, R. A. Harris and H. H. Zakon, *Science*, 2017, **357**, 1261–1266 (cause of frogs' resistance to epibatidine)

B. Salehi, S. Sestito, S. Rapposelli, G. Peron, D. Calina, M. Sharifi-Rad, F. Sharopov, N. Martins and J. Sharifi-Rad, *Biomolecules*, 2019, **9**, 6 (epibatidine review)

CHAPTER 8 ORGANOFLUORINE COMPOUNDS

A. Borodin, *Comptes Rendus Chimie*, 1862, **55**, 553–556 ('Facts for serve to the history of fluoride and preparation of benzoyl fluoride')

F. Swarts, *Academie Royale de Belgique*, 1892, **3**, 474–484 ('Étude sur le fluochloroforme')

R. E. Banks and J. C. Tatlow, Synthesis of C-F Bonds: The Pioneering Years, in: *Fluorine. The First Hundred Years*, R. E. Banks, D. W. A. Sharp and J. C. Tatlow (eds.), Lausanne and New York, Elsevier, 1986, pp. 71–108; *Journal of Fluorine Chemistry*, 1986, **33**, 71–108

P. Lebeau and A. Damiens, *Comptes rendus de l'Académie des Sciences*, 1926, **182**, 1340–1342; *Comptes rendus de l'Académie des Sciences*, 1930, **191**, 939

O. Ruff and R. Keim, *Zeitschrift für anorganische und allgemeine Chemie*, 1930, **192**, 249–256 (CF_4)

G. Balz and G. Schiemann, *Chemische Berichte*, 1927, **60**, 1186–1190 (synthesis of aromatic C-F bonds)

K. Müller, C. Faeh and F. Diederich, *Science*, 2007, **317**, 1881–1886 (fluorine in pharmaceuticals)

D. O'Hagan, *Chemical Society Reviews*, 2008, **37**, 308–319 (organofluorine chemistry and the C–F bond)

T. Okazoe, *Proceedings of the Japan Academy, Series B*, 2009, **85**, 276–289 (overview of organofluorine chemistry)

CFCS AND THEIR SUCCESSORS

T. Midgley and A. C. Henne, *Industrial & Engineering Chemistry Research*, 1930, **22**, 542–545 (organic fluorides as refrigerants, especially CF_2Cl_2)

R. J. Thompson, *Industrial & Engineering Chemistry Research*, 1932, **24**, 620–623 (Freon–a refrigerant)

T. Midgley, *Industrial & Engineering Chemistry*, 1937, **29**, 241–244 ('From the Periodic Table to Production')

C. J. Giunta, *Bulletin for the History of Chemistry*, 2006, **31**, 66–74 (Thomas Midgley Jr., and the invention of chlorofluorocarbon refrigerants)

J. E. Lovelock, *Nature*, 1971, **230**, 379 (atmospheric fluorine compounds as indicators of air movements)

J. E. Lovelock, R. J. Maggs and R. J. Wade, *Nature*, 1973, **241**, 194–196 (halogenated hydrocarbons in and over the Atlantic)

P. A. Crutzen, *Canadian Journal of Chemistry*, 1974, **52**, 1569–1581

M. J. Molina and F. S. Rowland, *Nature*, 1974, **249**, 810–812 (chlorine atom-catalysed destruction of ozone)

M. McFarland and J. Kaye, *Photochemistry and Photobiology*, 1992, **55**, 911–929 (CFC review)

W. Dekant, *Environmental Health Perspectives*, 1996, **104**, 75–83 (toxicology of CFCs; also greenhouse gases)

K. R. Williams, *Journal of Chemical Education*, 2000, **77**, 1540–1541 (refrigerator history)

P. A. Newman, L. D. Oman, A. R. Douglass, E. L. Fleming, S. M. Frith, M. M. Hurwitz, S. R. Kawa, C. H. Jackman, N. A. Krotkov, E. R. Nash, J. E. Nielsen, S. Pawson, R. S. Stolarski and G. J. M. Velders, *Atmospheric Chemistry and Physics*, 2009, **9**, 2113–2128 (what would have happened to the ozone layer if chlorofluorocarbons (CFCs) had not been regulated?)

K-H. Kim, Z-H. Shon, H. T. Nguyen and E-C. Jeon, *Atmospheric Environment*, 2011, **45**, 1369–1382 (CFCs and halocarbon alternatives reviewed)

M. I. Hegglin, *Nature*, 2018, **557**, 317–318; S. A. Montzka et al., *Nature*, 2018, **557**, 413–417 (unexpected CFC-11 emissions)

P. G. Simmonds et al., *Atmospheric Chemistry and Physics*, 2018, **18**, 4153–4169 (increase in HFC-23 (CHF_3) emissions linked to HCFC-22 ($CHClF_2$) production)

M. Rigby et al., *Nature*, 2019, **569**, 546–550 (increase in CFC-11 emissions from eastern China)

S. A. Montzka et al., *Nature*, 2021, **590**, 428–432 (decline in global CFC-11 emissions during 2018–2019)

S. Park et al., *Nature*, 2021, **590**, 433–437 (decline in emissions of CFC-11 from eastern China)

L. M Western, M. K. Vollmer, P. B. Krummel, K. E. Adcock, P. J. Fraser, C. M. Harth, R. L. Langenfelds, S. A. Montzka, J. Mühle, S. O'Doherty, D. E. Oram, S. Reimann, M. L. Rigby, I. Vimont, R. F. Weiss, T. D. S. Young and J. C. Laube, *Nature Geoscience*, 2023, **16**, 309–313 (increase in atmospheric CFC levels 2010–2020)

D. J. Sheldon and M. R. Crimmin, *Chemical Society Reviews*, 2022, **51**, 4977–4995 (repurposing of fluorinated gases)

H. Park, J. Kim, H. Choi, S. Geum, Y. Kim, R. L. Thompson, J. Mühle, P. K. Salameh, C. M. Harth, K. M. Stanley, S. O'Doherty, P. J. Fraser, P. G. Simmonds, P. B. Krummel, R. F. Weiss, R. G. Prinn and S. Park, *Atmospheric Chemistry and Physics*, 2023, **23**, 9401–9411 (rise in HFC-23 emissions from eastern Asia since 2015)

PFOA AND PFOS

L. Vierke, C. Staude, A. Biegel-Engler, W. Drost and C. Schulte, *Environmental Sciences Europe*, 2012, **24**, 16 (PFOA concerns)

S. E. Fenton, A. Ducatman, A. Boobis, J. C. DeWitt, C. Lau, C. Ng, J. S. Smith and S. M. Roberts, *Environmental Toxicology and Chemistry*, 2021, **40**, 606–630 (PFAs, toxicity and health)

J. P. Guin, J. A. Sullivan and K. R. Thampi, *ACS Engineering Au*, 2022, **2**, 134–150 (visible light induced degradation of PFAs)

X. Z. Lim, *Nature*, 2023, **620**, 24–27 (problems with 'Forever chemicals')

FLUORINE-CONTAINING NATURAL PRODUCTS

D. O'Hagan and D. B. Harper, *Journal of Fluorine Chemistry*, 1999, **100**, 127–133

G. W. Gribble, *'Naturally Occurring Organohalogen Compounds'*, *Progress in the Chemistry of Organic Natural Products*, Springer, Switzerland, 2023, **121**, pp. 150–152

ANAESTHETICS

K. L. Merrett and R. M. Jones, *British Journal of Hospital Medicine*, 1994, **52**, 260–263 (inhalational anaesthetic agents)

T. Hudlicky, C. Duan, J. W. Reed, F. Yan, M. Hudlicky, M. A. Endoma and E. I. Eger, *Journal of Fluorine Chemistry*, 2000, **102**, 363–367 (synthesis of fluorinated ethyl methyl ethers)

I. Smith, M. Nathanson and P. F. White, *British Journal of Anaesthesia*, 1996, **76**, 435–445

L. Delgado-Herrera, R. D. Ostroff and S. A. Rogers, *CNS Drug Reviews*, 2001, **7**(1), 48–120 (Sevoflurane as an ideal anesthetic)

R. K. Burdick, J. P. Villabona-Monsalve, G. A. Mashour and T. Goodson, *Scientific Reports*, 2019, **9**, 11351 (theories of anaesthesia)

S. Varughese and R. Ahmed, *Anesthesia and Analgesia*, 2021, **133**, 826–835 (environmental and occupational considerations of anesthesia)

K. Pożarowska, A. Rosińska, M. Orczykowski, M. Tyszkiewicz and M. Choina, *Journal of Education, Health and Sport*. 2022, **12**(11), 114–119 (diethyl ether as an anaesthetic)

SODIUM FLUOROACETATE POISONING

E. R. Kalmbach, *Science*, 1945, **102**, 232–233

A. T. Proudfoot, S. M. Bradberry and J. A. Vale *Toxicological Reviews*, 2006, **25**, 213–219

D. Á. Reyes, J. C. G. Mejía, J. F. G. González and M. A. Flórez, *La Revista Médica de Risaralda*, 2020, **26**, 166–171

J. Timbrell, *The Poison Paradox*, Oxford, OUP, 2005, p. 152

MEFLOQUINE/LARIAM

R. Speich and A. Haller, *New England Journal of Medicine*, 1994, **331**, 57–58 (Mefloquine side effects)

C. Hennequin, N. Bazin, F. Bisaro and A. Feline, *Archives of Internal Medicine*, 1994, **154**, 2360–2362 (severe psychiatric side effects with Mefloquine)

A. M. Croft, *Journal of the Royal Society of Medicine*, 2007, **100**, 170–174 (review of problems associated with Lariam)

P. Schlagenhauf, M. Adamcova, L. Regep, M. T. Schaerer and H-G. Rhein, *Malaria Journal*, 2010, **9**, 357 (mefloquine as a malaria prophylactic)

I. W. Sherman, *Magic Bullets to Conquer Malaria*, Washington, DC, ASM Press, 2011, pp. 137–149

K. M. Masterson, *The Malaria Project*, New York, New American Library, 2014, esp. pp. 338–339

5-FLUOROURACIL

C. Heidelberger, N. K. Chaudhuri, P. Danneberg, D. Mooren, L. Griesbach, R. Duchinsky, R. J. Schnitzer, E. Pleven and J. Scheiner, *Nature*, 1957, **179**, 663–666 (fluorinated pyrimidines as tumour inhibitors)

X. H. Xu, G. M. Yao, Y. M. Li, J. H. Lu, C. J. Lin, X. Wang and C. H. Kong, *Journal of Natural Products*, 2003, **66**, 285–288 (5-fluorouracil derivatives from the sponge *Phakellia fusca*)

V. C. Jordan, *Cancer Research*, 2016, **76**, 767–768 (background to the discovery of 5-Fluorouracil)

CIPROFLOXACIN

I. Brook, *Journal of Antimicrobial Agents*, 2002, **20**, 320–325 (cipro and anthrax)

P. C. Sharma, A. Jain, S. Jain, R. Pahwa and M. S. Yar, *Journal of Enzyme Inhibition and Medicinal Chemistry*, 2010, **25**, 577–589 (ciprofloxacin reviewed)

D. Sharma, R. P. Patel, S. T. R. Zaidi, M. M. R. Sarker, Q. Y. Lean and L. C. Ming, *Frontiers in Pharmacology*, 2017, **8**, 546 (ciprofloxacin and antibiotic resistance in developing countries)

Z. C. Conley, T. J. Bodine, A. Chou and L. Zechiedrich, *PLoS Pathogens*, 2018, **14**, e1006805 (ciprofloxacin and resistance)

A. Shariati, M. Arshadi, M. A. Khosrojerdi, M. Abedinzadeh, M. Ganjalishahi, A. Maleki, M. Heidary and S. Khoshnood, *Frontiers in Public Health*, 2022, **10**, 1025633 (bacterial resistance against ciprofloxacin and ways to improve efficacy)

ATORVASTATIN

B. D. Roth, *Progress in Medicinal Chemistry*, 2002, **40**, 1–22 (the discovery and development of atorvastatin)

J. J. Li, *Triumph of The Heart*, Oxford, OUP, 2009 (the story of statins)

A. Endo, *Proceedings of the Japan Academy, Series B*, 2010, **86**, 484–493 (history of the discovery of statins)

J. J. Li, *Top Drugs, Their History, Pharmacology and Syntheses*, Oxford, OUP, 2015, pp. 5–19 (Lipitor and statins)

R. Collins et al., *Lancet*, 2016, **388**, 2532–2561 (review of the efficacy and safety of statin therapy)

G. S. Costa, L. S. Julião-Silva, V. S. Belo, H. C. F. de Oliveira and V. E. Chaves, *European Heart Journal - Cardiovascular Pharmacotherapy*, 2023, **9**, 100–115 (effects of atorvastatin on blood pressure)

CELECOXIB

J. E. Frampton and G. M. Keating, *Drugs*, 2007, **67**, 2433–2472 (Celecoxib for arthritis and acute pain)

P. L. McCormack, *Drugs*, 2011, **71**, 2457–2489 (Celecoxib)

M. C. Gerald, *The Drug Book*, New York, Sterling, 2013, pp. 458–459 (Celecoxib)

J. J. Li, *Block-Buster Drugs*, Oxford, OUP, 2014, pp. 142–144 (Celecoxib, Vioxx and COX-2 inhibitors)

L. Puljak. A. Marin, D. Vrdoljak, F. Markotic, A. Utrobicic and P. Tugwell, *Cochrane Database of Systematic Reviews*. 2017, Issue 5, Art. No. CD009865. (Cochrane review of Celecoxib for osteoarthritis)

P. G. Shete, N. G. Shete, D. N. Kumbhakaran, N. S. Mane, V. S. Padole and R. P. Kalsait, *International Journal of Research in Pharmacy and Chemistry*, 2020, **10**, 382–389

B-R. Cheng, J-Q. Chen, X-W. Zhang, Q-Y. Gao, W-H. Li, Li-J. Yan, Y-Q. Zhang, C-J. Wu, J-L. Xing and J-P. Liu, *PLoS One*, 2021, **16**, e0261239 (systematic review and metaanalysis of celecoxib in rheumatoid arthritis and osteoarthritis)

OXYGEN CARRIERS

B. Remy, G. Deby-Dupont and M. Lamy, *British Medical Bulletin*, 1999, **55**, 277–298

J. Jägers, A. Wrobeln and K. B. Ferenz, *Pflügers Archiv - European Journal of Physiology*, 2021, **473**, 139–150 (perfluorocarbon-based oxygen carriers)

MEDICINES

F. M. D. Ismail, *Journal of Fluorine Chemistry*, 2002, **118**, 27–33 (important fluorinated drugs)

P. Shah and A. D. Westwell, *Journal of Enzyme Inhibition and Medicinal Chemistry*, 2007, **22**, 527–540 (fluorine in medicinal chemistry)

K. Müller, C. Faeh and F. Diederich, *Science*, 2007, **317**, 1881–1886 (fluorine in pharmaceuticals)

M. Inoue, Y. Sumii and N. Shibata, *ACS Omega*, 2020, **5**, 10633–10640 (organofluorine compounds as pharmaceuticals)

G. Shabir, A. Saeed, W. Zahid, F. Naseer, Z. Riaz, N. Khalil, Muneeba and F. Albericio, *Pharmaceuticals*, 2023, 16, 1162 (fluorinated drugs approved by the FDA (2016–2022))

G. Chandra, D. V. Singh, G. K. Mahato and S. Patel, *Chemical Papers*, 2023, **77**, 4985–4106 (fluorine - a small magic bullet atom in drugs)

PERFLUOROCUBANE

M. Sugiyama, M. Akiyama, Y. Yonezawa, K. Komaguchi, M. Higashi, K. Nozaki and T. Okazoe, *Science*, 2022, **377**, 756–759

CHAPTER 9 SMOKING AND VAPING

GENERAL

James I, *King of England, A Counter-Blaste to Tobacco*, imprinted at London by R. B., 1604 (London, the Rodale Press, 1954)

R. Doll and A. B. Hill, *BMJ*, 1954, **328**, 1451–1455 ('the mortality of doctors in relation to their smoking habits')

J. Goodman, *Tobacco in History: The Cultures of Dependence*, London, Routledge, 1993

T. Parker-Pope, *Cigarettes: Anatomy of an Industry from Seed to Smoke*, New York, New Press, 2000

K. Stratton, P. Shetty, R. Wallace and S. Bondurant, eds., *Clearing the Smoke: Assessing the Science Base for Tobacco Harm Reduction*, Washington, DC, National Academies Press, 2001

I. Gately, *La Diva Nicotina: The Story of How Tobacco Seduced the World*, London, Simon & Schuster, 2001

J. J. Buccafusco and A. V. Terry, *Life Science*, 2003, **72**, 2931–2942 (cotinine and Alzheimer's)

S. L. Gilman and Z. Xun (eds.), *Smoke. A Global History of Smoking*, London, Reaktion Books, 2004

A. Steppuhn, K. Gase, B. Krock, R. Halitschke and I. T. Baldwin, *PLoS Biology*, 2004, **2**, 1074–1080 (nicotine's defensive function in nature)

N. Minematsu, H. Nakamura, M. Furuuchi, T. Nakajima, S. Takahashi, H. Tateno and A. Ishizaka, *European Respiratory Journal*, 2006, **27**, 289–292 (genetics of CYP2A6 and smoking)

A. Brandt, *Cigarette Century: The Rise, Fall, and Deadly Persistence of the Product That Defined America*, New York, Basic Books, 2007

D. Kessler, K. Gase and I. T. Baldwin, *Science*, 2008, **321**, 1200–1202 (nicotine in floral scents)

X. Xiu, N. L. Puskar, J. A. P. Shanata, H. A. Lester and D. A. Dougherty, *Nature*, 2009, **458**, 534–537 (nicotine binding to brain receptors)

N. Oreskes and E. M. Conway, *Merchants of Doubt: How a Handful of Scientists Obscured the Truth on Issues from Tobacco Smoke to Global Warming*, New York, Bloomsbury Press, 2010

L. Etter, *The Devil's Playbook*, New York, Random House, 2021 (e-cigarettes and JUUL)

K. Wailoo, *Pushing Cool*, Chicago, University of Chicago Press, 2021 (menthol cigarettes and Afro-Americans)

NICOTINE PHARMACOLOGY AND ADDICTION

J. Cao, J. D. Belluzzi, S. E. Loughlin, D. E Keyler, P. R Pentel and F. M. Leslie, *Neuropsychopharmacology*, 2007, **32**, 2025–2035 (acetaldehyde increases nicotine self-administration in adolescent rats)

N. L. Benowitz, J. Hukkanen and P. Jacob, *Handbook of Experimental Pharmacology*, 2009, **192**, 29–60 (nicotine chemistry, metabolism, kinetics and biomarkers)

N. L. Benowitz, *Annual Review of Pharmacology and Toxicology*, 2009, **49**, 57–71 (pharmacology of nicotine, addiction and smoking-induced disease)

E. R. Kandel and D. B. Kandel, *New England Journal of Medicine*, 2014, **371**, 932–943 (nicotine as a gateway drug)

N. A. Goriounova and H. D. Mansvelder, *Cold Spring Harbor Perspectives in Medicine*, 2012, **2**, a012120; M. Yuan, S. J. Cross, S. E. Loughlin and F. M. Leslie, *Journal of Physiology*, 2015, **593**, 3397–3412 (nicotine and the adolescent brain)

F. Alasmari, L. E. C. Alexander, A. M. Hammad, C. M. Bojanowski, A. Moshensky and Y. Sari, *Frontiers in Pharmacology*, 2019, **10**, 885 (e-cigarettes and nicotine dependence)

M. R. Picciotto and P. J. Kenny, *Cold Spring Harbor Perspectives in Medicine*, 2021, **11**, a039610 (mechanisms of nicotine addiction)

A. Alhowail, *Molecular Medicine Reports*, 2021, **23**, 398 (nicotine, memory and cognition in Alzheimer's and Parkinson's, review)

E. M. Castro, S. Lotfipour and F. M. Leslie, *Pharmacological Research*, 2023, **190**, 106716 (nicotine on the developing brain)

G. Chen, S. Rahman and K. Lutfy, *Advances in Drug and Alcohol Research*, 2023, **3**, 11345 (e-cigarettes as a gateway)

J. O'Neill, M. P. Diaz, J. R. Alger, J-B. Pochon, D. Ghahremani, A. C. Dean, R. F. Tyndale, N. Petersen, S. Marohnic, A. Karaiskaki and E. D. London, *Molecular Psychiatry*, 2023, **28**, 4756–4765 (smoking and glutamate)

NICOTINE POISONING

J. M. Faulkner, *Journal of the American Medical Association*, 1933, **100**, 1664–1665 (nicotine poisoning by absorption through a florist's skin)

S. A. Cotton, *Every Molecule Tells a Story*, Boca Raton, CRC Press, 2011, pp. 106–107 (tabun as an alternative to nicotine)

B. Mayer, *Archives of Toxicology*, 2014, **88**, 5–7 (how much nicotine kills a human?)

GREEN TOBACCO SICKNESS

S-J. Park, H-S. Lim, K. Lee and S-J. Yoo, *Safety and Health at Work*, 2018, **9**, 71–74 (green tobacco sickness among tobacco harvesters)

J. Becam, E. Martin, G. Pouradier, N. Doudka, C. Solas, R. Guilhaumou and N. Fabresse, *Toxics*, 2023, **11**, 464 (transdermal nicotine poisoning in a man making e-liquids)

SMOKING AND DISEASE

F. L. Hoffman, *Annals of Surgery*, 1931, **93**, 50–67 (cancer and smoking habits)

F. H. Müller, *Zeitschrift für Krebsforschung*, 1939, **49**, 57–85 (smoking and lung cancer)

E. Schairer and E. Schöniger, *Zeitschrift für Krebsforschung*, 1943, **54**, 261–269 (lung cancer and smoking)

R. Schrek, L. A. Baker, G. P. Ballard and S. Dolgoff, *Cancer Research*, 1950, **10**, 49–58

M. L. Levin, H. Goldstein and P. R. Gerhardt, *Journal of the American Medical Association*, 1950, **143**, 336–338

C. A. Mills and M. M. Porter, *Cancer Research*, 1950, **10**, 539–542

E. L. Wynder and E. A. Graham, *Journal of the American Medical Association*, 1950, **143**, 329–336

R. Doll and A. B. Hill, *British Medical Journal*, 1950, 2, 739–748 (smoking and disease)

R. Doll and A. B. Hill, *British Medical Journal*, 1954, **1**, 145–155; *British Medical Journal*, 1956, **2**, 1071–1076 (smoking and lung cancer among British doctors)

E. C. Hammond and D. Horn, *Journal of the American Medical Association*, 1954, 155, 1316–1328 (smoking habits and death rates)

Smoking and *health. Report of the Advisory Committee to the Surgeon General of the Public Health Service*, US Department of Health, Education and Welfare. Washington DC, PHS Pub. No. 1103. 1964

R. Doll, *Statistical Methods in Medical Research*, 1998, **7**, 87–117 (history of evidence linking smoking with cancer)

A. Morabia, *American Journal of Public Health*, 2017, **107**, 1708–1710 (Nazi and Soviet views of smoking)

C. J. Murray, A. Y. Aravkin, P. Zheng, C. Abbafati, K. M. Abbas, M. Abbasi-Kangevari et al., *Lancet*, 2020, **396**, 1223–1249 (global smoking risk)

TOXIC CHEMICALS IN TOBACCO SMOKE

H. Witschi, I. Espiritu, J. L. Peake, K. Wu, R. R. Maronpot and K. E. Pinkerton, *Carcinogenesis*, 1997, **18**, 575–586 (the carcinogenicity of environmental tobacco smoke)

S. S. Hecht, *Journal of the National Cancer Institute*, 1999, **91**, 1194–1210 (tobacco smoke carcinogens and lung cancer)

D. Hoffmann, I. Hoffmann and K. El-Bayoumy, *Chemical Research in Toxicology*, 2001, **14**, 767–790 (carcinogens and tobacco smoke)

G. P. Pfeifer, M. F. Denissenko, M. Olivier, N. Tretyakova, S. S. Hecht and P. Hainaut, *Oncogene*, 2002, **21**, 7435–7451 (tobacco smoke carcinogens and smoking-associated cancers)

S. S. Hecht, *Nature Reviews Cancer*, 2003, **3**, 733–744 (tobacco carcinogens and biomarkers)

R. N. Proctor, *Tobacco Control*, 2012, **21**, 87–91 (the history of the discovery of the cigarette-lung cancer link)

Y. Li and S. S. Hecht, *Food and Chemical Toxicology*, 2022, **165**, 113179 (carcinogenic components of tobacco and tobacco smoke: a 2022 update)

COMPOUNDS IN TOBACCO SMOKE

R. L. Cooper, A. J. Lindsey and R. E. Waller, *Chemistry & Industry*, 1954, **46**, 1418 (3,4–benzopyrene in cigarette smoke)

S. S. Hecht and D. Hoffmann, *Carcinogenesis*, 1988, **9**, 875–884 (tobacco-specific nitrosamines)

A. Rodgman and T. A. Perfetti, *Beiträge zur Tabakforschung International / Contributions to Tobacco Research*, (a) 2006, **22**, 13–69; (b) 2006, **22**, 208–254 (polycyclic aromatic hydrocarbons in tobacco smoke)

A. Rodgman and T. A. Perfetti, *The Chemical Components of Tobacco and Tobacco Smoke*, 2nd edition, Boca Raton, FL, CRC Press, 2013

NITROSAMINES

S. S. Hecht, *Chemical Research in Toxicology*, 1998, **11**, 559–603 (biochemistry, biology, and carcinogenicity of tobacco-specific nitrosamines)

B. Siminszky, L. Gavilano, S. W. Bowen and R. E. Dewey, *Proceedings of the National Academy of Sciences of the United States of America*, 2005, **102**, 14919–14924 (conversion of nicotine to nornicotine)

A. T. Zarth, P. Upadhyaya, J. Yang and S. S. Hecht, *Chemical Research in Toxicology*, 2016, **29**, 380–389 (DNA adduct formation by N'-nitrosonornicotine (NNN))

E. Yalcin and S. de la Monte, *Journal of Physiology and Biochemistry*, 2016, **72**, 107–120 (tobacco nitrosamines as culprits in disease)

G. Bustamante, B. Ma, G. Yakovlev, K. Yershova, C. Le, J. Jensen, D. K. Hatsukami and I. Stepanov, *Chemical Research in Toxicology*, 2018, **31**, 731–738 (detection of N′-nitrosonornicotine in the saliva of e-cigarette users)

Y. Li and S. S. Hecht, *International Journal of Molecular Sciences*, 2022, **23**, 5109 (DNA adduct formation by N-nitrosamines)

VAPING

The best single sources of information on e-cigarettes is: National Academies of Sciences, Engineering, and Medicine. *Public Health Consequences of E-Cigarettes*, 2018, Washington, DC, The National Academies Press. https://doi.org/10.17226/24952 https://nap.nationalacademies.org/catalog/24952/public-health-consequences-of-e-cigarettes

J. E. Gotts, S-E. Jordt, R. McConnell and R. Tarran, *BMJ*, 2019, **366**, 15275 (respiratory effects of e-cigarettes reviewed)

P. Laucks and G. A. Salzman, *Missouri Medicine*, 2020, **117**, 159–164 (history of vaping)

E. A. Eshraghian and W. K. Al-Delaimy, *Tobacco Prevention & Cessation*, 2021, **7**(2), 10 (review, constituents of e-cigarette liquids and aerosols, to 2020)

C. Esquer, O. Echeagaray, F. Firouzi, C. Savko, G. Shain, P. Bose, A. Rieder, S. Rokaw, A. Witon-Paulo, N. Gude and M. A Sussman, *Life Science Alliance*, 2021, **22**, e202101246 (review, vaping and pulmonary hazards)

P. Marques, L. Piqueras and M-J. Sanz, *Respiratory Research*, 2021, **22**, 151 (e-cigarettes and health reviewed)

F. Effah, B. Taiwo, D. Baines, A. Bailey and T. Marczylo, *Journal of Toxicology and Environmental Health, Part B*, 2022, **25**, 343–371 (pulmonary effects of e-liquid flavours reviewed)

R. Sahu, K. Shah, R. Malviya, D. Paliwal, S. Sagar, S. Singh, B. G. Prajapati and S. Bhattacharya, *Advances in Respiratory Medicine*, 2023, **91**, 516–531 (e-cigarettes and associated health risks: an update on cancer potential)

J. J. Rose, S. Krishnan-Sarin, V. J. Exil, N. M. Hamburg, J. L. Fetterman, F. Ichinose, M. A. Perez-Pinzon, M. Rezk-Hanna and E. Williamson, *Circulation*, 2023, **148**, 703–728 (review, cardiopulmonary impact of e-cigarettes and vaping products)

H. Zong, Z. Hu, W. Li, M. Wang, Q. Zhou, X. Li and H. Liu, *Pflügers Archiv - European Journal of Physiology*, 2024, **476**, 875–888 (review, electronic cigarettes and cardiovascular disease)

STUDENT VAPING

T. W. Wang, A. Gentzke, S. Sharapova, K. A. Cullen, B. K. Ambrose and A. Jamal, *MMWR Morbidity and Mortality Weekly Report is a Weekly*, 2018, **67**, 629–633 (vaping in US students 2011–2017)

K. Jones and G. A. Salzman, *Missouri Medicine*, 2020, **117**, 56–58 (student vaping to 2018)

A. L. Groom, T-H. T. Vu, A. Kesh, J. L. Hart, K. L. Walker, A. L. Giachello, C. G. Sears, L. K. Tompkins, D. T. Mattingly, R. L. Landry, R. M. Robertson and T. J. Payne, *Preventive Medicine Reports*, 2020, **18**, 101094 (youth vaping flavour preferences)

S. M. Gaiha, J. Cheng and B. Halpern-Felsher, *Journal of Adolescent Health*, 2020, **67**, 519–523 (young vapers predisposed to COVID)

J. Birdsey, M. Cornelius, A. Jamal, E. Park-Lee, M. R. Cooper, J. Wang, M. D. Sawdey, K. A. Cullen and L. Neff, *Morbidity and Mortality Weekly Report*, 2023, **72**, 1173–1182 (tobacco use among U.S. middle and high school servants 2023)

S. E. Jackson, H. Tattan-Birch, L. Shahab and J. Brown, *BMJ*, 2024, **386**, e079016 (long term vaping trends among adults in England, 2013–23)

NICOTINE

E. R. Kandel and D. B. Kandel, *New England Journal of Medicine*, 2014, **371**, 932–943 (nicotine as a gateway drug)

M. Yuan, S. J. Cross, S. E. Loughlin and F. M. Leslie, *Journal of Physiology*, 2015, **593**, 3397–3412 (nicotine and the adolescent brain)

L. Shahab, M. L. Goniewicz, B. C. Blount, J. Brown, A. McNeill, K. U. Alwis, J. Feng, L. Wang and R. West, *Annals of Internal Medicine*, 2017, **166**, 390–400 (nicotine levels in smokers of e-cigarettes)

A. K. Duell, J. F. Pankow and D. H. Peyton, *Chemical Research in Toxicology*, 2018, **31**, 431–434 (free nicotine levels in e-liquids)

S. Talih, R. Salman, R. El-Hage, E. Karam, N. Karaoghlanian, A. El-Hellani, N. Saliba and A. Shihadeh, *Tobacco Control*, 2019, **28**, 678–680 (nicotine levels in JUUL e-cigarettes)

S. Talih, R. Salman, R. El-Hage, E. Karam, S. Salam, N. Karaoghlanian, A. El-Hellani, N. Saliba and A. Shihadeh, *Scientific Reports*, 2020, **10**, 7322 (nicotine levels in JUUL e-cigarettes in USA and UK compared)

TSNAS

H. J. Kim and H. S. Shin, *Journal of Chromatography A*, 2013, 1291, 48–55 (TSNAs in e-liquids in Korea)

K. E. Farsalinos, I. G. Gillman, M. S. Melvin, A. R. Paolantonio, W. J. Gardow, K. E. Humphries, S. E. Brown, K. Poulas and V. Voudris. *International Journal of Environmental Research and Public Health*, 2015, **12**, 3439–3452 (nicotine and TSNAs in e-cig refills)

K. E. Farsalinos, G. Gillman, K. Poulas and V. Voudris, *International Journal of Environmental Research and Public Health*, 2015, **12**, 9046–9053 (TSNAs compared in liquids and aerosols)

W. E. Stephens, *Tobacco Control*, 2018, **27**, 10–17 (comparing emissions from e-cigarettes with tobacco smoke)

METAL HEATING ELEMENTS AND METAL PIECES

C. A. Hess, P. Olmedo, A. Navas-Acien, W. Goessler, J. E. Cohen and A. M. Rule, *Environmental Research*, 2017, **152**, 221–225 (e-cigarettes as a source of toxic and potentially carcinogenic metals)

M. Williams, K. Bozhilov, S. Ghai and P. Talbot, *PLoS One*, 2017, **12**, e0175430 (metals in the atomizer and aerosol of disposable electronic cigarettes and electronic hookahs)

P. Olmedo, W. Goessler, S. Tanda, M. Grau-Perez, S. Jarmul, A. Aherrera, R. Chen, M. Hilpert, J. E. Cohen, A. Navas-Acien and A. M. Rule, *Environmental Health Perspectives*, 2018, **126**, 027010 (metal concentrations in e-cigarette liquid and aerosol samples)

C-J. Na, S-H. Jo, K-H. Kim, J-R. Sohn and Y-S. Son, *Environmental Research*, 2019, **174**, 152–159 (heavy metals in e-cigarette liquid)

D. R. Fels Eliott, R. Shah, C. A. Hess, B. Elicker, T. S. Henry, A. M. Rule, R. Chen, M. Golozar and K. D. Jones, *European Respiratory Journal*, 2019, **54**, 1901922 (giant cell interstitial pneumonia secondary to cobalt exposure from e-cigarette use)

S. Goto, R. M. H. Grange, R. Pinciroli, I. A. Rosales, R. Li, S. L. Boerboom, K. F. Ostrom, E. Marutani, H. V. Wanderley, A. Bagchi, R. B. Colvin, L. Berra, O. Minaeva, L. E. Goldstein, R. Malhotra, W. M. Zapol, F. Ichinose and B. Yu, *Archives of Toxicology*, 2022, **96**, 3363–3371 (vaping with aged coils increases emission of toxic aldehydes, and lung injury in mice)

SOLVENTS AND ACETALS

B. E. Erickson, *Chemical & Engineering News*, 2015, **93**, 10–13 (decomposition products of glycerol and propylene glycol)

L. Jin, J. Lynch, A. Richardson, P. Lorkiewicz, S. Srivastava, W. Theis, G. Shirk, A. Hand, A. Bhatnagar, S. Srivastava and D. J. Conklin, *American Journal of Physiology-Heart and Circulatory Physiology*, 2021, **320**, H1510–H1525 (e-cig solvents and the role of methanol and ethanol in pulmonary irritation and endothelial dysfunction)

N. R. Jaegers, W. Hu, T. J. Weber and J. Z. Hu, *Scientific Reports*, 2021, **11**, 7800 (decomp. of propylene glycol and glycerol below 200°C)

H. C. Erythropel, S. V. Jabba, T. M. DeWinter, M. Mendizabal, P. T. Anastas, S. E. Jordt and J. B. Zimmerman, *Nicotine & Tobacco Research*, 2019, **21**, 1248–1258 (propylene glycol adducts with benzaldehyde, cinnamaldehyde, citral, ethylvanillin, and vanillin activate TRPA1 receptors)

S. V. Jabba, A. N. Diaz, H. C. Erythropel, J. B. Zimmerman and S-E. Jordt, *Nicotine & Tobacco Research*, 2020, **22**, S25–S34 (acetals of vanillin, ethyl vanillin (vanilla), and benzaldehyde with propylene glycol and glycerol are cytotoxic and inhibit mitochondrial function)

H. C. Erythropel, L. M. Davis, T. M. DeWinter, S. E. Jordt, P. T. Anastas, S. S. O'Malley, S. Krishnan-Sarin and J. B. Zimmerman, *American Journal of Preventive Medicine*, 2019, **57**, 425–427 (vanillin acetals in JUUL e-liquids and aerosols)

FLAVOURINGS

A. L. Groom, T-H. T. Vu, A. Kesh, J. L. Hart, K. L. Walker, A. L. Giachello, C. G. Sears, L. K. Tompkins, D. T. Mattingly, R. L. Landry, R. M. Robertson and T. J. Payne, *Preventive Medicine Reports*, 2020, **18**, 101094 (youth vaping flavour preferences)

ALDEHYDES

R. P. Jensen, W. Luo, J. F. Pankow, R. M. Strongin and D. H. Peyton, *New England Journal of Medicine*, 2015, **372**, 392–394 (formaldehyde in e-cigarette aerosols)

P. A. Tierney, C. D. Karpinski, J. E. Brown, W. Luo and J. F. Pankow, *Tobacco Control*, 2016, **25**, e10–e15 (analysis of flavourings in 30 e-liquids)

A. Khlystov and V. Samburova, *Environmental Science & Technology*, 2016, **50**, 13080–13085 (aldehydes formed by decomposition of flavourings)

M. A. Ogunwale, M. Li, M. V. R. Raju, Y. Chen, M. H. Nantz, D. J. Conklin and X-A. Fu, *ACS Omega*, 2017, **2**, 1207–1214 (aldehydes as a potential risk to the cardiovascular system)

P. Fagan, P. Pokhrel, T. A. Herzog, E. T. Moolchan, K. D. Cassel, A. A. Franke, X. Li, I. Pagano, D. R. Trinidad, K. K. Sakuma, K. Sterling, D. Jorgensen, T. Lynch, C. Kawamoto, M. C. Guy, I. Lagua, S. Hanes, L. A. Alexander, M. S. Clanton, C. Graham-Tutt, T. Eissenberg et al., *Nicotine & Tobacco Research*, 2018, **20**, 985–992 (sugars and aldehydes in e-liquids)

N. Beauval, M. Verrièle, A Garat, I. Fronval, R. Dusautoir, S. Anthérieu, G. Garçon, J-M. Lo-Guidice, D. Allorge and N. Locoge, *International Journal of Hygiene and Environmental Health*, 2019, **222**, 136–146 (puffing conditions affect aldehyde levels in e-cigarette aerosols)

L. Kosmider, S. Cox, M. Zaciera, J. Kurek, M. L. Goniewicz, H. McRobbie, C. Kimber and L. Dawkins, *Scientific Reports*, 2020, **10**, 6546 (nicotine levels and exposure to methanal and ethanal)

DIACETYL

K. Kreiss, A. Gomaa, G. Kullman, K. Fedan, E. J. Simoes and P. L. Enright, *New England Journal of Medicine*, 2002, **347**, 330–338 (workers at popcorn plant with severe bronchiolitis obliterans)

R. Kanwal, G. Kullman, C. Piacitelli, R. Boylstein, N. Sahakian, S. Martin, et al. *Journal of Occupational and Environmental Medicine*, 2006, **48**, 149–157 (flavourings-related lung disease risk at six microwave popcorn plants)

S. Clark and C. K. Winter, *Comprehensive Reviews in Food Science and Food Safety*, 2015, **14**, 634–643 (review, diacetyl in foods)

J. G. Allen, S. S. Flanigan, M. LeBlanc, J. Vallarino, P. MacNaughton, J. H. Stewart and D. C. Christiani, *Environmental Health Perspectives*, 2016, **124**, 733–739 (flavouring chemicals in flavoured e-cigarettes, particularly diacetyl, 2,3-pentanedione, and acetoin)

K. Kreiss, *Toxicology*, 2017, **388**, 48–54 (occupational effects of diacetyl reviewed)

A. F. Hubbs, K. Kreiss, K. J. Cummings, K. L. Fluharty, Ryan O'Connell, A. Cole, T. M. Dodd, S. M. Clingerman, J. R. Flesher, R. Lee, S. Pagel, L. A. Battelli, A. Cumpston, M. Jackson, M. Kashon, M. S. Orandle, J. S. Fedan and K. Sriram, *Toxicologic Pathology*, 2019, **47**, 1012–1026 (flavourings-related lung disease reviewed)

M. D. McGraw, M. Yee, S-Y. Kim, A. M. Dylag, B. P. Lawrence and M. A. O'Reilly, *American Journal of Physiology-Lung Cellular and Molecular Physiology*, 2022, **323**, L578–L592 (diacetyl impairs airway epithelial repair in mice infected with influenza A virus)

S. N. Langel, F. L. Kelly, D. M. Brass, A. E. Nagler, D. Carmack, J. J. Tu, T. Travieso, R. Goswami, S. R. Permar, M. Blasi and S. M. Palmer, *Cell Death Discovery*, 2022, **8**, 64 (diacetyl alters airway cell morphology, inflammatory and antiviral response, and susceptibility to SARS-CoV-2)

BENZALDEHYDE, CINNAMALDEHYDE AND VANILLIN

R. Z. Behar, W. Luo, S. C. Lin, Y. Wang, J. Valle, J. F. Pankow and P. Talbot, *Tobacco Control*, 2016, **25**(Suppl 2), ii94–ii102 (toxicity of cinnamaldehyde in e-cigarette refill fluids and aerosols)

P. W. Clapp, K. S. Lavrich, C. A. van Heusden, E. R. Lazarowski, J. L. Carson and I. Jaspers, *American Journal of Physiology-Lung Cellular and Molecular Physiology*, 2019, **316**, L470–L486 (cinnamaldehyde and respiratory function)

S. Salam, N. A. Saliba, A. Shihadeh, T. Eissenberg and A. El-Hellani, *Chemical Research in Toxicology*, 2020, **33**, 2932–2938 (flavour-toxicant correlation in e-cigarettes)

B. P. Rickard, H. Ho, J. B. Tiley, I. Jaspers and K. L. R. Brouwer, *ACS Omega*, 2021, **6**, 6708–6713 (vanillin and ethylvanillin induce cytotoxicity in HepG2 cells)

L. Kosmider, A. Sobczak, A. Prokopowicz, J. Kurek, M. Zaciera, J. Knysak, D. Smith and M. L. Goniewicz, *Thorax*, 2016, **71**, 376–377 (benzaldehyde in cherry-flavoured e-cigs)

A. Martin, C. Tempra, Y. Yu, J. Liekkinen, R. Thakker, H. Lee, B. de Santos Moreno, I. Vattulainen, C. Rossios, M. Javanainen and J. B. de la Serna, *Environmental Science & Technology*, 2024, **58**, 1495–1508 (benzaldehyde disrupts pulmonary surfactant biophysical function)

VITAMIN E ACETATE AND EVALI

A. K. Werner, E. H. Koumans, K. Chatham-Stephens, P. P. Salvatore, C. Armatas, P. Byers, C. R. Clark, I. Ghinai, S. M. Holzbauer, K. A. Navarette et al., *New England Journal of Medicine*, 2020, **382**, 1589–1598 (hospitalizations and deaths associated with EVALI)

S. G. Doukas, L. Kavali, R. S. Menon, B. N. Izotov and A. Bukhari, *Toxicology Reports*, 2020, **7**, 1381–1386 (review of EVALI)

B. C. Blount, M. P. Karwowski, P. G. Shields, M. Morel-Espinosa, L. Valentin-Blasini, M. Gardner, M. Braselton, C. R. Brosius, K. T. Caron, D. Chambers et al., *New England Journal of Medicine*, 2020, **382**, 697–705 (Vitamin E acetate associated with EVALI)

D. Wu and D. F. O'Shea, *Proceedings of the National Academy of Sciences of the United States of America*, 2020, **117**, 6349–6355

R. M. Strongin, *Proceedings of the National Academy of Sciences of the United States of America*, 2020, **117**, 7553–7554 (formation of ketene from Vitamin E acetate)

K. R. Attfield, W. Chen, K. J. Cummings, P. Jacob, D. F. O'Shea, J. Wagner, P. Wang and J. Fowles, *American Journal of Respiratory and Critical Care Medicine*, 2020, **202**, 1187–1188 (potential of ethenone (ketene) to contribute to EVALI)

K. R. Munger, R. P. Jensen and R. M. Strongin, *Chemical Research in Toxicology*, 2022, **35**, 1202–1205 (vaping cannabinoid acetates leads to ketene formation)

M. E. Rebuli, J. J. Rose, A. Noël, D. P. Croft, N. L. Benowitz, A. H. Cohen, M. L. Goniewicz, B. T. Larsen, N. Leigh, M. D. McGraw, A. C. Melzer, A. L. Penn, I. Rahman, D. Upson, L. E. Crotty Alexander, G. Ewart, I. Jaspers, S. E. Jordt, S. Kligerman, C. E. Loughlin, R. McConnell, E. R. Neptune, T. B. Nguyen, K. E. Pinkerton and T. J. Witek, *Annals of the American Thoracic Society*, 2023, **20**, 1–17 (EVALI and e-cigarette or vaping product use)

K. R. Munger, K. M. Anreise, R. P. Jensen, D. H. Peyton and R. M. Strongin, *JACS Au*, 2024, **4**, 2403–2410 (ketene formation from cannabinoid acetates)

NICOTINE ANALOGUES

H. C. Erythropel, S. V. Jabba, P. Silinski, P. T. Anastas, S. Krishnan-Sarin, J. B. Zimmerman and S. E. Jordt, *JAMA*, 2024, **332**, 753–755

S. E. Jordt, S. V. Jabba, P. J. Zettler and M. L. Berman, *Tobacco Control*, 2024, **0**, 1–5. In press doi: 10.1136/tc-2023-058469 (6-methyl nicotine in e-cigarettes) This last paper embargoed until September 2025

CHAPTER 10 ISOTOPES

Patrick Coffey, *Cathedrals of Science. The Personalities and Rivalries that made Modern Chemistry*, Oxford, OUP, 2008

J. B. West, G. J. Bowen, T. E. Dawson and K. P. Tu, *Isoscapes*, Dordrecht, Springer, 2010

J. F. Marra, *Hot Carbon. Carbon-14 and a Revolution in Science*, New York, Columbia University Press, 2019

D. I. Mendeleev, *Zeitschrift für Chemi*, 1869, **12**, 405–406 ('Ueber die Beziehungen der Eigenschaften zu den Atomgewichten der Elemente')

F. W. Soddy, *Nature*, 1913, **92**, 399–400 (the term isotope coined)

H. G. J. Moseley, *Philosophical Magazine*, 1913, **1913**, 1024 (X-ray spectra leading to Atomic Number concept)

J. W. van Spronsen, *The Periodic System of Chemical Elements*, Amsterdam, Elsevier, 1969 (pp. 46–47 for table showing values of RAM in the 19th c.)

HEAVY WATER

F. Franks, Water, *London, Royal Society of Chemistry*, first edition 1983 (pp. 10–11 for comparison of H_2O and D_2O)

J. Bigeleisen, *Journal of Chemical Education*, 1984, **61**, 108–116 (G.N. Lewis's work with isotopes)

N. Ben Abu, P. E. Mason, H. Klein, N. Dubovski, Y. Ben Shoshan-Galeczki, E. Malach, V. Pražienková, L. Maletínská, C. Tempra, V. C. Chamorro, J. Cvačka, M. Behrens , M. Y. Niv and P. Jungwirth, *Communications Biology*, 2021, **4**, 440 (sweet taste of heavy water)

For the effect of D_2O on the ^1H NMR spectrum of menthol, see: https://www.chm.bris.ac.uk/motm/menthol/mentholh.htm

ISOTOPES AND ESTERIFICATION

M. Polanyi and A. L. Szabo, *Transactions of the Faraday Society*, 1934, **30**, 508–512 (alkaline hydrolysis of amyl acetate)

I. Roberts and H. C. Urey, *Journal of the American Chemical Society*, 1938, **60**, 2391–2393

VANILLA AND FOOD FRAUD

D. A. Kruger and H. W. Kruger, *Journal of Agricultural and Food Chemistry*, 1985, **33**, 323–325 (NMR detection of carbon-13 labelled fraudulent vanillin)

G. S. Remaud, Y. L. Martin, G. G. Martin and G. J. Martin, *Journal of Agricultural and Food Chemistry*, 1997, **45**, 859–866 (use of SNIF-NMR method to detect adulterated vanillin)

A-M. Hansen, A. Fromberg and H. Frandsen, *Journal of Agricultural and Food Chemistry*, 2014, **62**, 10326–10331 (authentication of vanilla flavours using carbon and hydrogen isotopes)

N. J. Gallage and B. L. Moller, *Molecular Plant*, 2015, **8**, 40–57 (review of bioconversion routes for vanillin from corn, rice, clove and turmeric)

M. M. Bomgarder, *Chemical & Engineering News*, 2016, **94**, 38–42

M. Fache, B. Boutevin and S. Caillol, *ACS Sustainable Chemistry & Engineering*, 2016, **4**, 35–46 (vanillin from lignin)

A. S. Wilde, T. Strucko, C. R. Veje, U. H. Mortensen and L. Duedahl-Olesen, *Food Control*, 2022, **131**, 108389 (authentication of vanillin ex glucose using $\delta^{13}C$ and δ^2H values)

P. M. Le, E. Martineau, S. Akoka, G. Remaud, M. M. G. Chartrand, J. Meija and Z. Mester, *Analytical and Bioanalytical Chemistry*, 2022, **414**, 7153–7165 (site specific carbon isotope studies of vanillin)

ISOTOPES AND BOTTLED WATERS

G. J. Bowen, D. A. Winter, H. J. Spero, R. A. Zierenberg, M. D. Reeder, T. E. Cerling and J. R. Ehleringer, *Rapid Communications in Mass Spectrometry*, 2005, **19**, 3442–3450 (isotopic analyses of world bottled waters)

M. Brencic and P. Vreca, *Rapid Communications in Mass Spectrometry*, 2006, **20**, 3205–3212 (isotopic analyses of Central European bottle waters)

G-E. Kim, J-S. Ryu, W-J. Shin, Y-S. Bong, K-S. Lee and M-S. Choi, *Rapid Communications in Mass Spectrometry*, 2012, **26**, 25–31 (isotopic analysis of Korean bottled water)

TRUFFLES

D. Sciarrone, A. Schepis, M. Zoccali, P. Donato, F. Vita, D. Creti, A. Alpi and L. Mondello, *Analytical Chemistry*, 2018, **90**, 6610–6617 (fraudulent white truffles)

ISOTOPES AND DRUGS

J. R. Ehleringer, D. A. Cooper, M. J. Lott and C. S. Cook, *Forensic Science International*, 1999, **106**, 27–35

J. R. Ehleringer, J. F. Casale, M. J. Lott and V. L. Ford, *Nature*, 2000, **408**, 311–312

J. R. Mallette, J. F. Casale, J. Jordan, D. R. Morello and P. M. Beyer, *Scientific Reports*, 2016, **6**, 23520 (isoscapes and cocaine growing regions in South America)

J. R. Mallette and J. F. Casale, *Journal of Chromatography A*, 2014, **1364**, 234–240 (truxillines in cocaine)

J. B. West, J. M. Hurley, F. Ö. Dudás and J. R. Ehleringer, *Journal of Forensic Sciences*, 2009, **54**, 1261–1269 (Sr isotopes show geographical origin of marijuana)

S. Schneiders, T. Holdermann and R. Dahlenburg, *Science and Justice*, 2009, **49**, 94–101 (isotope ratios of amphetamines depend on the precursor and the synthetic pathway)

N. NicDaéid, S. Jayamana, W. J. Kerr, W. Meier-Augenstein and H. F. Kemp, *Analytical and Bioanalytical Chemistry*, 2013, **405**, 2931–2941 (stable isotope signatures of methylamphetamine prepared from OTC medicines by different routes)

METAL ISOTOPES

A-M. Desaulty, P. Telouk, E. Albalata and F. Albarède, *Proceedings of the National Academy of Sciences of the United States of America*, 2011, **108**, 9002–9007 (isotopic Ag–Cu–Pb record of silver circulation through 16th–18th century Spain)

A-M. Desaulty and F. Albarède, *Geology*, 2013, **41**, 135–138 (isotopic Ag–Cu–Pb record of Tudor and early Stuart Europe)

B. A. Hofmann, S. B. Schreyer, S. Biswas, L. Gerchow, D. Wiebe, M. Schumann, S. Lindemann, D. R. García, P. Lanari, F. Gfeller, C. Vigo, D. Das, F. Hotz, K. von Schoeler, K. Ninomiya, M. Niikura, N. Ritjoho and A. Amato, *Journal of Archaeological Science*, 2023, **157**, 105827 (source of Bronze Age arrowhead made of meteoritic iron)

LEAD ISOTOPES

H. Delile, J. Blichert-Toft, J-P. Goiran, S. Keay and F. Albarède, *Proceedings of the National Academy of Sciences of the United States of America*, 2014, **111**, 6594–6599 (Lead in ancient Rome's city waters)

H. Delile, D. Keenan-Jones, J. Blichert-Toft, J-P. Goiran, F. Arnaud-Godet, P. Romano and F. Albarède, *Proceedings of the National Academy of Sciences of the United States of America*, 2016, **113**, 6148–6153 (water supply of Naples after AD 79)

R. D. Russell, R. M. Farquhar and J. E. Hawley, *Transactions, American Geophysical Union*, 1957, **38**, 557–565 (ores from Broken Hill, NSW)

J. E. Landmeyer, P. M. Bradley and T. D. Bullen, *Environmental Geology*, 2003, **45**, 12–22 (lead isotopes in South Carolina groundwater, and tetraethyllead)
M. Komárek, V. Ettler, V. Chrastný and M. Mihaljevič, *Environment International*, 2008, **34**, 562–577 (Pb isotopes in ores and leaded petrol)

STRONTIUM ISOTOPES

J. B. West, J. M. Hurley, F. Ö. Dudás and J. R. Ehleringer, *Journal of Forensic Sciences*, 2009, **54**, 1261–1269 (Sr isotopes and where cannabis has been grown)
S. Voerkelius, G. D. Lorenz, S. Rummel, C. R. Quétel, G. Heiss, M. Baxter, C. Brach-Papa, P. Deters-Itzelsberge, S. Hoelzl, J. Hoogewerff, E. Ponzevera, M. Van Bocxstaele and H. Ueckermann, *Food Chemistry*, 2010, **118**, 933–940 (Sr isotopic signatures of mineral waters)

URANIUM ISOTOPES AND UF$_6$

L. Bickel, *The Deadly Element. The Story of Uranium*, London, Macmillan, 1980
A. D. Aczel, *Uranium Wars*, New York, Palgrave Macmillan, 2009

KING RICHARD III

R. Buckley, M. Morris, J. Appleby, T. King, D. O'Sullivan and L. Foxhall, *Antiquity*, 2013, **87**, 519–538 (burial of Richard III)
A. L. Lamb, J. E. Evans, R. Buckley and J. Appleby, *Journal of Archaeological Science*, 2014, **50**, 559–565 (Multi-isotope analysis demonstrates significant lifestyle changes in King Richard III)

CRIME

W. Meier-Augenstein and I. Fraser, *Science and Justice*, 2008, **48**, 153–159
H. F. Kemp and W. Meier-Augenstein, *Minerva Medicolegle*, 2009, **129**, 219–231 (forensic isotope analysis identifies an Irish murder victim, the 'Scissor Sisters' case)
W. Meier-Augenstein, H. F. Kemp, I. Brewster and G. Ronayne, in: *Stable Isotope Forensics*, W. Meier-Augenstein, Chichester, Wiley-Blackwell, 2010, pp. 197–201 (murder of Vietnamese migrant)
A. Fischer, K-G. Sjögren, TZT Jensen, ML Jørkov, P. Lysdahl, T. Vimala, et al. *PLoS One*, 2024, **19**, e0297032 (Vittrup Man–The life-history of a genetic foreigner in Neolithic Denmark)

CHAPTER 11 METHANE

B. Cooper and T. F. Gaskell, *North Sea Oil–The Great Gamble*, London, Heinemann, 1966
C. Callow, *Power from the Sea*, London, Victor Gollancz, 1973 (North Sea Gas)
P. McGeer and E. Durbin, eds., *Methane: Fuel for the Future*, New York and London, Plenum Press, 1982
D. Knight, *Humphry Davy: Science and Power*, Cambridge, Cambridge University Press, 1992 (Davy lamp)
National Academy of Sciences, *Methane Generation from Human, Animal, and Agricultural Wastes*, New York, Books for Business, 2001
J. Darley, *High Noon for Natural Gas*, Vermont, Chelsea Green Publishing, 2004

S. Shah, *Crude: The Story of Oil*, New York and London, Seven Stories Press, 2004

M. Yeomans, *Oil: Anatomy of An Industry*, New York and London, The New Press, 2004

R. M. Hazen, *Symphony in C*, London, William Collins, 2019

R. L. Myers, *The 100 Most Important Chemical Compounds. A Reference Guide*, Westport, Greenwood Press, 2007, pp. 171–173

Oil and gas production figures at https://www.ukeiti.org/oil-gas

A. Guisasola, D. de Haas, J. Keller and Z. Yuan, *Water Research*, 2008, **42**, 1421–1430 (methane formation in sewers)

S. Kirschke et al, *Nature Geoscience*, 2013, **6**, 813–823 ('Three decades of global methane sources and sinks')

M. Granda, C. Blanco, P. Alvarez, J. W. Patrick and R. Menéndez, *Chemical Reviews*, 2014, **114**, 1608–1636 (chemicals from coal coking)

C. D. Ruppel and J. D. Kessler, *Reviews of Geophysics*, 2017, **55**, 126–168 (methane hydrate emissions buffered by oxidation)

W. Steffen et al., *Proceedings of the National Academy of Sciences of the United States of America*, 2018, **115**, 8252–8259 (potential breakdown of climate stabilisation, including methane hydrates)

S. Shetty, J. Abraham, C. Balusami and R. Shehnaaz, *International Journal of Livestock Research*, 2020, **10**(8), 24–32 (methane production in biogas plant)

M. Nikolaisen, T. Cornulier, J. Hillier, P. Smith, F. Albanito and D. Nayak, *Journal of Cleaner Production*, 2023, **409**, 137245 (methane emissions from rice paddies)

L. Ernst, U. Barayeu, J. Hädeler, T. P. Dick, J. M. Klatt, F. Keppler and J. G. Rebelein, *Nature Communications*, 2023, **14**, 4364 (methane formation prior to the origin of life)

LONDON SMOGS

P. Brimblecombe, *The Big Smoke: A History of Air Pollution in London Since Medieval Times*, Methuen, London, 1987, esp. pp. 161–177

M. L. Bell and D. L. Davis, *Environmental Health Perspectives*, 2001, **109** (suppl. 3), 389–394 ("Reassessment of the lethal London fog of 1952")

R. Stone, *Science*, 2002, **298**, 2106–2107 ("Counting the Cost of London's Killer Smog")

M. L. Bell, D. L. Davis and T. Fletcher, *Environmental Health Perspectives*, 2004, **112**, 6–8 ("A Retrospective Assessment of Mortality from the London Smog Episode of 1952")

PERMAFROST EMISSIONS

A. A. Bloom, P. I. Palmer, A. Fraser, D. S. Reay and C. Frankenberg, *Science*, 2010, **327**, 322–325 (human influence in Arctic methane emissions)

C. Knoblauch, C. Beer, S. Liebner, M. N. Grigoriev and E-M. Pfeiffer, *Nature Climate Change*, 2018, **8**, 309–312 (methanogens stimulate CH_4 production in thawing permafrost)

M. N. Dyonisius et al., *Science*, 2020, **367**, 907–910 ('Old carbon reservoirs were not important in the deglacial methane budget')

C. D. Elder, D. R. Thompson, A. K. Thorpe, H. A. Chandanpurkar, P. J. Hanke, N. Hasson et al., *Global Biogeochemical Cycles*, 2021, **35**, e2020GB006922 (permafrost hotspots)

N. Rößger, T. Sachs, C. Wille, J. Boike and L. Kutzbach, *Nature Climate Change*, 2022, **12**, 1031–1036 (methane emissions from Siberian permafrost)

ATMOSPHERIC CHEMISTRY, METHANE AND GLOBAL WARMING

S. Arrhenius, *The London, Edinburgh, and Dublin Philosophical Magazine and Journal of Science (fifth series)*, 1896, **41**, 237–275 ('On the Influence of Carbonic Acid in the Air upon the Temperature of the Ground')

H. Craig and C. C. Chou, *Geophysical Research Letters*, 1982, **9**, 1221–1224

R. A. Rasmussen and M. A. K. Khalil, *Journal of Geophysical Research*, 1984, **89**, 11, 599–605

T. Blunier, J. A. Chappell, J. Schwander, J. M. Barnola, T. Desperts, B. Stauffer and D. Raynaud, *Geophysical Research Letters*, 1993, **20**, 219–222

D. Raynaud, J. Jouzel, J. M. Barnola, J. Chapellaz, R. J. Delmas and C. A. M. Lorius, *Science*, 1993, **259**, 926–934 (ice core analyses for atmospheric methane)

A. M. Holloway and R. P. Wayne, *Atmospheric Chemistry*, Cambridge, RSC Publishing, 2010.

Index

Note: **Bold** page numbers refer to tables and *italic* page numbers refer to figures.

acetate, amyl (pentyl) 211
    benzyl 136
    Δ⁹-THC 204–205
    geranyl 53
    hexyl 137, *139*
    menthyl 84, *85*
    myrtenyl 50, *51*
    phenyl 205
    vitamin E 73, 204–205
acetic acid 129
acetoin 202
acetylcholine 171, 193
acetyl coenzyme A 17, 47, 64
2-acetylpyrroline 12, 24, 25
acrolein 195, 200, 201
Actinoplanes teichomyceticus 169
ADB-FUBINACA molecule 111, 112
Aedes aegypti 131, 156
Aedes mosquito 155
Agaricus bisporus 33–36
ague 93
AK-47 24 Karat Gold 112
alcohol 120
    dehydrogenase enzyme 137, 139, 200
alcohol acyltransferase (AAT) enzyme 47, 50
aldehyde 12–14, 20–21, 24, 26–27, 30–32, 38–39,
        44, 54–55, 126, 200–203, 215–217
    short-chain 201
    toxic 201
    unsaturated 139–141
aldehyde methional, sulfur-containing 12–13,
        20–21, 27–28, 30–31, 35, 122
Alder Wright, Charles Romley 95
Aldrich, Thomas 123
Alliaria petiolata 91
allicin 37
allyl isothiocyanate 89–90
allylsulfenic acid (2-propenesulfenic acid) 37
α-amino acid 1, 8, 25, 30
α-amino acid proline 25
α-chlorocodide 107
α-D-fructopyranose 5
α-D-glucose 3
α-dicarbonyl compounds 1–2
α-glucose 6
α-1,4-glycosidic bonds 7
α-pinene 50–51, 149
α-sanshool 82–84
α-sinensal 54

aluminium, natural 213
AMB-FUBINACA molecule 111–112
Amblyomma americanum 165
American Civil War 94
American dog tick (*Dermacentor variabilis*) 165
amine transferase enzyme 26
amino acid 1–2, 8–10, 12, 15–16, 19–20, 26–28,
        30, 32–33, 35, 37, 42, 45, 65–67, 87,
        115, 117–118, 122, 126, 132–134,
        152, 171
ammonia 19, 115, 128, 174, 198–199, 225
Amorphophallus eichleri 119
amylase 6, 23
amylopectin 3, 6–7, 23
amylose 3, 6, 23
anaemia, pernicious 68
anaesthetics 164, 183–184
Anopheles gambiae 131, 155
Anopheles mosquitoes 131, 155, 161
anthocyanins 45
antibiotics, tetracycline 166–167
antidepressant 164, 184–185
    fluorine-containing 184–185
antioxidants 45
aromatic fluorine compounds, synthesis of 174
aromatic hydrocarbons, polycyclic
        195–196, 225
artificial sweetener 4, 164
arum (*Amorphophallus titanum*) 119–120, 134
L-ascorbic acid (vitamin C) 69–70
Aston, F. W. 208–209
atomic mass 208–209
atomic weights 208
atomiser 198
atorvastatin 187
aureomycin (chlorotetracycline) 166
avidin 66

baked potatoes 31–32
baking 24–28
Balz, Günther 174
Balz-Schiemann reaction 174
'basis of radiocarbon dating' 217
bath salts 110
Batis maritima 160
Beef 2, 8, 12–14
    raw 12
    roast 13
bees, Euglossine 136, 154

beetles (*Neopyrochroa flabellata*) 152
   blister 151–152
   bombardier 146–147
benzaldehyde 35, 150, 201, 203–204
benzopyrene 195, 196
3,4-benzopyrene 195
benzopyrene diol epoxide 196
benzoquinone 146
1,4-benzoquinone 147
2-benzylbenzimidazoles 113
benzylmercaptan 121
beta-carotene 39, 42, 43, 56–58
β-D-glucose 3
β-fructofuranose 5
β-galactosidase (enzyme lactase) 6
β-1,4-glycosidic bonds 7
β-sinensal 54
biotin (vitamin B$_7$) 66
black mustard (*Brassica nigra*) 89
black pepper (*Piper nigrum*) 76, 79
blister beetles 151–152
blood substitutes 188
blue cheeses 15–18
    Bleu des Causses 18
    Danish Blue 18
    Gorgonzola 16, 18
    Roquefort 15, 16, 18
    Stilton 15, 16, 18
Blyth, Alexander Wynter 60
body odour 120, 125, 129, 130
boiled potatoes 6, 21, 30–31, 35, 122
bombardier beetles 146–147
Borodin, Alexander Porfiryevich 137, 173
branched-chain acids 130
bread 2, 6, 8, 22–26, 28–29, 35, 59, 72
Brevibacterium linens bacteria 16
Brown dog tick (*Rhipicephalus sanguineus*) 165
Browne, Collis Dr. 94
brush-tailed possums 183
butanoic acid 48, 129, 131
S-2-butenyl thioacetate 124
butter (Latin *butyrum*) 129

C$_6$ acid 129
cadaverine 132–133
caffeine 90, 142, 191
calciferol (vitamin D) 70–72
calcitriol 71
Camembert cheese 16, 18–19, 21, 117, 126
cancer 104, 182, 186
   lung 194–195, 198, 206, 207
cannabinoid 108–109, 112, 205
   psychoactive 205
   receptor 108, 109
   synthetic 108, 112
Cannabis sativa 108
cantharidin 151–152

capsaicin 76–83, 89, 90, 144
capsaicinoids 76–83
capsaicin receptor 78
carbohydrates 2–4, 12, 16, 18, 23, 28, 56, 59, 61, 64, 132, 152
carbon 214
   atoms 214
   chiral 33, 84, 85, 87, 122, 130, 132, 185, 192
   in foods 219isotopes 209, 214–218, *215*
   radioactive 219
carbon monoxide (CO) 195
carboxylic acid 16, 17, 19, 21, 24, 31, 69, 115, 129, 131, 132, 137, 147, 219
carfentanil and powerful fentanyls 97, *100*, 100–102
carotenoids 42, 56–57
Carson, Rachel 161
carvone 87–88, 122, 158
catalytic hydrogenation 103, 167
celebrex 187
celecoxib 187
cellular respiration 3
cellulose 2, 3, 6, 7
C–F bond 173, 176, 181–183
CFCs *see* chlorofluorocarbons (CFCs)
Cheddar 8, 16, 21, 117, 122
cheese 2, 8, 15–22, 27, 58, 73, 117, 122, 126, 130
   blue 15–18
   Camembert 16, 18–19, 21, 117, 126
   Cheddar 16, 21, 117, 122
   Emmental 19
   Gruyère 19, 20
chicken
   cooked 8, 14–15
   fried 14–15, 55
chilli pepper, hotness of 79
chillis 75–80, 82, 89, 144
"Chinese five-spice" powder 83
Chinese peppers 82–83
Chinese prickly ash 82
chips, potatoes 29, 32
chiral carbon 33, 84, 85, 87, 122, 130, 132, 185, 192
chloramphenicol 170–171
chlorinated phenols 165–166
chlorine
   compounds 106, 159–172, 174–180
   elemental 159
chlorine radicals, ozone-depleting 160, 176
Chlorodyne 94
chlorofluorocarbons (CFCs) 159–160, 174–180, 182, 184, 227–228; *see also* hydrochlorofluorocarbons (HCFCs)
chloroform 164, 174, 183
chloromethane 159–161, 174
chloromycetin 170
chlorophyll 39

chlorotetracycline (aureomycin) 166–167
chloroxylenol 166
cholecalciferol 71
chromoproteins, light-sensitive 57
cigarette-making machine 191
cigarettes 86, 191–195; *see also* e-cigarette
    197–202, 204, 206–207
  machine-made 191
  mentholated 86
cigar smoking 190–191
cinnamaldehyde 82, 203–204
  in e-liquids 203–204
cinnamon (*Cinnamomum verum* or *Cinnamomum
    zeylandicum*) 75, 76, 82, 92, 203
ciprofloxacin (Cipro) 186–187
11-cis-retinal 57–58
citalopram 185
citral 51–53, 55, 152
citric acid (Krebs) cycle 64, 183
citriodiol 157
citronellal 52–53, 85–86, 157
Civil War 94–95
Clark, Jim 166
clove (*Syzygium aromaticum*) 75–76,
    80–82, 92
CoA *see* coenzyme A (CoA)
coal 225–227
coal gas 225–227
cobalamin (vitamin B_{12}) 56, 68–69
codeine 94, 102–103, *103*, 106–108
coenzyme A (CoA) 64
coinage and isotopes 213
Colman, Jeremiah 89
Colorado tick fever 165
cooked onions 39
cool and menthol 84–85
Corynebacteria 131
cotinine 192–193
COVID-19 198, 204, 206–207
COX-2 inhibitor 187
Crème de menthe 86
crime, isotopes in solving
  'scissor sisters' case 221–222
  Welsh case 221
crimes, unsolved 222–223
crops, domesticated 1
cucumber aldehyde 44
Culex mosquitoes 155
Culex quinquefasciatus mosquitoes 156
cyanadin 45
cyanide (CN) group 69, 150
cyanocobalamin 68–69
cyclic esters (lactones) 19, 48
cyclisation reactions of hydroxy acids 48
cysteine 8, **9**, 14–15, 90, 115
  degradation 12
cytochrome P450 enzyme 50–51, 196–197

Dalton, John 208, 224
Dam, Henrik 73
'dash to gas' 227
DDT *see* dichlorodiphenyltrichloroethane (DDT)
dead bodies 132–135
dead-horse arum 120
decarboxylation 17, 20, 26, 27, 30, 126, 127, 132
DEET 155–156, 158
degradation
  cysteine 12
  Maillard 1, 2, 12, 15, 24, 28, 31, 32
  Strecker 1, 2, 30, 32
dehydrogenase enzyme, alcohol 137, 139
dengue fever 131, 155, 162
De Quincey, Thomas 93
desocodeine 107
desomorphine 105–107, *107*
Dettol 166
deuterium oxide 209–210
D-glucose 2–4, *3*
diacetyl *18*, 19, 202, 207
diacetylmorphine 95, 104, 106; *see also* heroin
diastereoisomers of p-menthane-3,8-diol 157–158
diazotisation reaction 174
Dicerandra frutescens 88
dichlorodiphenyltrichloroethane (DDT) 154,
    161–163
2,4-dichlorophenol 165
Dieldrin and Aldrin 163
Diels-Alder reaction 163
N, N-diethyl-3-methyl-benzamide 155
diet, King Richard III 220–221
dimethyl sulfide (DMS) 116–119, *119*, 120, 134
dimethylsulfoniopropionate (DMSP) 118,
    119, *119*
disaccharide lactose 5
DNA synthesis 69
D_2O and H_2O 209–211
domesticated crops 1
dough, polypeptide chains in formation of *22*
doxycycline 167
Drosophila melanogaster flies 156
drugs, isotopes and 219, *220*
Dumas, Jean-Baptiste 173
durian fruit 121

e-cigarette 197–202, 204, 206–207
  liquids 192, 198–202, 204–205, 207
  refill fluids 204
e-cigarette or vaping product use-associated lung
    injury (EVALI) 204
Ecstasy (MDMA) 219
Eijkman, Christiaan 59
Eisner, Thomas 88
elemental chlorine 159
e-liquids 192, 198–202, 204–205, 207
  effect of cinnamaldehyde in 203–204

emissions of volatile organics 140
Emmental cheese 19
environment and smogs 226
enzyme; *see also specific types*
    alcohol acyltransferase 47, 50
    alcohol dehydrogenase 137, *139*
    amine transferase 26
    cytochrome P450 50, *51*, 196, 197
    hydrolase 29, 65, 137
    hydroperoxide lyase *34, 41, 137, 139, 140*
    isomerase 65
    'isomerisation factor' 139
    lipoxygenase 29, 34, *34*, 44, 137, *138, 139,
        139, 140*
    13-lipoxygenase 40, *41*
    lyase 34, 40, 65
    oxidoreductase 65
    transaminase *19*, 20
    transferase 65
enzyme lactase (β-galactosidase) 6
epibatidine 171
Epipedobates tricolor 171
epizingiberene 43
escitalopram 185
esterification 17, 19, *127*, 211
esters 11, 12, 16, 17, 19, 21, 24, 36, 38, 45–48, 84,
        95, 117, 127–129, 134–137, 139–140,
        144–145
ethanal *18*, 54, 140, 141, 195, 199, 200, *200*,
        201, *201*
etheneone (ketene) 205
ethers, fluorinated 183
ethoxyethane 183
2-ethyl-3,5-dimethylpyrazine 28, *28*
ethyl 3-mercaptopropionate 21, 126
ethyl vanillin 201, 204
Eufriesea purpurata 154
eugenol 50, 80–82
Euglossa viridissima 154
euglossine bees (euglossini) 136, 154, 162
EVALI *see* e-cigarette or vaping product
        use-associated lung injury (EVALI)
Evans, Herbert 72

fatal menthol intoxication 86
fatty acids, thermal oxidation of 12
fentanyl 96–99, *100*, 101, 102, 112–113
    carfentanil and 97, 100–102
fentanyl-related deaths 96–99, *100*
fever
    Colorado tick 165
    dengue 155, 162
    malarial (ague) 93
    Rift Valley 155
    Rocky Mountain spotted 165, 171
    yellow 131, 155
F–F bond energy 173

fish-breath syndrome 129
fish odour syndrome 129
flammable gas 225
flavin monoxygenases (FMOs) 128
flavour 12, 14–16, 18, 21, 22, 24, 29, 32, 35, 38,
        40, 44–46, 48, 131, 201, 203, 207
flavourings 1, 16, 18, 19, 80, 86, 87, 198, 201–202,
        207, 215
flesh-eating drug 105, 108
fluorinated ethers 183
fluorinated pharmaceuticals 183–188
fluorine (elemental) 173–174
fluorine-containing antidepressant 184, 185
fluoroacetate, toxic nature of 182–183
fluorocarbons 188
fluoroquinolones 186
5-fluorouracil 186
fluoxetine (Prozac) 164, 184
FMOs *see* flavin monoxygenases (FMOs)
folate 67
folic acid (vitamin B$_9$) 66–67, 69
Fonzarelli, Arthur Herbert 87
food
    carbon in 219
    fraud detection 215–218
forensics 219, 221–223
forest of walnuts 145
Forever Chemicals 182
Fragaria ananassa 50–51
'fraudulent' vanillin molecules 215–217
fraudulent white truffles 218
freons 175, 181
fresh onions 38–39
fried chicken 14–15, 55
fructose 3–5, 40, 45

garlic (*Allium sativum*) 37–38
gas
    coal 225–227
    flammable 225
    greenhouse 227–228
    marsh 224
    natural 226–227
    town 225
Gastrolobium 182
geranial 51–52, *52*, 55, 152
geranyl diphosphate 50, *51*, 85
ginger (*Zingiber officinale*) 75, 79–80
gliadins 22
global climate cycle 118
global warming 160, 178, 224, 227–228
glucose 1–7, *3*, 12, 18, *18*, 23, 24, 40, 45,
        151, 215
glutamine residues 131
glutenins 22
gluten proteins 24
GLV *see* green leaf volatiles (GLV)

glycerol 129, 198, 200–201, *200, 201*
   triester of 11, 16
glycogen 3, 6, 7, 12
grapefruit mercaptan 121–122, *122*
grapes (*Vitis vinifera*) 121
green grass volatiles 137, 139
greenhouse effect 176, 227
greenhouse gas 227
green leaf volatiles (GLV) 137, 140
green odour 137, 141
green tobacco sickness 192
Gruyère cheese 19, 20
György, Hungarian Paul 65, 66

halogenated compounds, from marine fungi 171
halothane 164, 183–184
Haworth projections 3
heavy water 210; *see also* deuterium oxide
Helicodiceros muscivorus 120
Heptan-2-one 17–18
heroin 95–99, 102, 104–108, 219
hexanal 13, 29, 31, 40, 41, 137–140, *138, 139*
hexanoic acid 48, 129, 146
(Z)-3-hexenal 139–141, *140*
hexuronic acid 69–70
HFCs *see* hydrofluorocarbons (HFCs)
HFOs *see* hydrofluoroolefins (HFOs)
high-performance liquid chromatography 79
Hillbilly heroin 105
histidine 8, 9, *9*
H₂O and D₂O 210
Hodgkin, Dorothy 68
hog-nosed skunk (*Conepatus mesoleucus*) 124
homocysteine 126, *127*
Homo sapiens 143
Hopkins, Frederick G. 56, 59
human marker molecules 134
hydrocarbons, polycyclic aromatic 195, 196, 225
hydrochlorofluorocarbons (HCFCs) 174–175,
   177–181
hydrocodone 105
hydrofluorocarbons (HFCs) 177–178, 180
hydrofluoroolefins (HFOs) 178
hydrogen 210
hydrogenation, catalytic 103, 167
hydrogen cyanide (HCN) 115, 150, 195
hydrogen peroxide 103, 146
hydrogen sulfide (H₂S) 115–116, 120
hydrolase enzyme 29, 65, 137
hydrolysis reactions 8, 16, *17*, 23, 85, 95, 124, 131,
   196, 211
9-hydroperoxidelyase 44
hydroperoxide lyase enzyme 137, 139, 140
hydroxy acids 19, 48
   cyclisation reactions of 48
hydroxy-alpha sanshool 83
14-hydroxycodeinone *104*

9- and 13-hydroxyoctadecadienoic acid
   (9- and 13-HODE) 78

indigenous energy source 227
indoleacetate 133
Industrial Revolution 93, 225, 228
infrared spectroscopy 210
inhibitor, COX-2 187
insects
   plants attracting 152–154
   repellents 52, 155–158
   weaponised 145–152
intoxication, fatal menthol 86
isoleucine 8, 9, 26, 42, 66
isomenthol 85, 87
isomerisation factor enzymes 139
isomers, optical 33–34, 53–54, 87, 130, 132, 192
isoscapes 218–223
isotopes 181, 208–223
   aluminium 213
   carbon 214–219, *215*
   coinage and 213
   and drugs 219–220, *220*
   lead 211–213
   oxygen 222
   radioactive 208, 209, 213, 214, 217
   strontium 219, 220
   uranium, separation 181, 213–214
isotopes in solving crimes
   'scissor sisters' case 221–222
   Welsh case 221

Janssen, Paul 97, 101
Japanese horseradish 90
Johnson, Samuel 92
JWH-018 molecules 109

Karrer, Paul 56
kerogen, waxy 224
ketene (etheneone) 73, 205, *205*
2-keto-3-methylpentanoic acid 26
ketone 12, 17–18, 24, 84, 117, 132, 205
King Richard III (1452–1485) 220–221
Koagulationsvitamin 73
Kornfield, E. C. 168
Krebs cycle 62, 64, 183
Krokodil 105–108
Kuhn, Richard 60, 65

lactones (cyclic esters) 19, 21, 48, 51
lactose 3, 5–6, 16, 18–20
   and Camembert 18
   disaccharide 5–6
lactose intolerance 6
laudanum 92–93
lead isotopes 211–212
leaf aldehyde 140

lemon 51–56, 59, 82
    odours 53–54
    oil 51, 53
    peel oil 53
lemongrass oil 53
Lentinula edodes 33, 35
leucine 8, 9, 20, *20*, 26, 42, 66
Lewis, Gilbert N. 209–210
Libby, Willard 217
lightning-triggered fires 160
light-sensitive chromoproteins 57
limonene 51–54, *52*, 85, 149
    isomers smell 53–54
Lind, James 56, 69
linoleic acid 34, 36, 40, 78, 137, *138, 139*
linolenic acid 11, 40, 137, 139, *140*
lipids 11, 12, 16, 17, 24, 26, 29, 31, 32, 72, 131,
    132, 137
    oxidation of 24, 26
    in sweat 129
Lipitor 187
lipolysis 16, 17, 19
lipoxygenase enzyme 29, 34, *34*, 40, 41, 44,
    137–140, *138*
9-lipoxygenase enzyme 44
13-lipoxygenase enzymes 41
L-lactic acid 131
Lovelock, James 118, 159, 176
lung cancer 194–195, 198, 206, 207
lyase enzyme 34
Lyme disease 165, 167
lysine 8, 9, 132, *133*

machine-made cigarettes 191
Maillard degradation 1
    of proline 12
Maillard, Louis Camille 1
Maillard reaction 1, 2, 12, 15, 24, 28, 31, 32
    pyrazine formation 24, 28, 29, 31, 32
malaria 93, 131, 155, 157, 161, 162, 167, 185, 190
maltase 23
maltose 3, 5, 23
malt sugar *see* maltose
mangroves 160
Manhattan Project 181
marine fungi, halogenated compounds from 171
marsh gas 224
McGee, Harold 1, 12, 15, 82
m-chlorobenzoic acid (m-CPBA) 103
MDMA (Ecstasy) 219
meat 1, 12–15, 39, 60, 62, 65, 69, 72, 75, 219, 222
mefloquine 185, 186
MeJA *see* methyl jasmonate (MeJA)
Meloidae 151
meningoencephalitis 165
Mentha arvensis 84
Mentha x piperita 84

menthol 78, 81, 84–87, 89, 191, 207, 211
    biosynthesis of 85
    cool and 84, 85, 87
    intoxication, fatal 86
mentholated cigarettes 86
menthone 84, 85
menthyl acetate 84, 85
menthyl lactate 84, 85
meperidine 97
Mephedrone 110
Mercaptans 120–123
3-mercaptohexan-1-ol 125, 127
3-mercaptohexyl acetate 127
3-mercapto-3-methylbutan-1-ol 127, 128
3-mercapto-2-methylpentan-1-ol 126
metabolites, secondary 45, 76, 84, 89, 142, 191
metal heating elements 199
metal pieces 199
Metatine 206
methanal 195, 199–201, *200, 201*
methane 224–228
methanethiol 21, 118, 120, 121, 160
methanogenic (methane-making) bacteria 129
methanoic acid 129, 145
methanol 140, 141, 211
methicillin-resistant *Staphylococcus aureus*
    (MRSA) 167
methional 12, 13, 20, 21, *21*, 27, 29–31, 35, 122
    sulfur-containing aldehyde 27, 120
methionine 8, 9, 12, 21, 27, 28, 35, 115–118, *119*,
    120, 122, 126, 134
methoxyflurane 183
methoxypyrazines *21*, 29, 31
3-methylbenzoic acid 155
2-methylbutanal 1, 13, 20, 21, 26, 32, 36, 41–42
3-methylbutanal 1, 13, 20, 21, 26, 31–32, 36,
    41–42
2-methylbutanoic acid 20, 36, 48, 130
3-methylbutanoic acid 19–20, 36, 130–131
(*R*)-2-methylbutanoic acid cheesy, sweaty 130
(*S*)-2-methylbutanoic acid fruity, sweet 130
S-3-methylbutyl-1-thioacetate 124
methyl derivative of cobalamin 69
methyleugenol 81
α-methylfentanyl 98
3-methylfentanyl 98–99
2-methyl-3-furanthiol 14
methyl jasmonate (MeJA) 145
S-methylmethionine 116–117, *117*
6-methyl nicotine 205–206
8-methyl-6-nonenoic acid 77
2-methylpropanal 1, 2, *19*, 20, 21, 26, 32
2-methylquinoline 123
methyl salicylate (MeSA) 42, *42*, 136, 144–145,
    152, 154
3-methyl-3-sulfanylhexan-1-ol 132
(*R*)-3-methyl-3-sulfanylhexan-1-ol 132

(*S*)-3-methyl-3-sulfanylhexan-1-ol 132
Midgley, Thomas Jr. 174–176, 212
Minot, George 68
minty molecule 84, 87–88
Moissan, Henri 173
6-monoacetylmorphine (6-MAM) 95
Montreal Protocol 176, 178, 181
morphine 92–95, 97–98, 101, 102, 104–106, 113,
    142, 171, 191
mosquitoes 130–131, 155–158, 190
mosquito magnets 131
MRSA *see* methicillin-resistant *Staphylococcus*
    aureus (MRSA)
Murphy, William P. 68
mushroom alcohol 33
mushrooms 31, 33–37
    puffball 36–37
    smell of 33, 35
mustard 82, 89–91
myrtenol 50–51
myrtenyl acetate 50

NAD *see* nicotinamide adenine dinucleotide
    (NAD)
naloxone 101, 102
naltrexone 101, 102
Nasonov pheromone 53
natural aluminium isotope 213
natural amino acids 8–10
natural carboxylic acids 17
natural chlorine compounds 159, 161, 166, 171
natural fluorine compound 182–183
natural gas and methane 226–227
natural germ-killers from Earth 166–170
natural repellent 157
natural vanilla 215
neoisomenthol 85, 87
neomenthol 85, 87
neon isotopes 208
neral 51, 52, 54, 55
neural tube defects (NTDs) 67
neurotransmitter 65, 193–194
neutrons 213, 214
new psychoactive substances (NPS) 109, 113
niacin (vitamin B$_3$) 61–62
Nicotiana rustica 190
Nicotiana tabacum 142, 190–191
nicotinamide 56, 205, 206
nicotinamide adenine dinucleotide (NAD) 62, 63
nicotine 86, 90, 142, 155, 171–172, 190–195,
    197–199, 201
    analogues 205–206
    issues with 198–199
    lethal dose 193
    6-methyl 205–206
    poisoning 191–192
    protonated 194, 199

nicotinic receptors 194
nitazenes 112–113
nitrogen-containing compounds 133
nitrosamines 195, 199
    tobacco-specific 195, 197, 199
Noyori synthesis of menthol 85
NPS *see* new psychoactive substances (NPS)
NTDs *see* neural tube defects (NTDs)
Nubian beer 167
nuclear magnetic resonance (NMR) spectroscopy
    123, 210–211, 214–215, 217
nucleophilic substitution 161
nutmeg (*Myristica fragrans*) 75, 76, 81–82, 92

1-octen-3-ol 18, 33–36, 156
1-octen-3-ol, biosynthesis of 34
1-octen-3-one 29, 31, 33, 35–36
oil
    lemon 51, 53
    lemongrass 53
    lemon peel 53
    orange 53, 54
    peppermint 84, 86
onions (*Allium cepa*) 2, 37–39, 126
    cooked 39
    fresh 38–39
    lachrymatory factor 38
opium 92–94, 96, 102, 103, 109, 191
opium poppy (*Papaver somniferum*) 92, 102
opsins 57–58
optical isomers 33–34, 53, 84, 85, 87, 130,
    132, 192
oranges 51–53
    juice 51, 53–55
organic chlorine compounds 159–172
organic sulfur compounds 118, 120
organochlorine compounds 159–172
organofluorine compounds 173–189
orthinine 132, *133*
oxidation of lipids 24, 26, 32
2-oxopropanal 25
10-oxo-*trans*-8-decenoic acid (ODA) 34
oxycodone 102–105, *103*
OxyContin® 102–105
oxygen isotope 211, 218, 220–222
ozone 119, 160, 176–177, 184, 226–228
ozone-depleting chlorine radicals 160

painkiller addiction 105
pain-killing effects 106
pantothenic acid (vitamin B$_5$) 56, 64
Paracelsus 92
patent medicines *see* proprietary medicines
Pauling scale 183
peat bogs 160
pelargonidin 45
Péligot, Eugène 173

pellagra 62
penicillin 68, 166, 168, 187
penicillin-resistant staphylococci 168
Penicillium camemberti 16, 18
Penicillium roqueforti 16
2,3-pentanedione 202
pepper
    black pepper (*Piper nigrum*) 76, 79
    chilli pepper 76–79, 82
    Chinese peppers 82–84
    Szechuan (Sichuan) peppers 82–84
peppermint oil 84, 86, 94
Percocet® 105
perfluorocubane 189
perfluorodecalin 188
Periodic Table (1869) 209
pernicious anaemia 68
personal hygiene 125–128
pesticides 161–162
pethidine 97
petrochemicals 216
PG *see* propylene glycol (PG)
pharmaceuticals, fluorinated 183–189
phenols, chlorinated 165–166
phenyl acetate, pyrolysis of 205
phenylalanine 8, 9, 27, *42*, 45, 70
2-phenylethanal 27, *27*, 35
2-phenylethanol 27, *27*, 42, *42*, 152
phenylmethanethiol 121
pheromone 117, 149, 165
    Nasonov 53
pheromone 2,6-dichlorophenol 165
phosphoglycerides 11
phospholipids 11
phytoplankton 118, *119*
pipe smoking 190–191
plant defence 141–145
plant food 110
plant hormone 43, 145
plants attracting insects 144–145, 152–155
Plasmodium falciparum 155, 186
PLP *see* pyridoxal 5′-phosphate (PLP)
Plunkett, Roy J. 181
PMD *see* p-menthane-3,8-diol (PMD)
p-menthane-3,8-diol (PMD) 157–158
    diastereoisomers of 157–158, *158*
1-p-menthene-8-thiol 122
poison ivy (*Toxicodendron radicans*) 143
Poisons and Pharmacy Act (1868) 94
poly- and perfluoroalkyls (PFAs) 182
polycyclic aromatic hydrocarbons 195–196, 225
polypeptide chains, role of water in alignment
    of 22
polypeptides 10
polysaccharides 6–7
polysubstance abuse 99
pond cleaner 110

potatoes 2, 6, 27–32
    baked 31–32
    boiled 30–31
    chips 32
problematic organofluorine compounds 182
proline 8–10, 25, 35
    Maillard degradation of 12
propane-1,2-diol 201
propane-1,2,3-triol 200
proprietary medicines 94
propylene glycol (PG) (propane-1,2-diol) 198,
    200–201, *201*, 204
proteins 8–12, 15, 16, 19, 22–24, 28, 33, 56–58,
    61, 62, 64, 66, 70–73, 78, 115, 117,
    132, 145, 146, 170
    gluten 22–24
proteolysis 19
protonated nicotine 193–194
Prozac (fluoxetine) 164, 184
psychoactive cannabinoid 205
psychoactive molecules 108–110, 113
PTFE 180–181
puffball mushrooms (*Calvatia gigantea*) 36–37
pulmonary illness, vaping-associated 204–205
putrescine 132, *133*
pyrazine 28
    formation 2, 28
pyrethrins 143
pyridoxal 5′-phosphate (PLP) 65
pyridoxine (vitamin B_6) 56, 64–65
pyrolysis of phenyl acetate 205
pyruvaldehyde (2-oxopropanal) 201

quinine 93, 155, 161

radioactive isotope 208–209, 211, 213–214
radiocarbon 216–217
    dating 75, 217, 222
raw beef 12
repellents
    insect 89, 90, 131, 155–158
    natural 89, 90, 131, 157–158
    synthetic 155–156
retinyl palmitate 58
rhodopsins 57
riboflavin (vitamin B_2) 56, 60–61
Richard III's diet 220–221
rickets 62, 71–72
Rift Valley fever 155
roast beef 13, 25
Rocky Mountain spotted fever 165, 171
Rocky Mountain tick (*Dermacentor
    andersoni*) 165

salicylic acid 145
S-allylcysteine sulfoxide 37
salt marshes 119, 160

sanshool
    hydroxy-alpha sanshool 83–84
    synthetic 84
saponification 16
scented molecules 152
scent of death 132–136
Schieberle, Peter 46
Schieman, Günther 174
scurvy 56, 59, 62, 69–70
secondary metabolites 45, 76, 84, 89, 142, 191
semisynthesis 106
semisynthetic 96, 99, 102
sensations, taste 12
serotonin 65, 164, 184, 185, 193
sertraline 164, *165*
shiitake mushroom (*Lentinula edodes*) 33, 35–36
short-chain aldehydes 201
Siddall, Elizabeth 93
sinalbin 90
sinigrin 89
site-sensitive ^{13}C NMR techniques
        (SNIF-NMR) 217
skatole 133, 134, 136, 154
skunks 123–125
smell
    2-acetylpyrroline 12, 25
    beef 12–14
    bread 23–27
    cheese 15–21
    chicken 14–15, 55
    lemon 51–55
    limonene isomer 53–54
    of living and dead 129–136
    mushrooms 33–36
    onion 37–39
    orange 51–54
    orange juice 53–54
    potato 29–32
    strawberry 46–51
    sulfur compounds 115–128, 132, 134
    tomato 40–44
smogs, environment and 226
smoke 226
smokers 86, 142, 191–195
smoking 190–193
    cigar 190–191
    health effects of 194–195, 206–207
    pipe 190–191
    tobacco, health risks of conventional 194–195
Soddy, Frederick 208
Solanum tuberosum 28
Spanish fly 151–152
spice 75–84
    and hot 75–79
'Spice' (drug) 108–113
spotted skunk (*Spilogale gracilis*) 124
staphylococci, penicillin-resistant 168

starch 2, 3, 6, 7, 23, 28, 32
statins 187–188
strawberries 45–51
    wild 45, 48–51
Strecker degradation 1, 2, 30, 32
Streptomyces venezuelae 170
striped skunk (*Mephitis mephitis*) 123–124
strontium 219, 222
    isotope 219, 220, 222
student vaping 206
sucrose 2–5, 25, 45
sufentanil 97, 101
sugarcane 4
sulcatone 132, 156
sulfides 38–39, 115–120, 134
sulfur-containing aldehyde methional 12, 13, 20,
        21, 27, *28*, 30, 31, 35, 122
sulfur-containing aldehyde
        3-methylthio-1-propanal 127; *see also*
        methional
sulfur cycle 118, *119*
Swarts, Frédéric 174
Swarts reaction 174
sweat glands 125
sweat, lipids in 129
sweeteners
    artificial 4, 164
Sydenham, Thomas 92
synapse 194
synthetic cannabinoids 108
synthetic psychoactive molecules 109
synthetic repellent 155
synthetic sanshools 84
synthetic sweeteners 164
synthetic vanillin 215–217
Szechuan (Sichuan) peppers 82–84
Szent-Gyorgyi, Albert 69–70

taste
    sensations 12
    of tomato 40
TBM *see* tert-butyl mercaptan (TBM)
TCP stood for trichlorophenol 166
Teflon 180
teicoplanin 169–170
terpenoids 43, 46, 50, 51, 146, 152
tert-butyl mercaptan (TBM) 120
tetracycline 166–167
tetracycline antibiotics 166–167
tetrahydrothiophen (THT) 120
thebaine 94, 103, *104*
thermal oxidation of fatty acids 12
thiamin (vitamin B_1) 56, 58–60
thiols 14, 21, 38, 118, 120–128
Thomson, J. J. 208
THT *see* tetrahydrothiophen (THT)
tigecycline 167

tissue decay 132
titan arum (*Amorphophallus titanum*)
    119–120, 134
tobacco 141–142, 145, 190–201, 205–207
    plant 141–142, 145, 191–192
    seeds 190
    toxic chemicals in 191–193, 195–199
tobacco mosaic virus 145
tobacco smoking, health risks of conventional
    194–196, 198, 206
tobacco-specific nitrosamines (TSNAs)
    195–197, 199
tocopherols 72, 73
tomatoes 39–44, 57, 64, 72, 141
town gas 225
toxic aldehydes 199–201
toxic chemicals in tobacco 191–193, 195–199
toxicity of heavy water 210
toxic nature of fluoroacetate 182–183
transaminase enzyme 20
transamination 20, 26, 27, 126, 127
Transient Receptor Potential (TRP) cation
    channels 90
all-trans-retinal 57–58
trichloromethane 164, 174, 183
trichlorophenol, TCP stood for 166
trimethylamine 128–129
Triticum aestivum 22
TRPA1 receptors 80, 82, 84, 90, 201
TRP cation channels *see* Transient Receptor
    Potential (TRP) cation channels
TRPC5 channels 80
TRPM8 receptor (transient receptor potential
    cation channel subfamily M member 8
    receptor) 78, 81, 87
TRPV1 receptor (transient receptor potential
    cation channel, vanilloid subfamily
    member 1) 77–81, 84, 90, 201
truffles 117, 118, *118*
    fraudulent white 218
    synthesis of $(CH_3)_2S$ 118
tryptophan 8–10, 62, 70, 133, 134
tryptophan deaminase 133
TSNAs *see* tobacco-specific nitrosamines
    (TSNAs)
Tuber melanosporum 117, 118, 218
turmeric 75
typhus 161, 162, 165, 167

unsaturated aldehydes 12, 32, 38, 39, 54, 55, 141
unsaturated thiols 125
uranium
    hexafluoride 213–214
    isotopes and separation 213–214
urushiols 143
US Tobacco Control Act 205

vancomycin 168, 169, *169*
vanilla flavour 203, 215, 216
vanillin 154, 201, 203–204, 214–217
    synthetic 215
vanillylamine 77, *77*
vaping 198–207
    student 206
vaping-associated pulmonary illness (EVALI)
    204–205
Vicodin® 105
Vioxx (Merck) 187
vitamins 56–74
vitamin A 56–59
vitamin B_1 (thiamin) 58–60
vitamin B_2 (riboflavin) 60–61
vitamin B_3 (niacin) 61–63
vitamin B_5 (pantothenic acid) 64
vitamin B_6 (pyridoxine) 64–65
vitamin B_7 (biotin) 66
vitamin B_9 (folic acid) 66–67
vitamin B_{12} (cobalamin) 68–69
vitamin C (L-ascorbic acid) 69–70
vitamin D (calciferol) 70–72
vitamin E 72–73
vitamin E acetate 73, 204–205
vitamin K 73–74
vitamin K_1 73–74
volatile compound of uranium 213–214
    green grass 137–140
volatile organics, emissions of 140
Volta, Alessandro 224

walnuts (*Juglans californica* × *Juglans regia*) 145
water (H_2O) 4, 10, 11, 16, 18, 22, 23, 37, 45, 56,
    59, 64, 70, 78, 92, 94, 106, 115–117,
    119, 124, 146, 159, 162, 181, 182,
    202, 209–212, 214, 218, 219,
    222, 227
water, heavy (D_2O) 209–211
waxy kerogen 224
weaponised insects 145–150
western grey kangaroos 183
Whipple, George 68
Wilberforce, William 92–93
wild strawberries 45, 48–51
Williams, Robert R. 59
Wills, Lucy 67
wines 117, 121, 125–127
wintergreen embrocation 144
wood 154, 160, 174, 215, 217, 225–226

yellow fever 131, 155

zooplankton 118
zwiebelanes 38
zymase 24

For Product Safety Concerns and Information please contact our EU
representative GPSR@taylorandfrancis.com
Taylor & Francis Verlag GmbH, Kaufingerstraße 24, 80331 München, Germany

* 9 7 8 1 0 4 1 1 1 0 6 2 0 *